D1713911

PALEOSOLS
THEIR RECOGNITION AND INTERPRETATION

PALEOSOLS
THEIR RECOGNITION AND INTERPRETATION

V. PAUL WRIGHT, EDITOR

PRINCETON UNIVERSITY PRESS
PRINCETON, NEW JERSEY

Published by Princeton University Press, 41 William Street,
Princeton, New Jersey 08540

Copyright © 1986 by Blackwell Scientific Publications
All rights reserved

Library of Congress Cataloging in Publication Data

Main entry under title:

Paleosols: their recognition and interpretation.

   Includes index.
   1. Paleopedology. I. Wright, V. Paul, 1953–
QE473.P35 1986   552'.5   85–43201
ISBN 0–691–08405–X (alk. paper)

Clothbound editions of Princeton University Press books are printed on acid-free paper, and binding materials are chosen for strength and durability. Paperbacks, while satisfactory for personal collections, are not usually suitable for library rebinding.

CONTENTS

CONTRIBUTORS, viii

PREFACE, ix

INTRODUCTION, x

Chapter 1
THE FOSSIL RECORD OF SOILS by G.J. Retallack, 1

Introduction, 1
Earth's Early Atmosphere, 2
Early Life on Land, 13
Large Plants and Animals on Land, 17
Afforestation of the Land, 23
Grasses in Dry Continental Interiors, 30
Paleosols and Human Evolution, 35
Conclusions, 41

Chapter 2
PEDOGENIC CALCRETES IN THE OLD RED SANDSTONE FACIES (LATE SILURIAN—EARLY CARBONIFEROUS) OF THE ANGLO-WELSH AREA, SOUTHERN BRITAIN by J.R.L. Alien, 58

Introduction, 58
Stratigraphical Context of the Calcretes, 59
Field Characteristics of the Calcretes, 62
Microscopic Features of the Calcretes, 74
The Pedogenic Model, 75
Significance of the Calcretes, 78
Conclusions, 82

Chapter 3
PALEOSOLS CONTAINING AN ALBIC HORIZON: EXAMPLES FROM THE UPPER CARBONIFEROUS OF NORTHERN BRITAIN by C.J. Percival, 87

Introduction, 87
The Upper Carboniferous Environment of Northern England, 87

Factors Affecting Soil Formation, 87
Modern Soils Containing an Albic Horizon, 88
Paleosols Containing and Albic Horizon from the Upper Carboniferous of Northern England, 91
Conclusions, 109

Chapter 4
THE CALCAREOUS PALEOSOLS OF THE BASAL PURBECK FORMATION (UPPER JURASSIC), SOUTHERN ENGLAND by Jane E. Francis, 112

Introduction, 112
Geological Setting, 112
Palaeobotany of the Paleosols, 116
Descriptions of the Paleosols, 117
Interpretation of the Paleosols, 122
The Pebbles of the Great Dirt Bed, 126
Discussion, 130
Summary, 134

Chapter 5
TECTONIC CONTROL ON ALLUVIAL SEDIMENTATION AS REVEALED BY AN ANCIENT CATENA IN THE CAPELLA FORMATION (EOCENE) OF NORTHERN SPAIN by Christopher D. Atkinson, 139

Introduction, 139
Geological Setting, 140
Stratigraphic Setting, 144
Location of the Studied Sections, 147
Field and Laboratory Procedures, 147
Capella Formation: Sedimentation and Paleosol Development, 147
Origin and Causes of Paleosol Variation, 167
Origin of Differential Sedimentation Rates Across the Basin, 172
Conclusions, 174

Chapter 6
PALEOSOLS AND TIME RESOLUTION IN ALLUVIAL STRATIGRAPHY by Mary J. Kraus and Thomas M. Bown, 180

Introduction, 180
Stratigraphic Concepts, 183
Palaeotopography and the Alluvial Record, 188
Section Completeness, Paleosols and the Fossil Record, 194
Paleosols and Sediment Accumulation Rates, 199
Summary, 201

Chapter 7
RECOGNITION OF PALEOSOLS IN QUATERNARY PERIGLACIAL AND VOLCANIC ENVIRONMENTS IN NEW ZEALAND by I.B. Campbell, 208

Introduction, 208
Paleosols of Glacially and Periglacially Related Deposits, 209
Paleosols of Volcanic Deposits, 222
Paleosol Indentification and Paleosol Stratigraphy, 233
Summary, 238

Chapter 8
PRE-FLANDRIAN QUATERNARY SOILS AND PEDOGENIC PROCESSES IN BRITAIN by R.A. Kemp, 242

Introduction, 242
Quaternary Stratigraphy in Britain, 243
Buried Pre-Flandrian Soils, 245
Pre-Flandrian Relict Features in Non-Buried Soils, 252
Conclusions, 257

Chapter 9
PALEOSOLS IN ARCHAEOLOGY: THEIR ROLE IN UNDERSTANDING FLANDRIAN PEDOGENESIS by Richard I Macphail, 263

Introduction, 263
History, 263
Methodology, 266
The Processes of Decalcification and Lessivage, and Man's Effects Upon Them, 271
Agriculture, Erosion and their Effects on Pedogenic Trends, 275
Podzolization and Upland Hydromorphism, 276
Conclusions, 283

Glossary, 291

Author Index, 297

Subject Index, 307

CONTRIBUTORS

J.R.L. ALLEN FRS *Department of Geology, University of Reading, Whiteknights, Reading RG6 AAB, UK*

C.D. ATKINSON *ARCO Exploration and Technology Company, Exploration and Production Research, 2300 West Plano Parkway, Plana, Texas 75075, USA*

T.M. BOWN *US Geological Survey, United States Department of the Interior, Box 25046, MS 919, Denver Federal Center, Denver, Colorado, USA*

I.B. CAMPBELL *New Zealand Soil Bureau, Private Bag, Nelson, New Zealand*

J.E. FRANCIS *Department of Geology and Geophysics, The University of Adelaide, Box 498, GPO, Adelaide, South Australia 5001*

R.A. KEMP *Department of Soil Science, Lincoln College, Canterbury, New Zealand*

M.J. KRAUS *Department of Geological Sciences, University of Colorado, Campus Box 250, Boulder, Colorado 80309, USA*

R.I. MacPHAIL *Institute of Archaeology, University of London, 31–34 Gordon Square, London WC1H 0PY, UK*

C.J. PERCIVAL *BP Alaska Exploration Inc, 100 Pine Street, San Francisco, California, USA*

C.J. RETALLACK *Department of Geology, University of Oregon, Eugene, Oregon 97403, USA*

V.P. Wright *Department of Geology, Wills Memorial Building, Queen's Road, Bristol BS8 1RJ, UK*

PREFACE

This book was conceived because of the interest being shown in pre-Quaternary paleosols. The book aims to provide a focus for this interest and to provide a data-base and a stimulus for future work. Even though the subject as a whole is still in the early stages of development it proved very difficult to cover every aspect of the rapidly growing sub-discipline. Books already exist which are devoted to Quaternary paleosols but it was felt that specific reviews on such paleosols should be included in this book so as to provide a yardstick with which to compare other work. An appreciation of the methodology and conceptual background used in studying Quaternary paleosols is important for the future development of studies on pre-Quaternary paleosols.

It was decided that the aims of the book would best be met by a mixture of reviews and case-studies covering various aspects and various types of paleosols. The book also serves to highlight gaps in our knowledge such as paleosol diagenesis, clay mineralogy and palaeoecology. The chapters are arranged in stratigraphic order. Other aspects of paleosols are covered in the book *Residual Deposits: Surface Related Processes and Materials*, edited by R.C.L. Wilson (Special Publication of Geological Society of London, number 11, 1983), published by Blackwell Scientific Publications, Oxford.

The term paleosol as used in this book follows the recommended spelling of the INQUA/ISSS Paleopedology Commission.

The chapters have benefited from the criticisms and suggestions made by the reviewers, whom I thank very much for their help.

Bristol.   V. Paul Wright.

INTRODUCTION

A paleosol is a soil that formed on a landscape of the past (Valentine and Dalrymple 1976). In the study of Quaternary sequences paleosols have become a standard tool and as such are well documented. It is now appreciated that paleosols are abundant in the pre-Quaternary geological record and even Precambrian types are now widely recognized. The study of these 'older' paleosols is still in its infancy but already such studies have caught the imagination of many, and this compilation of papers is aimed at providing both a data-base and a stimulus for future work.

The fact that paleosols are abundant in the geological record may come as a surprise to some for soils seem ephemeral features, mainly formed in areas undergoing active erosion, and so would have a low preservation potential. However, in areas undergoing active sedimentation such as floodplains, soils have high preservation potentials. Mary Kraus and Thomas Bown (Chapter 6) note that the development of soils in alluvial sequences is a normal phenomenon for the rates of sedimentation on floodplains are generally slow enough to allow sediments to be pedogenically modified. Alluvial sequences are very common in the geological record and are the main repositories for paleosols. Kraus and Bown note that the Eocene Willwood Formation of Wyoming contains from 500 to 1200 superposed paleosols. Professor John Allen (Chapter 2) describes the classic calcrete paleosols from the Old Red Sandstone (Devonian) alluvial sequences of southern Britain which also contain hundreds of profiles.

Paleosols are sufficiently well known even at this stage to allow some general comments to be made about their evolution through time. Professor Greg Retallack (Chapter 1) gives a stimulating review of the geological history of soils beginning with a discussion of the early evolution of the Earth with speculations on its earliest soils in comparison to those on the Moon, Venus, and Mars. The study of these Precambrian soils is a particularly fascinating aspect and provides information on the development of the early atmosphere. His chapter outlines some of the major events which occurred during the evolution of the biosphere which had major impacts on soil development. Even though a biologically active soil cover almost certainly existed long before the late Silurian, the rise of the vascular plants in the middle Palaeozoic must have had an enormous impact on soils. In the Upper Palaeozoic extensive afforestation occurred which also led to an increased variety of soils, and the appearance of grasses in the Tertiary further led to a diversification of soil types. It is clear from such a review that soil forming conditions have varied dramatically through geological time, especially as regards the biological influences on pedogenesis. This must be borne in mind when looking at the older paleosols, especially those in the Lower Palaeozoic and Precambrian. It may be necessary to take a non-uniformitarian approach and look at these paleosols from a process-based viewpoint rather than trying to fit them into the classification of present day soils. Perhaps what is striking is that many older paleosols are remarkably similar to present day soils even though the fauna and flora must have been different. Real advances will come when we are able to detect and interpret differences between present day soils and their apparent ancient analogues.

How then can we recognize paleosols, particularly in the older sedimentary sequences where diagenesis will have taken its toll? There is no set of simple criteria which can be used. It is more useful to ask what is a soil before we try to recognize fossil forms. At the simplest level a soil can be regarded as a natural body formed by the accumulation of organic and inorganic materials at the Earth's surface which differs from the underlying material in its morphology and its physical, chemical and biological properties. Such soils typically, but not always, contain horizons with different structures which often enables the recognition of fossil forms to be made. A common fallacy I have encountered is that geologists think that paleosols must have roots in them to show that they functioned as 'soils'. The preservation potential of roots or rootlets is low in many soil types and biological processes only form part of the spectrum of soil forming processes. To recognize paleosols it is necessary to show that the material has been modified by near surface, pedogenic processes. These processes include eluviation and illuviation, rubifaction and gleying to name a few, and the recognition of many of these processes is discussed in various chapters. Iain Campbell (Chapter 7) reviews the Quaternary paleosol record of New

Zealand and discusses the techniques and problems associated with paleosol recognition. Colin Percival (Chapter 3) uses vertical profile changes to illustrate alteration by pedogenesis in Carboniferous paleosols. His study provides a useful guide to the sorts of criteria used in studying older paleosols such as down-profile changes in grain-size and mineral composition, illuviation features, loss of primary sedimentary structures and the nature of contacts between horizons.

Several authors use micromorphological criteria to identify pedogenetic alteration but a sharp contrast can be seen between the sophisticated studies of Quaternary paleosols by Robert Kemp (Chapter 8) and Richard Macphail (Chapter 9) and those on older paleosols. Such micromorphological techniques may prove to be the most useful tools for recognizing and interpreting paleosols.

Classifying paleosols can be very difficult and a variety of classifications of present day soils are available. Some authors in this book have used the United States Department of Agriculture Soil Survey Staff system while others have used the FAO/UNESCO system. This may seem confusing but each system has advantages and disadvantages. This is especially the case with paleosols where many of the criteria used in classifying present day soils are not preserved, for example base status and thermal regime. As stated above when looking at very ancient paleosols it is unwise to try to fit the soil into one of the present day categories. It is much more sensible to take a process-based, rather than a model-fitting, approach.

The concepts and basic terminology of paleosol stratigraphy are discussed by Kraus and Bown (Chapter 6). In studying paleosols a common problem which is encountered is that many profiles consist of superimposed soils, that is, different soils overlap and overprint one another during vertical accretion. Such paleosols are said to be composite in nature and indeed such profiles may be the norm in many ancient alluvial sequences. Other paleosols are polygenetic, that is they have developed under conditions which have changed during pedogenesis. Overprinting is the typical result of this. Robert Kemp (Chapter 8) discusses pre-Flandrian Quaternary pedogenesis and describes paleosols with paleoargillic horizons which represent relict features formed during interglacial periods, later modified by cold-stage pedogenic processes. Detailed micromorphological studies are needed to recognize such relict features as is also shown in Macphail's study of Flandrian pedogenesis. Many paleosols in the geological record are developed at major stratigraphic breaks and may represent very long periods of pedogenesis. Polygenetic paleosols should be expected at such levels and deliberately searched for. These can provide a wealth of information, but at this early stage in the development of the subject many paleosols are described simply with the aim of proving that pedogenesis actually occurred rather than interpreting the features present in their own right. The occurrence of overprinting, in composite and polygenetic profiles, also makes classification difficult.

Catenas are groups of soils, with similar parent materials, developed under similar climates but with different characteristics related to variations in relief and drainage. Their recognition in paleosols is particularly useful and can be related to the environmental setting provided by sedimentological modelling. Christopher Atkinson (Chapter 5) provides a case-study of this approach based on the paleosols in the Eocene Capella Formation of north Spain. He recognizes a variety of composite profiles with Entisols including Aquents, Fluvents and Torrerts (USDA system). Variations in these paleosol types occur across the Tremp-Gaus basin in north Spain which represent a paleo-catena related to topography and sedimentation rate reflecting the overall tectonic setting.

The Capella Formation study serves to illustrate how important it is to relate paleosols to their sedimentological and broader setting. Colin Percival also illustrates this approach with his Carboniferous paleosols. Other types of data can also be used in interpreting paleosol environments. Jane Francis (Chapter 4) describes the well known Upper Jurassic Purbeck Dirt Bed paleosols of southern England and integrates palaeobotanical information into her study. Of historical interest is the fact that the Dirt Beds were recognized as fossil soils as early as 1826. The paleosols represent rendzinas with accumulations of calcium carbonate in the lower parts of the profiles. The vegetation is represented by spectacular tree roots and stumps of low branched, shallow rooted conifers, as well as the remains of cycads, ferns and lycopods. The flora, in conjunction with sedimentological data, indicate that these rendzinas formed in a semi-arid, strongly seasonal, Mediterranean-type climate. The tree-rings in the fossil plants are also used to give a minimum estimate of the length of pedogenesis. Similar integrated studies of paleosols and the vegetation are likely to be very successful in reconstructing palaeoenvironments.

Paleosols provide powerful tools for a variety of purposes. They have been widely used for stratigraphic correlation in the Quaternary. This approach has many problems which are discussed by Robert Kemp for the pre-Flandrian paleosols of Britain and by Iain Campbell in his discussion of New Zealand Quaternary paleosols. Paleosols can be used in stratigraphy in other ways and Mary Kraus and Thomas Bown describe their use in time resolution in alluvial sequences. By this method the completeness of a particular stratigraphic record can be assessed and this can be particularly important in sequences which have been used for evolutionary studies. Paleosols provide information on sedimentation rates which can be integrated into studies of palaeogeomorphology, alluvial architecture and basin analysis as is illustrated in the chapters by John Allen and Chris Atkinson. Paleosols are frequently used to assess palaeoclimates with the calcrete paleosol horizons of red-bed sequences being one of the best known examples (Chapter 2). Paleosols have been widely used in archaeological studies and are reviewed by Richard Macphail who also shows how such studies provide feedback enabling changes in soil processes to be studied using archaeological information.

The study of paleosols now involves two 'camps', the Quaternary and pre-Quaternary groups. Research into pre-Quaternary paleosols is only just beginning and much can be learned from studying the methodology of those working on Quaternary paleosols. The amount of material available for study is staggering and I once calculated that within two hours drive of my home in South Wales there were at least *one thousand* Devonian, Carboniferous and Triassic paleosol profiles exposed.

Paleosols provide an enormously challenging field of research which we have only just begun to investigate.

REFERENCE

Valentine, K.W.G. and Dalrymple, J.B. 1976. Quaternary buried paleosols; a critical review. *Quaternary Research* **6**, 209–20.

Chapter 1

THE FOSSIL RECORD OF SOILS

G. J. RETALLACK
Department of Geology, University of Oregon, USA

INTRODUCTION

The record of organisms and environments of the past in sedimentary rocks is largely biased towards depositional areas, such as lakes and seas. What of non-depositional areas, such as the landscapes on which we humans play out our lives? Are they lost forever? They have left a different kind of record—one of physical, chemical and biological alteration of sediments and other materials during the quiet times between floods, landslides and other geological events. The resulting fossil soils (paleosols) are found at the hiatuses, diastems, disconformities and unconformities of the non-marine rock record. Those found at major unconformities representing millions of years of non-deposition, are the most problematic. They may have begun forming under very different climatic, or other soil forming factors, than those prevailing when the landscape was buried by younger deposits. It is easier to reconstruct paleosols and factors in their formation when the time over which they formed is short enough that conditions are unlikely to have changed. Burial of land surfaces on this kind of time-scale is found in river valleys and coastal plains. It is the thick sequences of red beds and coal measures of the world which have the best fossil record of soils.

The fossil record of soils is long. The oldest paleosol recognized here is 3100 million years old. There is no reason to doubt that, with increasing awareness of paleosols and increasingly sophisticated methods of detecting them, they will be found in rocks as old as the present sedimentary rock record, some 3800 million years before present.

Over this long sweep of geological time, soils have changed, just as there have been changes in continents, sediments, plants and animals. Some fossil soils of the distant past appear so different from modern soils, that they may represent extinct kinds of soil. It is difficult to be certain, because of the alteration of these paleosols by burial and metamorphism. The interpretation of fossil soils is in some ways similar to reconstructing past vegetation, animals and sedimentary environments from fossils and rocks. Nikiforoff (1943) has compared a fossil soil to the skeleton of a fossil animal. There are uncertainties in reconstructing the past, and many aspects of fossil soils may never be fully understood. Nevertheless, much current research is contributing to a history of soils on Earth. This essay is my attempt to review some major events in that history. I acknowledge a debt of inspiration to my predecessors in this effort (Yaalon, 1971; Hunt, 1972; Ortlam, 1980; Catt and Weir, 1981), and to numerous scientists cited here, who have made my recent review (Retallack 1981a) obsolete.

EARTH'S EARLY ATMOSPHERE

There are so many peculiar features of rocks older than 600 million years that the Earth is suspected to have been a rather different place at that time. Many of these features are better understood if free oxygen were present in much lower amounts than today, its place taken by other gases such as carbon dioxide and methane (Holland, 1984). Oxygen is generally thought to have built up to present levels in the atmosphere because of the activity of photosynthetic microbes (Schopf *et al.*, 1983). Although this view is widely accepted, other scientists have argued for appreciable Precambrian atmospheric oxygen (Dimroth and Kimberley, 1976; Clemmey and Badham, 1982). As for many aspects of Precambrian palaeoenvironments, evidence for atmospheric composition is hard to find and seldom compelling. Paleosols provide important clues.

Although many Precambrian paleosols are different from younger ones, they are not as different as they could be, judging from the kinds of soils now forming on the Moon and planets of our solar system. Soil formation on the Moon is most unlike anything on Earth, now or in the known past. In the absence of air and water there, micrometeorites are the major agent of soil formation (Lindsay, 1976). Impacts of large asteroids and meteorites are also important, but are less significant in the long run because they are much less frequent. Impact breaks down rock fragments and causes local melting. This results in the surficial accumulation of pulverized rock, composite particles welded by glass (agglutinates) and pure particles of glass. On Venus, possible duricrusts have been seen in images from Soviet probes (Florensky *et al.*, 1983). Considering likely atmospheric conditions on Venus, such duricrusts are probably most similar to what on Earth would be regarded as metamorphic rocks formed under very high temperature and low pressure (Nozette and Lewis, 1982). Paleosols of the Lunar and Venusian type have not yet been found on Earth, but it is unlikely anyone has seriously searched for them. They probably will not be found in rocks younger than 3800 million years. Sedimentary rocks of that age are an indication of free liquid water at the Earth's surface. Very ancient gypsum (3500 million years old) is evidence of surface temperatures below 58°C since that time (Walker *et al.*, 1983). Mars has the most similar of the known planetary surfaces to that of Earth. Images from Viking landers show basaltic boulders, loose sand, silt and clay, as well as partly buried, cracked surfaces (Strickland, 1979). These cracked surfaces may be evaporitic duricrusts, like those forming in desert soils and playas on Earth (Clark and van Hart, 1981). This could be interpreted as evidence of substantial reorganization of some constituents of the planetary surface, but the degree of weathering now demonstrated on Mars is less than in the deserts of Earth.

Precambrian paleosols on Earth clearly formed in a world of running water and atmosphere, and may provide evidence of variation in their fluid environment with time (Holland, 1984). Atmospheric oxygen is the main oxidizing agent in the present atmosphere and its effect on weathering is profound. Iron liberated by dissolution of minerals is oxidized to ferric oxides (Fe_2O_3 or haematite), hydroxides or oxyhydrates, in which iron is in the trivalent state (Fe^{3+}). These minerals are yellow, brown and red in colour, and are responsible for the warm hues of most modern soils. Under reducing conditions, as in swamps, iron released from minerals remains in its reduced state (Fe^{2+}), within drab coloured silicates (such as chlorite), sulphides (pyrite) or carbonates (siderite). Since bivalent iron is much more soluble than trivalent, iron tends to be lost from soils formed under reducing conditions. The iron content and colour of paleosols are potentially useful guides to the changing oxygen content of past atmospheres. However, a number of other soil forming factors must also be considered in interpreting any particular paleosol.

Among the oldest known paleosols and the most thoroughly studied, are profiles developed on Archaean granite and greenstone at the unconformity below 2300 million year old fluvial sediments of the Huronian Supergroup, north of Lake Huron, Ontario, Canada (Gay and Grandstaff, 1980; Kimberley *et al.*, 1984). A profile developed on greenstone has been called the Denison paleosol (Figure 1). This consists of green sericitic rock, lighter in colour and finer grained than

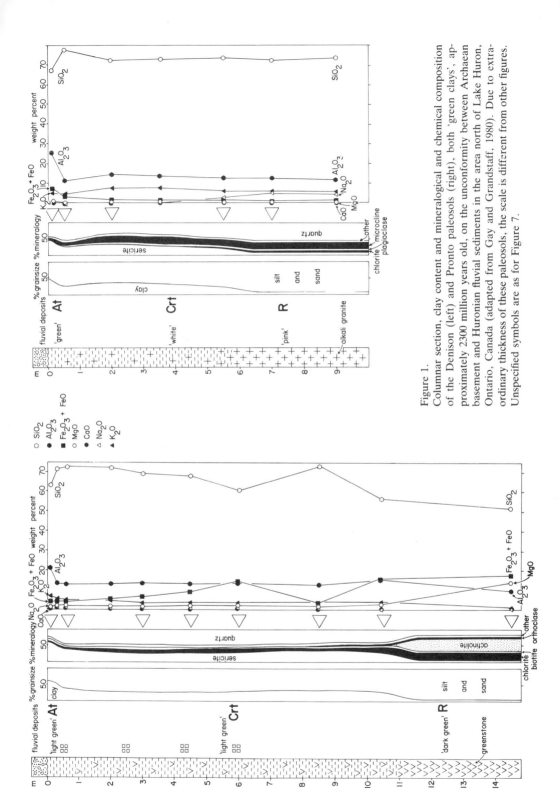

Figure 1.
Columnar section, clay content and mineralogical and chemical composition of the Denison (left) and Pronto paleosols (right), both 'green clays', approximately 2300 million years old, on the unconformity between Archaean basement and Huronian fluvial sediments in the area north of Lake Huron, Ontario, Canada (adapted from Gay and Grandstaff, 1980). Due to extraordinary thickness of these paleosols, the scale is different from other figures. Unspecified symbols are as for Figure 7.

the parent greenstone. Clay content (now sericite) was greatest near the surface (now to a depth of 0.6 m), moderately high to a considerable depth (now 10.5 m) and negligible within the parent greenstone. This very clayey surface layer is interpreted as an A horizon, in view of its contained organic carbon (Gay and Grandstaff, 1980). The less clayey subsurface layer is here identified as a C horizon (contrary to Gay and Grandstaff, 1980), comparable to the saprolite or Cr horizon of modern deeply weathered soils (as characterized by Birkeland, 1984). This layer is much thicker and shows more original rock fabric than is usual in the B horizons of modern soils. The mottled and pallid zones beneath soils on very ancient (millions of years old) landscapes (McFarlane, 1976) may be analogous to this extraordinarily thick, clayey layer.

Formation of clay (now sericite) from actinolite and feldspar, and to a lesser extent from chlorite, appears to have been the major soil forming process. There is also an increase in the amount of quartz at the expense of the same minerals. Since some of the quartz grains are larger than those in the parent greenstone, some quartz may have been newly formed in the paleosol rather than a residual concentrate. Alumina and silica vary in abundance with clay and quartz content, just as they would in a modern soil. The leaching of MgO from the profile is much as would be expected in modern soils of humid climates. The same is also true of CaO, although there was very little of this element in the parent material originally. There is even less Na_2O in the parent material, and it varies little within the profile. Surprisingly, K_2O appears to have accumulated. Total iron is strongly depleted from the profile. There is a little more trivalent iron than bivalent iron near the surface of the profile (above 0.6 m), but below that iron is largely present in the bivalent state.

Another profile along the same pre-Huronian unconformity, developed on pink alkali granite, has been called the Pronto paleosol (Figure 1). This has an especially clayey, light green surface (A horizon, now above 0.5 m) and a deep (now down to 5.5 m), clayey, white, subsurface (Crt) horizon. The main soil forming process appears to have been formation of clay (now sericite) from chlorite and feldspar. Chemical variation in Al_2O_3, SiO_2 and MgO is similar to that observed in the Denison paleosol. The amount of K_2O is rather constant, with slight subsurface (horizon Crt) accumulation and slight surficial (horizon A) depletion. There are very low levels of CaO throughout the Pronto paleosol, and Na_2O shows some surficial depletion, like that found in soils of humid climates. A notable feature of the Pronto paleosol is the surficial enrichment of total iron, largely in the trivalent state, in the upper part of the paleosol (above 6 m).

These two paleosols on the same ancient land surface are both drab coloured, as is typical of paleosols older than 2000 million years (Button and Tyler, 1979). However, they are surprisingly different in their degree of oxidation and amount of total iron. Just as modern soils are known to be products of many interacting factors, the interpretation of Precambrian paleosols must take into account complications such as metamorphism, clay diagenesis, original permeability, organic matter content, palaeodrainage and parent material.

Metamorphic environments are mostly reducing (Winkler, 1979), so it is possible that the lack of oxidized red beds older than about 2600 million years (Walker et al., 1983) is because most rocks of such great age are metamorphosed. However, red oxidized paleosols have persisted in the Silurian Bloomsburg Formation of Pennyslvania (Retallack, 1985), despite metamorphism to lower greenschist facies (Epstein et al., 1974). This is a much higher grade of metamorphism than suffered by the pre-Huronian paleosols (Roscoe, 1968). Furthermore, the existence of both weakly oxidized and unoxidized paleosols on this same ancient land surface is evidence against pervasive metamorphic alteration.

Changes in the composition of clays during burial may further obscure the original nature of paleosols. There is, for example, some reason to suspect that the anomalous K_2O enrichment seen in nearly all Precambrian (Williams, 1968; Button and Tyler, 1979; Gay and Grandstaff, 1980; Sokolov and Heiskanen, 1984) and many Palaeozoic paleosols (Retallack, 1985) may be diagenetic in origin. Comparable illitization of clays with depth has been observed in boreholes in many different parts of the world (Dunoyer de Segonzac, 1970; Van Moort, 1971; Heling, 1974; Hower et al., 1976). A second alternative for the enrichment of K_2O in Precambrian paleosols is that this element was less depleted in soils lacking vascular land plants (Weaver, 1969). The efficacy of land plants in removing potassium is well known. In one set of experiments 15 successive crops of Ladino clover (*Trifolium repens*) removed 410 kg/hectare of K_2O from an illitic-montmorillonitic soil and 356 kg/hectare from a kaolinitic soil (Mehlich and Drake, 1955). In uncropped natural vegetation there is some recycling of potassium (Graustein and Velbel, 1981), but in many situations, especially in humid forested soils, it is ultimately washed out of the soil (Basu, 1981). Out of 128 analyses of North American soils cited by Marbut (1935), 55% show surficial depletion of K_2O. Only 33% show enrichment, and 74% of these are aridland soils. Thus a third explanation for potassium enrichment in paleosols is a dry, evaporitic climate in which alkali and alkaline earth elements accumulate in the soil (Duchafour, 1982). Although this last explanation stretches one's credulity the least, it seems unlikely considering the abundant clay, depth of weathering, lack of evaporitic minerals or crystal casts, and depletion of Na_2O in pre-Huronian and in many other Precambrian paleosols. At present it is difficult to choose between a residual and metasomatic explanation for the enrichment of K_2O in Precambrian paleosols. If there has been widespread remobilization of potassium during burial, then analyses of other elements of Precambrian paleosols may also be suspect. Some encouragement can be gained from the composition of Precambrian and early Palaeozoic shales, which show enrichment in adjusted ratios of K_2O over $FeO + Fe_2O_3$ compared to igneous parent material, but no consistent trend of their own (Holland, 1984). This may mean that total iron content at least was not affected by potassium metasomatism, if indeed it occurred!

Drab coloured soils are found today in low lying parts of the landscape where atmospheric oxygen is excluded by waterlogging (Coultas, 1980). Such differences in palaeodrainage have been used to explain the different oxidation states of the Denison and Pronto paleosols (Gay and Grandstaff, 1980). This explanation has general appeal for the interpretation of Precambrian paleosols, because waterlogged soils are more abundant than oxidized ones in many sedimentary environments, and so would be more likely to be preserved in the rock record. In the case of the pre-Huronian paleosols, however, poor drainage seems unlikely. Their depth and degree of development is well in excess of modern waterlogged soils. In the Denison paleosol (thought to be waterlogged by Gay and Grandstaff, 1980) clay skins are evidence of a very low water table (below 6 m in the compacted profile). It is possible that the clay and soil structure of both Denison and Pronto paleosols formed under freely drained conditions, but were reduced under waterlogged conditions shortly before burial by overlying fluvial deposits. If this were the case though, it would have affected both paleosols more equally.

It is also possible that the degree of oxidation of Precambrian paleosols was related to their texture. This is an important limitation on depth of weathering of modern soils (Birkeland, 1984). In the case of the pre-Huronian paleosols the finer grained parent material (greenstone) is weathered to a greater depth than the coarse material (granite), the reverse of the usual relationship seen in modern soils. Although previous studies of pre-Huronian paleosols (Roscoe, 1968; Frarey & Roscoe, 1970) noted that greenstone profiles were thinner than granitic ones, the presence of exceptions such as the Denison paleosol, is an indication that grainsize was not of overriding importance. Nor was the texture of the resulting soil, because the Denison and Pronto paleosols do not differ appreciably in the amount of clay despite their different parent materials.

Even in the present oxidizing atmosphere, complexes of clay and stable organic matter in some well drained soils, such as Vertisols, can impart a drab colour (Duchafour, 1982). The effects of organic matter as an oxygen sink are difficult to assess for paleosols, because they are known to have consistently less organic matter than similar modern soils (Stevenson, 1969). Very low amounts of organic carbon have been found in the Denison paleosol (Gay and Grandstaff, 1980), but it is not dark coloured like modern organic soils. The organic matter content of the Pronto paleosol is not available for comparison.

A final consideration is the amount of iron present in the parent material. Under present oxygenic conditions, most iron released by weathering is oxidized. Thus, the reddest soils form on iron-rich parent materials such as basalt and greenstone. Curiously, the situation is reversed with the pre-Huronian and many other Precambrian paleosols: those developed on iron-rich rocks are less oxidized and have lost iron compared to those developed on iron-poor rocks. Holland (1984) has proposed a simplified model for weathering of silicate rocks, which explains this anomaly and allows calculation of some limits to the amount of oxygen in the atmosphere. Under conditions of low amounts of available oxygen within the soil there may not be enough to oxidize the large amounts of iron released from iron—rich rocks by the action of carbonic acid derived from CO_2. Thus much iron remains reduced and is washed out of the profile. Even low amounts of oxygen may be sufficient to oxidize slowly released iron from iron-poor parent materials. Holland's calculations based on a number of Precambrian paleosols (including these pre-Huronian profiles) reveal dramatic changes in the ratio of the partial pressure of oxygen over the partial pressure of carbon dioxide from 3000 to 1500 million years ago ($P_{O_2}/P_{CO_2} \leq 1.3 \pm 0.5$) to the present ($P_{O_2}/P_{CO_2} \doteq 600$). Although very early Precambrian atmospheres now appear to be a little more oxidizing than many scientists would have expected a few years ago, they were much less oxidizing than the modern atmosphere. All things considered, the pre-Huronian paleosols are in accord with the conventional view of very low amounts of oxygen in the atmosphere during the very early Precambrian.

The oldest paleosol yet recognized is some 3100 million years old and is developed on granitic basement under the Pongola System, near Amsterdam, Transvaal, South Africa (Edelman *et al.*, 1983; Holland, 1984). There are several other drab paleosols of similar antiquity (Rankama, 1955; Eskola, 1963; Vogel, 1975; Herd *et al.*, 1976; Button and Tyler, 1979; Schau and Henderson, 1983; Edelman *et al.*, 1983; Holland, 1984). The youngest of these peculiar drab paleosols may be an altered basalt on the western flank of the Anabar Shield, north central Siberia, a little more than 1650 million years old (Chayka and

Zaviyaka, 1968; Chayka, 1970; Salop, 1973). Identification of these paleosols within modern soil classifications is difficult. They were certainly not Spodosols (of USDA; Soil Survey Staff, 1975), because they are much too thick, clayey and lack an eluvial horizon. They are also too thick and well developed for Gleysols (of FAO: Fitzpatrick, 1980), another suggestion of Gay and Grandstaff (1980). The only early Precambrian paleosol which can be identified with confidence is the 2200 million year old Waterval Onder paleosol from Transvaal, South Africa (Figures 4 and 5; Button, 1979). Its clear deep clastic dikes (former cracks) and hummocky surface (gilgai structure) mark it as the oldest known Vertisol (of Soil Survey Staff, 1975). Comparable structures have not yet been observed in the other early Precambrian paleosols, but this may be because their outcrop in mines and boreholes is limited. It also could be that these other drab paleosols represent an extinct kind of deep weathering. For the moment, the author proposes informally to call them 'green clays.'

On some very stable, ancient land surfaces it appears that clay formation and alumina enrichment, characteristic of green clay paleosols, continued to the extent that bauxites formed. The Namies Schist of the Bushmanland Sequence, in the Cape Province of South Africa, contains as much as 74 weight percent Al_2O_3 (Button and Tyler, 1979). The age and stratigraphic relationships of these metamorphic rocks are uncertain, but they are likely to be between 1000 and 2500 million years old (Kent, 1980). This makes them possibly the oldest of the described Precambrian bauxites (Rozen, 1967; Valeton, 1972; Button and Tyler, 1979). Bauxites are thought to form by long term weathering of silicate minerals at moderate pH in humid, non-seasonal climates, on stable land surfaces (Valeton, 1972). Bauxites also form as a residuum or deposit on limestone in regions of low sediment influx. Numerous examples of karst bauxites are known ranging back in age to Devonian (Nicholas and Bildgen, 1979). The oldest example of this kind of bauxite is developed on karst topography on the 2300 million year old Malmani Dolomite in Transvaal, South Africa (Button and Tyler, 1979). An older example of karst, at least 2700 million years old, is known from the Steeprock Group west of Thunder Bay, Ontario, Canada (Schau and Henderson, 1983). This unconformity is overlain by concentrations of manganese and chert, like those found also in parts of the karst on the Malmani Dolomite.

By about 1000 million years ago the atmosphere was discernably more oxidizing. The Sheigra paleosol of about this age in northwest Scotland (Williams, 1968) was developed on a topographic surface of Lewisian biotite gneiss, amphibolite and microcline pegmatite, with a relief of at least 600 m, and covered by Torridonian alluvial fan deposits (Figures 2 and 3). In profiles developed on biotite gneiss, the surface (A) horizon (30 cm in the compacted paleosol) is stained red. Underlying this is a zone of mottled and light coloured rock including large round corestones of little weathered parent material (C horizon). Alteration extends more deeply into the parent material (up to 6 m from the surface) along joint planes. The profile is more clayey toward the surface but has a crystalline texture throughout. Quartz and microcline persist in the weathered part of the profile, but biotite and plagioclase of the parent gneiss have been extensively weathered to clay (now sericite). Clay formation is reflected in the decreased SiO_2 and increased Al_2O_3 toward the surface. Oxides of labile cations, such as CaO and Na_2O are depleted toward the surface, but K_2O appears to have accumulated, as is usual in Precambrian paleosols. Total iron is enriched in the surface horizon, and slightly depleted in the mottled and light coloured zone compared to parent material. Ferric iron (Fe_2O_3) is well in excess of ferrous iron (FeO) throughout the paleosol. This relationship is reversed in the parent material, as is usual for igneous and metamorphic rocks.

Figure 2. The Sheigra paleosol (bleached and reddened zone about 1 m thick), approximately 1000 million years old, developed on Lewisian amphibolite (left hand side) and biotite gneiss (to right) at the unconformity below Torridonian alluvial fan deposits, near the hamlet of Sheigra, north-west Scotland, UK. The white tape extending down from the unconformity, left of centre, is 2 m long.

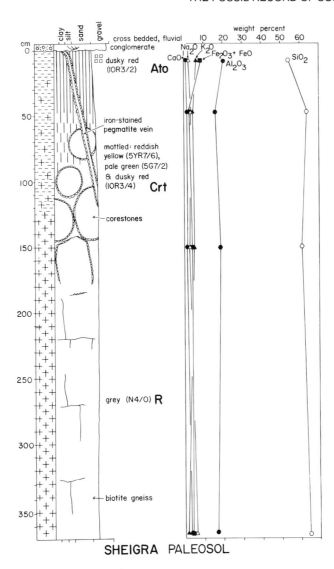

Figure 3. Columnar section and chemical composition of the Sheigra paleosol, probably an Inceptisol (adapted from Williams 1968, and personal field observations). Unspecified symbols are as for Figure 6.

Clay formation and ferruginization appear to have been the principal processes involved in the formation of this paleosol and it developed on a stable, ancient land surface. The bending of weather-resistant pegmatite veins at the surface of the paleosol has been interpreted (Williams, 1968) as evidence of Precambrian soil creep. This, and the depth of weathering of this paleosol are indications that it was well drained. Incomplete leaching of Na_2O and CaO, and the presence of dolomite veins in paleosols developed on amphibolite on the same ancient land surface, are evidence of a subhumid palaeoclimate. The mild metamorphic alteration, good former drainage and coarse texture of this paleosol would all have aided oxidation. The parent rock had a high demand for oxygen, closer to that of the Denison paleosol than that of the Pronto paleosol (as calculated from its amount of iron and other oxygen consuming elements by Holland, 1984). The clear oxidation and ferruginization of this paleosol is thus evidence of a significant advance in the amount of oxygen in the atmosphere since the early Precambrian.

The oldest paleosol yet reported to be ferruginized is developed on granite underlying the Lower Sosan Group in the Simpson Islands, Great Slave Lake, Canada, thought to be older than 2200 million years (Stanworth and Badham, 1984). Ferruginized paleosols are more common in younger Precambrian rocks (Sharp, 1940; Blades and Bickford, 1976; Elston and Scott, 1976; Grabert, 1976; Kroonenberg, 1978; Knight and Morgan, 1981; Stanworth and Badham, 1984). Even though many of these are on major unconformities, their degree of weathering (where data are available) correspond with Inceptisols (of Soil Survey, Staff 1975). Contrary to Williams (1968), they are unlikely to be 'Podzols' (USDA Alfisols, Ultisols and Spodosols), because they lack recognizable eluvial and illuvial horizons.

When examined more closely, some of these ferruginized paleosols may turn out to have been more deeply weathered, and perhaps referrable to Oxisols (of Soil Survey Staff, 1975). This is especially suggested by the 1660 million year old, deep weathering recognized beneath the Thelon Sandstone in the Thelon Basin, North West Territories, Canada (Chiarenzelli, 1983; Chiarenzelli et al., 1983a,b; Miller, 1983) and 1500 million year old weathering beneath the Athabaska Group in the Athabaska Basin in nearby northern Saskatchewan (Ramaekers, 1981; Tremblay, 1983; Holland, 1984). Kaolinite, an indicator of extreme weathering, has persisted in these profiles despite subsequent diagenetic alteration. These paleosols show a haematite zone (up to 24 m thick) at the surface, overlying a transitional horizon (up to 24 m), and then a chlorite-rich horizon (up to 30 m) over fresh bedrock (Chiarenzelli, 1983). These zones are comparable to incipient lateritic zones, mottled zones and pallid zones found beneath deep tropical soils (McFarlane, 1976). No actual laterite has yet been found, although it could have been eroded or otherwise altered. The profiles show strongly eroded surfaces, silcrete caps, and subsurface dolomitization, as well as metamorphic alteration. Until laterite or lateritic detritus is actually found, the oldest likely laterite remains the Cambrian or Ordovician Chester Emery of Massachusetts, USA (Norton, 1969). It could be that oxygen had not yet accumulated in the atmosphere to the extent needed to form laterite 1660 million years ago. Nevertheless, these deeply weathered Precambrian paleosols may represent the earliest known Oxisols.

Silcretes make their appearance in the rock record by about 1800 million years ago at the base of the Pitz Formation in the Thelon Basin (Chiarenzelli *et al.*, 1983a,b) in the North West Territories of Canada. Other Precambrian silcretes have also been reported (Donaldson and Ricketts, 1978). Silcretes are silica-rich rocks formed on stable cratonic land surfaces where the silica cement may be derived either from deep tropical weathering of silicates at moderate pH or from shallow, highly alkaline weathering, in desert soils (Summerfield, 1983).

Calcrete also appears in the rock record by about 1900 million years before present, in the form of pisolites in the Mara Formation, Goulburn Group, Bathurst Inlet, North West Territories, Canada (Campbell and Cecile, 1981; for age compare Easton, 1981). Numerous other Precambrian calcretes are known (Williams, 1968; Koryakin, 1971, 1977; Kalliokosi, 1975, 1977; Sochava *et al.*, 1975; Lewan, 1977; Donaldson and Ricketts, 1978; Bertrand-Sarfati and Moussine-Pouchkine, 1983; Chown and Caty, 1983). Many of these early calcretes are dolomitic in composition, like carbonate in modern soils of high base status (Doner and Lynn, 1977). Like the calcretes of modern Aridisols (Soil Survey Staff, 1975), they are thought to have formed in dry climates, in which soil solutions were not sufficiently copious or acidic to remove carbonate. The appearance of calcrete in the rock record may be related to the development of dry climates with increased size of continental nuclei and mountains, or perhaps to decreased amounts of atmospheric CO_2 and acidity of rainwater. Calcretes may be another indication of changes in terrestrial weathering roughly 2000 million years ago.

EARLY LIFE ON LAND

The microfossil record of life in marine cherts has now been found to extend back at least 3500 million years (Schopf *et al.*, 1983). Did a similar green slime also form the scum of the Earth? Theoretically, not only Precambrian soil microbiota, but even the origin of life in soil is likely (Nussinov and Vehkov, 1978; Bohn *et al.*, 1979). Unlike oceans and ponds, soils are intermittently wet and often contain concentrated solutions of salts, phosphates, metals and organic matter. The four most common cations in soils (Ca^{2+}, Mg^{2+}, Na^+, K^+) are closer in relative abundance to animals (dominated by Ca^{2+}) than the sea (dominated by Na^+). Soils are also less homogeneous than bodies of standing water. A variety of rare molecules may have been protected from hydrolysis in the nooks and crannies of mineral grains partly pulverized by micrometeorite bombardment, or breached by chemical or physical weathering. Soils are also the principal sites of clay formation on Earth, and clays may have been templates or catalysts for the formation of complex organic molecules. Hydrolysis-resistant short RNA molecules rich in the organic bases guanine and cytosine are viewed as possible molecular beginnings of life (Eigen et al., 1981). In solutions rich in clays, metals, organic bases and a single enzyme, such RNA has been experimentally shown to 'reproduce.' Complementary 'reproductive' reactions could maintain RNA 'quasispecies,' and ultimately, energy transfer pathways and organisms. Life in these formative stages within the soil would be less threatened by competition from other 'quasispecies,' because of reduced mobility compared to open water. Furthermore, microbes and molecules in the soil would have greater protection from ultraviolet and other cosmic radiation than in open water. Although there are some theoretical reasons to prefer a 'primordial sludge' over a 'primordial soup' as the source of life on Earth, these competing hypotheses have barely been tested by hard evidence.

One suggestive line of evidence for early life on land is the modern affinities of some Precambrian marine microfossils, which are similar to modern soil microbes (Campbell, 1979). *Eoastrion*, a stellate microfossil with an opaque central body, is abundant in parts of the 2000 million year old, marine, Gunflint Chert of south-west Ontario, Canada. It is identical to modern manganese-fixing bacterium or microbial trace fossil *Metallogenium* (Barghoorn, 1977; Staley et al., 1982), which is found in soils and rock varnish of modern deserts (Dorn and Oberlander, 1981). *Kakabekia* is another microfossil from the shallow marine, Gunflint Chert, which is similar to modern soil microbes (Siegel, 1977).

Some Precambrian microfossils have been found in situations which may have been terrestrial. Unfortunately, the palaeoenvironments of all of these are presently ambiguous. Microfossil nostocalean cyanobacteria have been found in cherts filling cracks into basement rocks underlying the Pokegama Quartzite in Minnesota, USA (Cloud, 1976; Hofmann and Schopf, 1983). Although this could have been a crevice fauna within a soil, it is likely that these microbes lived in the intertidal or subtidal zone of the sea in which the overlying Pokegama Quartzite was deposited. From a similarly ambiguous near-shore setting are supposed fossil fungi from the 2300 to 2700 million year Witwatersrand Group, South Africa (Hallbauer and van Warmelo, 1974). Although the form of these 'fossils' are probably artefacts of preparation procedures, and so not fungi, their carbon is probably biogenic (Cloud, 1976; Barghoorn, 1981). It is hoped that future micromorphological studies of Precambrian paleosols will be more revealing.

Evidence for Precambrian soil microbes may also be gained from chemical analysis for organic carbon. The 2300 million year old Denison paleosol (Figure 1) is overlain by fluvial deposits and contains 0.014 to 0.25 weight percent of reduced organic carbon (Gay and Grandstaff, 1980). Although this is a very small amount, it is known from studies of Quaternary paleosols and their equivalent modern soils, that a great deal of organic matter is lost from soils soon after burial (Stevenson, 1969). Very few other Precambrian paleosols appear to have been analysed for organic carbon. This is likely to be a productive and quick source of information on the antiquity of life on land.

Another line of evidence for life in soil is the development of certain kinds of soil structure. Soils appear quite different from blocky cracked clay, of the kind seen in modern brick pits and desert badlands. They have a characteristic appearance and structure, which can be called tilth (Russell, 1973). This appearance is especially well demonstrated by the 2200 million year old Waterval Onder paleosol, from Transvaal, South Africa (Figures 4 and 5). Compared to the yellow and green clayey portion (C horizon) of the paleosol, the dark surface (A) horizon in the swales of the Vertisolic gilgai structure, appears well structured. The transitional horizon (AC) between yellow and dark parts of the profile is cracked into subangular, blocky clods (former soil peds) separated by contorted veins of dark material washed in from the surface (deformed illuviation ferrimangano-argillans in the terminology of Brewer (1976)). The surface (A) horizon has some relict bedding, and probably continued to accumulate sediment during development of the paleosol (as a cumulative horizon in the sense of Birkeland (1984)). This horizon also contains rounded clay granules (crumb peds), which are encrusted by a thin rind of iron and manganese (diffusion ferrimangan) better developed on the top than the bottom of the ped (Retallack *et al.*, 1984, Figure 5). This is very like modern rock varnish, thought to be fixed from atmospheric dust by fungi and bacteria (Dorn and Oberlander, 1981; Staley *et al.*, 1982). As for other lines of evidence for life on land discussed, this is unlikely to be the oldest because so few Precambrian paleosols have been examined carefully from this point of view.

A final circumstantial line of evidence for the advent of life in soils is related to the increased resistance to erosion associated with soil microbes (Campbell, 1979). Armouring of desert landscapes with rock

16 CHAPTER 1

Figure 4. The Waterval Onder clay paleosol, approximately 2200 million years old, developed on clayey sediments overlying the Hekpoort Basalt and overlain by the fluvial Dwaal Heuvel Formation, in a road cutting east of Waterval Onder, Transvaal, South Africa. Thickening of the dark surface (A horizon as in Figure 5) to the right hand side, and clastic dyke (at arrow) below the thinner dark layer (A horizon) to the left, is gilgai structure. Hammer on light coloured rock (of C2 horizon to right) gives scale.

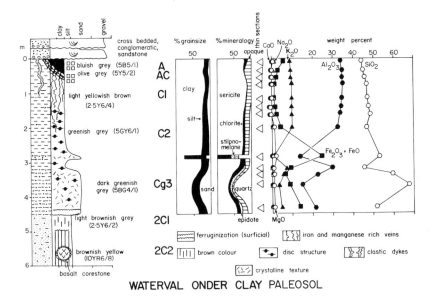

Figure 5. Columnar section, grainsize and mineralogical and chemical composition of the Waterval Onder clay paleosol, a Vertisol (adapted from Button 1979 and personal field observations and grainsize data). Scale is as for Figure 1 and unspecified symbols after Figure 6.

varnish (Dorn and Oberlander, 1981) and carbonate crusts (Krumbein and Geile, 1979) are well known. Microbial mats alone may appreciably stabilize desert soils (Booth, 1941). By contrast, in an hypothetical prebiotic landscape grains loosened from bedrock by physical or chemical weathering would tend to be washed away rather than weather in place to form clay (Schumm, 1968, 1977). Precambrian paleosols as old as 3100 million years are more clayey and deeply weathered than this hypothetical initial state. They represent land surfaces which were probably stable for many thousands, if not millions, of years.

In some cases (Gay and Grandstaff, 1980; Edelman et al., 1983), very ancient soils are more oxidized than would have been expected from other lines of evidence for atmospheric composition, such as the nature of marine rocks and geochemical models for formation of the Earth (Holland, 1984). Could it be that oxygen producing organisms appeared earlier in the soil than in the sea? Was the present situation of an oxidized atmosphere and more reducing CO_2- rich soil reversed during the earliest Precambrian? Could the redistribution of iron in Precambrian soils have been controlled more by microbial chelation than levels of atmospheric oxidants? These and other consequences of the evolution of soil biota remain to be explored.

LARGE PLANTS AND ANIMALS ON LAND

Although there is reason to believe that the greening of the land may have occurred well back in geological time, the appearance of multicellular land plants was a major advance in the complexity and biomass of terrestrial ecosystems, with consequences for soil formation. A few studies of paleosols relevant to this event have been published. They provide some constraints on when, which and how large plants and animals came to live on land.

Considering the very different lines of evidence used, it is not surprising that there has been some controversy concerning the antiquity of large plants and animals on land. Plants with a distinct vascular strand almost certainly lived on land and the oldest megafossils of these are mid-Silurian (Wenlockian) in age (Boucot and Gray, 1982). The extinct nematophytes also had tubular conducting tissues (Niklas and Smocovitis, 1983), although this differs from that of vascular plants. These fossils are poorly understood and peculiar in other ways, so that there is some doubt whether they were aquatic, semi-aquatic or land plants (Lang, 1937; Niklas, 1982). Nematophytes may be as old as Early Silurian (Llandoverian: Pratt et al., 1978; Strother and Traverse, 1979). All of these remains are restricted to sedimentary environments where aerobic decay of organic matter was limited by waterlogging. Fossil spores of plants are preserved in similar environments, but are more widespread, extending also into shallow marine environments. Although there has been some dispute over which of the many organic-walled palynomorphs extracted from early Palaeozoic rocks can be regarded as belonging to land plants (Banks, 1975; Gray and Boucot, 1977), this line of evidence for land plants extends back to the Late Ordovician (Caradocian: Gray et al., 1982). Late Silurian (Ludlovian and Pridolian) fossil millipedes are the oldest well accepted evidence of large land animals (Rolfe, 1980; Shear et al., 1984). Early land animals have an even more patchy fossil record than plants and spores, because they were rare, delicate, and destroyed in acidic as well as oxidizing environments.

Compared to this poor fossil record, paleosols are preserved in a wider range of environmental conditions and will probably provide more copious evidence of early multicellular organisms on land. To date, however, only two paleosols pertinent to this question have been reported. One of these is in the Late Ordovician (Ashgillian) Juniata Formation near Potters Mills, Pennsylvania, USA (Retallack, 1985). This is a red, clayey profile interbedded with other, similar paleosols and fluvial sandstones (Figure 6). The paleosol is riddled with burrows of large (3 to 16 mm diameter) invertebrates, which are subhorizontal in the surface of the paleosol (up to 4 cm from the top) and subvertical below that (to depths of 50 cm in the compacted paleosol). The burrows are thought to be original parts of the soil because calcareous nodules are preferentially distributed around them and because the increase in clay at the expense of mica toward the surface corresponds with increased density of burrows. The soil is not thought to have been marine influenced because palaeogeographic reconstructions place the sea 200 km to the west at the time it formed, and because there are no marine fossils (including palynomorphs) in these rocks. Nor are there any associated fossils or sedimentary structures which could be construed as evidence of lakes (Retallack, 1985). It thus seems likely that relatively large animals had colonized dry soils by Late Ordovician (Ashgillian) time. What kind of animals these could have been is uncertain, but it is difficult to imagine how they could have survived without plants for food and shelter.

Additional evidence for Late Ordovician (Caradocian or Ashgillian) land plants is provided by a paleosol from the Dunn Point Formation of the Arisaig area, Nova Scotia, Canada (Dewey in Boucot et al., 1974). This consists of 1.3 m of red, calcareous claystone on top of weathered corestones and columnar jointed flows of andesite. Near the surface of the paleosol are irregular pockets, about 1 m wide and

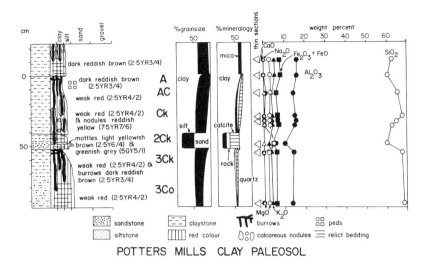

Figure 6. Columnar section, grainsize distribution and mineralogical and chemical composition of the Potters Mills clay paleosol, an Oxic Ustropept (Inceptisol), in the Late Ordovician (Ashgillian), Juniata Formation, near Potters Mills, Pennsylvania, USA (from Retallack 1985).

20 cm deep, filled with red shale redeposited from erosion of the paleosol. Small white reduction spots within the mounds between erosional pockets, are similar to drab mottles forming during diagenetic alteration of organic matter buried in red paleosols (Retallack, 1983b). Perhaps the mounds were stabilized against erosion by clumps of non-vascular land plants which lacked rooting structures substantial enough to leave obvious traces.

These two indications of relatively large, Late Ordovician, animals and plants on land may turn out to be among the oldest, because there are no large terrestrial fossils likely to be much older. Many more paleosols need to be examined, however, before the matter is closed.

Which large organisms were among the first to live on land is a problem for which the evidence of paleosols provides only the broadest of constraints. No obvious traces of rhizomes or roots were seen in either of the Late Ordovician paleosols. Considering the low preservation potential of organic matter in oxidized paleosols (Retallack, 1984a), there may have been a substantial biomass of multicellular filamentous or thallose plants and microbes growing in them. This may have included algae, lichens, bryophytes or completely extinct plants (Gray and Boucot, 1977). This earliest multicellular flora is unlikely to have included the enigmatic *Eohostimella* or vascular land plants, such as rhyniophytes. *Eohostimella* is preserved erect in the Early Silurian

(Llandoverian) Frenchville Formation of Maine, USA (Schopf et al., 1966). This could be regarded as a poorly developed coastal paleosol (such as an Aquent of Soil Survey Staff, 1975). It could also have been sediment at the bottom of a lake or tidal pool, although this seems less likely in view of the chemical affinities of *Eohostimella* with vascular land plants (Niklas, 1982). Wispy, tubular bioturbation has been found in a well drained, red, calcareous Late Silurian (Ludlovian) paleosol, near Palmerton, Pennsylvania, USA (Figure 7). This is the oldest likely trace of early vascular land plants, such as rhyniophytes, on dry land yet recognized. This paleosol is much better developed, and has better differentiated horizons, than the otherwise comparable Late Ordovician paleosol from Pennsylvania (Figure 6).

Although evidence is sparse at present, it appears that primitive vascular plants were preceded on land by a variety of poorly known non-vascular plants. The nature of the earliest large soil animals remains uncertain. The only generally accepted terrestrial fossils from rocks older than Devonian are millipedes, known definitely from rocks as old as Late Silurian (Ludlovian and Pridolian), but represented by very doubtful specimens from older Silurian (Llandoverian) rocks (Rolfe, 1980; Shear et al., 1984). Silurian scorpions have been widely considered as among the oldest land animals, but from careful assessment of their fossilized respiratory organs it appears that these were aquatic creatures until at least Late Carboniferous time (Kjellesvig-Waering, 1966). The extinct, largely aquatic eurypterids are known from rocks as old as Ordovician (Størmer, 1955). Respiratory organs of some especially well preserved Silurian eurypterids are indications that some of them may have been amphibious (Størmer, 1977). A case for the partly terrestrial habitat of some Late Carboniferous king crabs (xiphosurs) has been made but most of these creatures, extending back to Early Cambrian times, appear to have been marine (Størmer, 1955). Velvet worms (onychophorans) and earthworms (oligochaetes) are two common living terrestrial organisms with a virtually non-existent fossil record in non-marine rocks. Curiously they have been found in marine rocks as old as Middle Cambrian and Middle Ordovician, respectively (Thompson & Jones, 1980; Morris et al., 1982). Although the fossil record of water bears (tardigrades) does not extend back beyond the Cretaceous, they are an evolutionarily isolated group, are small, easily dispersed and remarkably resistant to desiccation (Rolfe, 1980). Among these possible early inhabitants of land, only a few are unlikely to have made the burrows observed in the Late Ordovician paleosol from Pennsylvania. Both earthworms and velvet worms are soft bodied, and probably would have dehydrated in these permanent dwelling burrows. Tardigrades are much too small. Accepting the

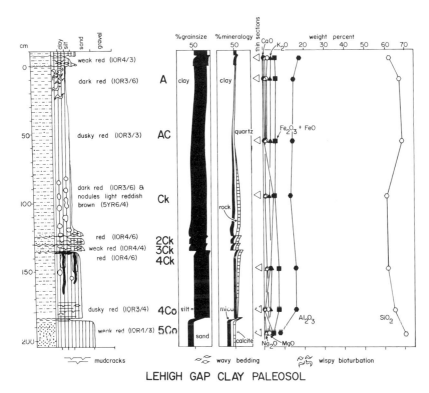

Figure 7. Columnar section, grainsize distribution and mineralogical and chemical composition of the Lehigh Gap clay paleosol, an Oxic Ustropept (Inceptisol), in the Late Silurian (Ludlovian), Bloomsburg Formation, near Palmerton, Pennsylvania, USA (from Retallack 1985). Unspecified symbols are as for Figure 6.

stratigraphic range of fossil millipedes at face value, this only narrows the field to amphibious chelicerates. Perhaps more detailed studies of burrows in early Palaeozoic paleosols will give a better idea of the earliest large animals on land.

Early Palaeozoic paleosols also provide some indications of how large plants and animals may have come on land. As presently understood, there is some support for the concept that the land was colonized first by microbes, then non-vascular plants, and then vascular plants. Terrestrial ecosystems appear to have been built up by degrees, becoming increasingly complex with the passage of time. This is compatible with the view that large land plants evolved from unicellular soil algae, independent of multicellular aquatic algae (Stebbins and Hill, 1980). It is also compatible with the alternative view that some aquatic plants may have invaded the land where it was prepared by microbial or non-vascular plants. Only the idea that bare, sterile earth was colonized by large plants now seems unlikely.

The advent of large animals on land may have been similar to the extent that they did not appear there until suitable food and shelter were available. If some of the earliest animals on land were as small as modern tardigrades, they would be very difficult to detect in paleosols. In the invasion of the land by large amphibious animals, pre-existing burrows may have been important microhabitats, in addition to abundant herbiage. In a similar way, rodent burrows in modern deserts may be small, semi-autonomous communities, including also algae, fungi and beetles, all protected within the moist, cool burrow from the harsh external environment (Halffter and Matthews, 1966; Martin and Bennett, 1977).

Increased biomass of terrestrial ecosystems in waterlogged terrain resulted in the appearance of a new kind of soil, peaty soils or Histosols (of Soil Survey Staff, 1975). The oldest known Histosol is the Early Devonian (Siegenian) Rhynie Chert of Scotland, which is a petrified peat with remains of vascular land plants in growth position (Kidston and Lang, 1921). Coaly layers of large, non-vascular plants in Silurian (Llandoverian to Ludlovian) rocks of the eastern United States (Willard, 1938; Pratt *et al.*, 1978; Strother and Traverse, 1979) do not meet the organic matter content and compacted thickness required of Histosols. They should be re-examined from this point of view.

Early Palaeozoic paleosols at major unconformities do not appear fundamentally different from Late Precambrian ones (Sharp, 1940; James *et al.*, 1961; Gariel *et al.*, 1968; Morey, 1972; Blaxland, 1973; Patel, 1977; Cummings and Scrivner, 1980). These also may have been Inceptisols and Oxisols (of Soil Survey Staff, 1975). Even less well developed Inceptisols and Entisols have been identified from Early Palaeozoic alluvial sequences (Retallack 1985). These are poorly developed soils and the existence of these orders is implied by better developed Precambrian paleosols. They are more easily recognized in Ordovician and younger rocks because of the presence of burrows and root and rhizome traces.

A cover of large land plants would have mitigated soil erosion compared to pre-existing microbial communities. The reduction spotted mounds in the surface of the Late Ordovician paleosol from Nova Scotia (Dewey in Boucot *et al.*, 1974) may represent direct evidence of this. There were probably also general effects of increased landscape stability, for example, in the increased abundance of high sinuosity, suspended load streams compared to low sinuosity, bedload streams (Schumm, 1968, 1977). The alluvial architectures of Ordovician to Permian red beds of the eastern United States show features, such as increased abundance of clay and fining upwards cycles, which may reflect just such changes in fluvial style with time (Cotter, 1978).

AFORESTATION OF THE LAND

Many of the changes in terrestrial ecosystems initiated with the early Palaeozoic appearance of large plants on land would have been further accentuated upon the evolution of trees. Evidence of these large plants and associated animals in paleosols can be used to assess the antiquity and nature of early woodland and forest ecosystems.

Presumably forests could have arisen during the Middle Devonian (Givetian), as large stumps of this age have been found in New York state, USA (Banks, 1980). More massive secondary wood may be almost this old (late Givetian), but it is especially common in Late Devonian (Frasnian) rocks of New York (Beck, 1964). Some of these massive woody trunks attained a diameter of 1.6 m (Beck, 1971). The actual time of origin of woodlands and forests may be hard to pinpoint, because measurement of stems of Palaeozoic fossil plants has revealed a steady increase in girth with time (Chaloner and Sheerin, 1979).

As would be expected from known Late Devonian fossil plants of New York and Pennsylvania, USA, there are abundant large root traces in rocks of that age there (Figure 8; other examples illustrated by Barrell, 1913; Walker and Harms, 1971). One of these horizons with large fossil root traces near the town of Hancock, New York, has recently been studied in detail (Figure 9: Retallack, 1985). This paleosol is very weakly developed and has conspicuous relict bedding throughout the profile. This is especially clear at the surface, which probably accumulated from floodwaters restrained by vegetation as the soil continued to form (a cumulative horizon of Birkeland, 1984). The large root traces and a fossil litter or partly decomposed leaves of *Archaeopteris halliana* within the cumulative horizon, provide evidence that this paleosol was forested. Other evidence for a forest cover is the differentiation of a laterally continuous, less clayey and iron poor surface (A) horizon over a more clayey, purple and iron rich, subsurface (B) horizon. This horizonation is so weakly expressed that the paleosol is identified as an Inceptisol (Soil Survey Staff, 1975). Translocation of clay and iron to subsurface horizons (lessivage) is partly mechanical, related to the penetration of roots and activity of soil fauna. Soil materials such as iron can also be translocated by the chemical action of rainfall leachates from leaves and of organic acids generated by decay of leaf litter (Fisher and Yam, 1984). Some of the most effective substances in this respect are phenols, manufactured by plants for defence against herbivores and pathogens. Such substances

24   CHAPTER 1

Figure 8. Large fossil root traces (disturbed clayed areas) in a paleosol in the Late Devonian (Frasnian) Oneonta Formation, in a roadcutting on Interstate Highway 88, 1 km west of the Unadilla exit, New York State, USA. Hammer is for scale.

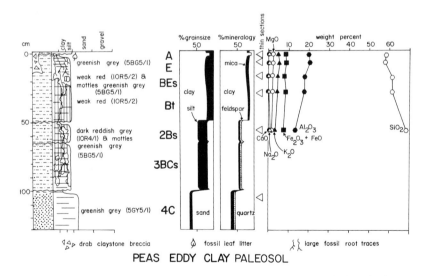

Figure 9. Columnar section, grainsize distribution and mineralogical and chemical composition of the Peas Eddy clay paleosol, a Tropaquept (Inceptisol), in the Late Devonian (Frasnian) Walton Formation, near Hancock, New York State, USA (from Retallack 1985). Unspecified symbols are as for Figure 6.

were probably widespread among early land plants, including *Archaeopteris*, just as they are in living plants (Swain and Cooper-Driver, 1981). This and other Late Devonian paleosols with large root traces in New York and Pennsylvania are non-calcareous and associated with sandstone of former stream channels. In contrast, red nodular calcareous paleosols away from palaeochannel deposits, lack large root traces and differentiation of A and B horizons. Thus, these early forests appear to have been restricted to streamside galleries, dissecting low shrubby or herbaceous vegetation of interfluves (as long ago proposed by Barrell, 1913). This distribution of early trees cannot be entirely attributed to their pteridophytic reproduction (Beck, 1971), because seed plants had not yet evolved (Gillespie *et al.*, 1981). Perhaps their demand for water and nutrients was not satisfied by soil microbiota or water in soils away from streams.

This early forested paleosol appears to be an Inceptisol, but well developed forested paleosols, such an Alfisols, Ultisols and Spodosols (of Soil Survey Staff, 1975) probably also date back to late Palaeozoic time. These each have laterally continuous A and B horizons, and include soils loosely called 'podzolic' in prior classifications (for example, Marbut, 1935; Kubiena, 1953). The USDA orders of forested soils are not always easy to distinguish in paleosols, because they are based on features such as pH and base status which must be inferred rather than measured (Retallack, 1981a, 1983a).

Alfisols are high base status, forested soils and are easily identified when calcareous and clayey. Sandy and non-calcareous Alfisols are known, and paleosols like this must be judged by the abundance of easily weathered minerals remaining in the profile, the nature of likely original clays and the abundance of elements likely to have been dominant cations (K_2O, Na_2O, CaO and MgO). Although not identified as such, it is likely that Alfisols are present among strongly developed, calcareous paleosols described from the Late Devonian (Fammenian), Aztec Siltstone, in Victoria Land, Antarctica (McPherson, 1979). Other possible fossil Alfisols, not yet studied in detail, are below the Early Permian (Leonardian) Minnekahta Limestone, South Dakota (Retallack, 1981a), within the Late Triassic (Carnian-Norian) Chinle Formation of Arizona (Retallack, 1981a), and in the Early Eocene (Wasatchian) Willwood Formation of Wyoming (Bown and Kraus, 1981a,b). The oldest fossil Alfisol studied adequately and identified as such, appears to be the Paleudalf (Interior clay paleosol) in the late Eocene (early Chadronian), Chadron Formation, Badlands National Park, South Dakota (Retallack, 1983a,b).

Ultisols are base poor, clayey, forested soils. The author suspects that very early Ultisols may be found in clayey sediments in the unconformity between the Middle Ordovician (Whiterockian) Jefferson City Dolomite and the Late Carboniferous (Desmoinesian or Westphalian) Fort Scott Limestone in the 'diaspore clay region' of south-eastern Missouri, USA (Keller et al., 1954). A variety of paleosols and sediments were recognized on this ancient karsted landscape. In places there are leached, grey clays (possible A horizons) grading down into reddish clays (possible B horizons). The clays contain abundant diaspore, and some boehmite, goethite and lepidocrocite. Alkalis and alkaline earths ($Na_2O + K_2O + MgO + CaO$) in this material total only 1.15 weight percent. This and the mineralogy of the clays, provides evidence of prolonged deep weathering of this material and very low base status. Early Triassic 'paleoplanosols' of southern France (Lucas, 1976) may also have been Ultisols. The oldest Ultisol actually identified and described as such is a partly redeposited profile, the Yellow Mounds paleosol, developed between the Late Eocene (Duchesnean) Slim Buttes Formation and Late Eocene to Early Oligocene (Chadronian) Chadron Formation, in Badlands National Park, South Dakota, USA (Retallack, 1983a,b).

Spodosols are acidic sandy soils, with a B horizon cemented by sesquioxides and organic matter (spodic horizon of Soil Survey Staff, 1975). The most promising place to search for the oldest Spodosols is among British Carboniferous ganisters: silicified sandstones mined for silica bricks and furnace linings. To date, however, there have been no detailed reports of iron or organic cemented horizons within ganisters which would qualify as spodic horizons in the USDA classification. Local ferruginized and carbonaceous zones have been noted in the Late Carboniferous (Westphalian) Sheffield Blue Ganister, near Sheffield, England (Searle, 1930; Percival, 1983 and this volume). This is tentatively accepted as the oldest likely Spodosol. The Avalon Series are comparable ganister-bearing paleosols from the Early to Middle Triassic (Scytho-Anisian) Newport Formation, near Sydney, Australia (Retallack, 1976, 1977). These have well differentiated sandy eluvial horizons (ganisters), but clear spodic horizons have not been recognized other than weakly enriched zones of amorphous organic matter within underlying siltstones. The author originally regarded prominent sub-surface horizons of iron carbonate (siderite) nodules in underlying siltstones as B horizons of these profiles. These are not spodic horizons, and there do not appear to be similar features in any reported modern Spodosols. Siderite implies a pH rather more neutral to alkaline than is typical for Spodosols. The nodules may be better regarded as products of waterlogging below or subsequent to burial of the original soil (deep gley or diagenetic pseudogley). Perhaps the British and Australian ganisters represent an extinct kind of soil formed under less acidifying vegetation than modern Spodosols. There are other problems with the Long Reef clay paleosol, within the Early Triassic (Scythian) Bald Hill Claystone in the same area (Retallack, 1976, 1977). In petrographic thin sections of the B horizon of the type profile an opaque, sesquioxide stain cements sand grains, as it should in a spodic horizon. However, the grains are for the most part weathered volcanic rock fragments, and the A horizon is much more clayey than usual for Spodosols. More detailed analytical work is warranted on both of these possible Early Triassic Spodosols. The oldest unquestionable Spodosols are between the Late Eocene (Bartonian) Sables de Beachamps and Formation d'Ezanville, near Paris, France (Pomerol, 1964, 1982). The profile at Ermenonville (Pomerol, 1982, Figure 3.24) appears to be an Orthod (of Soil Survey Staff, 1975).

From this preliminary outline of the appearance of the main kinds of well-drained forested soils on Earth, it appears that base-rich, alkaline, forested soils may have predated acidic, highly weathered ones. Whether this is a reflection of increasingly stable forest communities, or of palaeoclimatic and other peculiarities of the few areas best studied, remains to be established by further work. Fortunately, there is no shortage of well developed, red, Late Palaeozoic and Mesozoic paleosols (Hubert, 1960; Danilov, 1968; Kabata-Pendias and Ryka, 1968; Chalyshev, 1969; Power, 1969; Ortlam, 1971, 1980; McBride, 1974; Meyer, 1976; Sturt et al., 1979; Annovi et al., 1980; Franks, 1980; Retallack and Dilcher, 1981a,b; Ortlam and Zimmerle, 1982; Besly and Turner, 1983).

The advent of trees had substantial consequences for Histosols, the peaty soils of swamps and marshes (Soil Survey Staff, 1975). Moderately thick (10 cm) coal overlying claystone with large root traces has been found as old as Late Devonian (Fammenian) in Virginia, USA (Gillespie et al., 1981). Much thicker organic horizons (later altered to coal) of wetland soils were widespread during Carboniferous and later time. Drab paleosols of the Euramerican Carboniferous coal measures have been given a number of local names, such as underclay, seat earth, fire clay, tonstein and ganister (Huddle and Patterson, 1961; Roeschmann, 1971; Feofilova, 1973, 1977; Feofilova and Rekshinskaya, 1973). A number of Permian and Mesozoic fossil Histosols have also been reported (Retallack, 1976, 1977, 1980; Retallack and Dilcher, 1981a,b).

In addition to gains in biomass of terrestrial ecosystems there were also gains in complexity and diversity of life associated with woodlands and forests. There is a distinct Late Carboniferous to Early Permian peak in the diversity of fossil land plants (Tiffney, 1981). By this time also, many of the most important soil animals had evolved, including mites, collembollans, centipedes, insects, land snails, amphibians and reptiles (Rolfe, 1980; Milner, 1980). Increased diversity of soil fauna with time is also indicated by studies of trace fossils in paleosols. A low diversity assemblage of burrows and faecal pellets has been found in a thin paleosol A horizon in the Early Carboniferous (Viséan), Llanelly Formation, of South Wales, UK (Wright, 1983). Burrows of three kinds, including those of earthworms, cicada-like insects and vertebrates were recognized in paleosols formed under woodland and heath in the Early to Middle Triassic (Scytho-Anisian), Newport Formation, near Sydney, Australia (Retallack, 1976). Nine kinds of trace fossils have been found in woodland paleosols of the Early Eocene (Wasatchian), Willwood Formation, of Wyoming (Bown and Kraus, 1983). Beautifully preserved nests of termites and 15 other kinds of animal burrows were found in forested paleosols of the Oligocene, Jebel Quatrani Formation, in the Fayum Depression of Egypt (Bown, 1982). Five kinds of trace fossils, including indications of dung beetles and sweat bees were recognized in paleosols of woodland, savanna and grassland in the Oligocene White River and basal Arikaree Groups, in Badlands National Park, South Dakota (Retallack, 1983b, 1984b). In addition to the three kinds of vertebrate burrows found in Early Miocene desert paleosols of the Harrison Formation of north-eastern Nebraska (Martin and Bennett, 1977), the author can personally testify to the existence of an equivalent number of invertebrate burrows in these paleosols. These few accounts represent nearly all the descriptions of trace fossils explicitly recorded from paleosols. This promising source of information on the evolution of soil fauna has been sadly neglected, compared to trace fossils of other environments (Ratcliffe and Fagerstrom, 1980).

Forests would also have had a greater stabilizing effect on the landscape than pre-existing vegetation. Erosion of land protected only by primitive land plants may be the reason why Silurian marine red beds are so common, whereas there are few red beds with marine fossils in younger rocks (Ziegler and McKerrow, 1975). The author has seen very few sequences of Palaeozoic paleosols (setting aside the spectacular Devonian '*Psammosteus*' Limestones of the Welsh Borderland: Allen, 1974b) which have comparably well developed profiles to the red and variegated badlands of non-marine early Tertiary rocks in western North America (Figure 12: Bown and Kraus, 1981a,b; Retallack, 1983a,b). There are very few pre-Carboniferous coals, and even the Euramerican Carboniferous coal measures have more and thicker clastic partings than the spectacularly thick early Tertiary coals of North America, East Germany and south-eastern Australia (Schumm, 1968). Although the depth of weathering observed on Precambrian and early Palaeozoic unconformities may be impressive (Morey, 1972; Gay and Grandstaff, 1980), it does not rival some late Mesozoic and early Tertiary lateritic weathering profiles (McFarlane, 1976).

Even today the land is not completely forested, and more ancient kinds of soils persist in other parts of the landscape. Weakly developed paleosols (Entisols and Inceptisols of Soil Survey Staff, 1975), recognized mainly by their root traces, are found in many late Palaeozoic and Mesozoic alluvial sequences (Allen, 1947, 1959, 1976; Batten, 1973; Grigor'ev, 1973; Hlustik, 1974; Retallack, 1976, 1977, 1979, 1980, 1983c; Retallack & Dilcher, 1981a,b). Late Palaeozoic and Mesozoic calcretes, formed within a variety of aridland soils such as Inceptisols, Aridisols and Alfisols, are largely calcitic, rather than dolomitic like older calcretes (Blom, 1970; Allen, 1973, 1974a,b,c; Freytet, 1973; Steel, 1974; West, 1975, 1979; Folk and McBride, 1976; Lucas, 1976; Watts, 1976, 1978; Hubert, 1977a,b; Sochava, 1979; Adams, 1980; Adams and Cossey, 1981; Boucot et al., 1982; Freytet and Plaziat, 1982; Wright, 1982; Parnell, 1983; Blodgett, 1984). Changes in water table and in rainfall have been interpreted from the morphology and depth of karst topography and residual paleosols on late Palaeozoic and Mesozoic limestones (Keller et al., 1954; Dunham, 1969; Faugeres and Robert, 1969; Bernoulli and Wagner, 1971; Maiklem, 1971; Wardlaw and Reinson, 1971; Bosellini and Rossi, 1974; Walkden, 1974; Walls et al., 1975; Goldbery, 1979; Poty, 1980; Wright, 1981, 1982; Buchbinder et al., 1983). Bauxites are indicators of continuously wet tropical climates and laterites of wet seasonal climates, and both are widespread on late Palaeozoic and Mesozoic land surfaces (Goldich, 1938; Sloan, 1964; Sombroek, 1971; Valeton, 1972; Loughnan, 1975; Philobbos and Hassan, 1975; Singer, 1975; Blank, 1978; Goldbery, 1979; Abed, 1979; Nicholas and Bildgen, 1979). Silcretes may be associated with bauxitic or lateritic deep weathering or form in deserts, and these also are known in Palaeozoic, Mesozoic and Tertiary rocks (James et al., 1968; Dury and Habermann, 1978; Selleck, 1978; Wopfner, 1978; Rubin and Friedman, 1981; Isaac, 1983b). Strong seasonality is indicated by Vertisols, which also have been found in non-marine rocks of various ages through Phanerozoic time (McBride et al., 1968; Jungerius and Mücher, 1969; Allen, 1973, 1974b; Galloway, 1978; Goldbery, 1982a,b). Persistence of these ancient kinds of paleosols and soil features along with forested soils is evidence that the diversification of soils with time was more a process of addition of new kinds of soils, than of replacement of pre-existing kinds of soils.

GRASSES IN DRY CONTINENTAL INTERIORS

Before the advent of grasses, dry regions of the world were probably vegetated by a variety of woody plants. Many of these may have been even more bizarre than the Joshua tree (*Yucca brevifolia*), Boojum (*Idria columnaris*) and Saguaro cactus (*Cereus giganteus*) of the North American Southwest. Although large plants and animals play a conspicuous role in such desert ecosystems, there is much bare earth exposed. Soils forming in such environments (Aridisols of Soil Survey Staff, 1975) do not appear greatly different from calcareous paleosols as old as 1900 million years. The advent of grasses in subhumid to semi-arid plains of continental interiors, and in physiologically dry mountain regions, signalled the beginning of a new kind of ecosystem and new kinds of soils, the Mollisols. These are soils with nutrient rich, well structured surface horizons (mollic epipedon of Soil Survey Staff, 1975). Nutrients are not leached from these soils as is usual in humid climates, nor slowly redistributed into duricrusts as in very dry climates. Instead, a dense, low growth of herbaceous, annual grasses, abundant soil invertebrates and a variety of large mammals recycle nutrients to such an extent that these are some of the most productive ecosystems on land.

Evidence for grasslands among plant and animal fossils is poor and indirect. Grasses are not fossilized in the dry, well-drained environments of most grasslands. The direct fossil record of grasses is confined to swampy and near-stream habitats. Fossil grass pollen can be traced securely back to Palaeocene time (Muller, 1981) and megafossil grasses are known as old as Eocene (Daghlian, 1981). Grass fossils found within Late Miocene and Pliocene paleosols are largely biogenic silica encrustations (Thomassen, 1979) which may have coevolved with mammalian grazers long after the appearance of grasslands (Stebbins, 1981). There are similar problems with other adaptations thought to indicate former grasslands, such as the high-crowned cheek teeth (hypsodonty) and elegant slender limbs (cursoriality) of Tertiary mammals. Such skeletal evidence is an indication that grasslands were already established in South America by Eocene time (Webb, 1978), in North America by early Miocene (Webb, 1977), and in Africa, Eurasia and Australia by Late Miocene to Pliocene (Van Couvering, 1980; Sanson, 1982; Flannery, 1982). As a completely independent line of evidence for grasslands, paleosols may reveal not only the antiquity of grasslands, but how they developed.

There is abundant paleopedological evidence that woodland and forest was much more widespread than at present during late Mesozoic and early Tertiary time. This can be seen especially from the wide distribution of laterites and deeply weathered paleosols of this age

(Pettyjohn, 1966; Sombroek, 1971; Thompson *et al.*, 1972; Chernyakhovskii and Khosbayar, 1973; Abbott *et al.*, 1976; McFarlane, 1976; Nilsen, 1978; Nilsen and Kerr, 1978; McGowran, 1979; Singer and Nkedi-Kizza, 1980; Abbott, 1981; Thompson *et al.*, 1982; Isaac, 1983a). In Tertiary alluvial sediments, also, paleosols formed under woodlands and forests are abundant and widespread (Pomerol, 1964, 1982; Morand *et al.*, 1968; Reffay and Ricq-Debouard, 1970; Ritzkowski, 1973; Braunagel and Stanley, 1977; Buurman, 1980; Bown and Kraus, 1981a,b; Retallack, 1981b,c; Winkler, 1983).

A transition from forest to woodland, savannah and open grassland has been documented in volcaniclastic Late Eocene to Oligocene alluvium of the White River and lower Arikaree Groups in Badlands National Park, South Dakota, USA (Figure 10: Retallack, 1983a,b). Under an initially humid Late Eocene climate a forested Ultisol (of Soil Survey Staff, 1975) developed on an unconformity of smectitic Late Cretaceous (Maastrichtian) marine shale and thin residuals of Late Eocene (Duchesnean) alluvial sediments. In subhumid climates of the Eocene–Oligocene transition (early Chadronian), the first soil formed on the volcaniclastic alluvium was a strongly developed Paleudalf. This was followed by numerous superimposed Early

Figure 10. The Pinnacles area, Badlands National Park, South Dakota, USA. Above the weathered Cretaceous marine shale (lower foreground) are 87 superimposed paleosols of Late Eocene and Oligocene age within 143 m of alluvial deposits. These spectacular, colourful outcrops show well the field appearance of sequences of clayey paleosols.

Oligocene (Chadronian) Paleustalfs (for example, Figure 11). Judging from the laterally continuous, drab, surface (E) horizons and abundant large, drab-haloed root traces in these fossil Alfisols, they supported woodland vegetation. By early Late Oligocene (Orellan and Whitneyan) time there were a variety of soils, and the climate appears to have been distinctly drier and more seasonal. Paleustalf paleosols at this stratigraphic level have iron stained calcareous stringers (petrocalcic horizons) and are closely associated with sandstone palaeochannels. Red Fluvents (Entisols), showing relict bedding and only small root traces, are interbedded with stream deposits. Away from palaeochannels are Andic Ustochrept (Inceptisol) paleosols, in which there are limited development of A and B horizons, abundant small root traces, fine granular structure (peds), but only scattered large, drab-haloed root traces (Figure 11B). Early Late Oligocene vegetation is interpreted as a mosaic of early successional herbaceous vegetation in streamside swales, streamside gallery woodland and widespread savannah on the interfluves. In late Late Oligocene (Arikareean) rocks only one kind of paleosol, identified as Fluvaquentic Eutrochrept (Inceptisol), has large drab haloed root traces. These are interbedded with the sandstones of deeply incised streams. Other paleosols at this stratigraphic level all have simple (A-Ck) profiles and abundant small root traces. Some of these are light coloured Calciorthids (Aridisols: Figure 11), and others with darker surface horizons were identified as Ustollic Eutrandepts (Inceptisols). During late Late Oligocene time, trees appear to have been confined to stream margins within deep erosional gullies, whereas the open floodplain depressions and dry areas supported open grassland. Sediments at this stratigraphic level in the badlands are more calcareous and less clayey than older sediments. The climate was becoming increasingly dry, and was probably semi-arid by Late Oligocene time. Climatic drying can be attributed to the lengthening rain shadow cast by the Rocky Mountains to the west, but was also related to global climatic changes at this time (Wolfe, 1978; Kennett, 1982).

This sequence of paleosols can be used to assess a number of factors proposed for the maintenance of grasslands over other kinds of vegetation (Vogl, 1974; Walker et al., 1981). For example, grasses flourish over trees in dry or otherwise unfavourable, highly seasonal and unpredictable environments. This is because of their small stature, unspecialized pollination and dispersal mechanisms, and protection of most of their tissues in rhizomes and other underground structures. Climatic drying is a plausible explanation for the origin of grasslands in Badlands National Park. There is good evidence for increasingly dry climate with the appearance of grasslands, as far as can be judged from

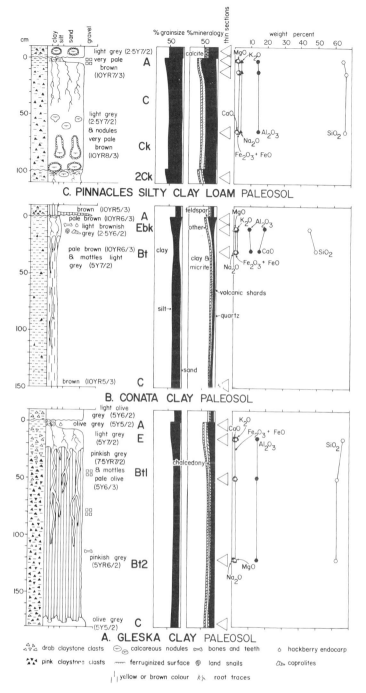

Figure 11. Columnar sections grainsize distributions and mineralogical and chemical compositions of C, the type Pinnacles silty loam paleosol, a Calciorthid (Aridisol) in the late Late Oligocene (Arikareean) Sharps Formation; B, the type Conata clay paleosol, an Andic Ustochrept (Inceptisol) in the early Late Oligocene (Orellan) Scenic Member of the Brule Formation; and A, the type Gleska clay paleosol, an Udic Paleustalf (Alfisol) in the early Oligocene (Chadronian) Chadron Formation, in Badlands National Park, South Dakota, USA (adapted from Retallack 1983b). Unspecified symbols are as for Figure 6.

the less severe weathering of volcaniclastic parent material and the shallower depth of calcareous horizons in paleosols with time.

Grasses also recover more quickly from fire than trees, and many modern grasslands are thought to be maintained by periodic fires. This is a difficult feature of ancient environments to interpret from grassland paleosols, because phytoliths and ash left by grass fires may appear no different in the soil than organic and mineral matter left by ordinary decay of grasses. The burning of trees, on the other hand, creates charcoal which can be distinguished from coalified wood, and is known to persist in paleosols under a wide variety of conditions (Retallack, 1984a). No charcoal was seen in savannah and woodland paleosols of the sequence in Badlands National Park. Although fire may have played a role in the maintenance of grassland once it appeared, it does not appear to have been important in the development of grassland from pre-existing savannah and woodland. The author suspects that fires were much less frequent and important to grasslands before the advent of humans.

Animal activity may also play a role in the maintenance of modern grasslands, but there is little evidence of this among fossil mammals of Badlands National Park. A new mammalian fauna appreciably better adapted to open country than Eocene or Paleocene Faunas of North America, appeared during Early Oligocene time, when the area was mostly wooded, judging from fossil soils. This evidence of woodland, independent of fossil mammals, confirms long-held suspicions that these creatures invaded from savannahs elsewhere, rather than evolving locally (Clark *et al.*, 1967; Emry, 1981). These faunas thrived in savannah environments which appeared during early Late Oligocene time. Surprisingly, they also persisted into late Late Oligocene time, when trees were very restricted in distribution, and most of the area was open grassland. At this time there were many mammals smaller than their presumed ancestors and more burrowing animals, so the observed environmental deterioration did have some effect on fauna. However, most endemic evolutionary lineages persisted, diversity remained much higher than usual for open grasslands and their tooth and limb structure was not appreciably better adapted to open country than before (MacDonald, 1963, 1970). This fauna was eventually replaced by a different assemblage of mammals, including a number of new evolutionary lineages with clear open country adaptations, during the very latest Oligocene, a time of extreme desertification in the Great Plains (Webb, 1977). Since open grasslands appeared several million years before mammalian faunas were especially well adapted to them, it is unlikely that grazing pressure was the most important factor in the origin of open grasslands in Badlands National Park.

A final factor encouraging development of grasslands is competition between plants. Although trees may have an advantage in shading out and chemically poisoning lower growing plants, grasslands can still persist by choking out small seedlings of trees. Plant competition is unlikely to have been a significant factor in the Badlands, because the oldest grassland soils there formed in very dry climates. These would have been marginal even for modern grasses, as in much of the intermontane rangelands of the present North American West. Well structured portions of these Late Oligocene paleosols are variable in thickness. This may be an indication that grasses and forbs of these early prairies were clumped, and that plant competition was less important than being able to tolerate dry conditions.

It remains to be seen whether grassland in other parts of the world appeared under similar circumstances. There is a comparable sequence of paleosols in Eocene and Oligocene volcaniclastic alluvial deposits of Argentina. These contain abundant fossil mammals, phytoliths and trace fossils of invertebrates, including dung beetles (Frenguelli, 1939; Andreis, 1972; Spalleti and Mazzoni, 1978). An early Oligocene (Deseadean) paleosol near Paso Flores, Neuquèn, has been briefly characterized (Frenguelli, 1939, p. 344). It may have been an Inceptisol or Alfisol (of Soil Survey Staff, 1975). Other paleosols of about the same age near Lago Colhue Huapi, Chubut, have been identified as Udolls and Aquolls (Spalletti and Mazzoni, 1978). These are the oldest Mollisols yet recorded. Since high crowned teeth and slender limbs are found in Eocene mammals of Argentina (Webb, 1978), it is likely that older Mollisols will be reported from there.

Perhaps, as appears to have been the case in South Dakota, the oldest grassland soils in Argentina will turn out to belong to different soil orders, such as Inceptisols. Similarly, the oldest forested paleosol now recognized cannot be attributed to one of the modern orders of forested soils. The early stages in the evolution of ecosystems may be very different, perhaps less integrated than their modern equivalents. The general question of coevolution within ecosystems is another issue for which paleosols may provide useful information.

Grasslands are only one of a number of plant communities poorly represented in the record of fossil plants, because they grow far from waterlogged sedimentary environments favourable for plant preservation (Retallack, 1984a). Clues to the geological history of chaparral, desert and alpine vegetation eventually may be gleaned from the study of paleosols.

PALEOSOLS AND HUMAN EVOLUTION

Even if our bones and the debris of civilization soon become diagnostic fossils of one of the briefest biostratigraphic zones in geological history, our effect on land surfaces of the world is already conspicuous and irreversible. Modern cities, dams, parking lots and highways are reshaping the landscape (Bidwell and Hole, 1964). Introduction of carbon that accumulated in the Earth over millions of years into the atmosphere as carbon dioxide, spreading of peculiar local phosphate deposits over fertile bottomlands and burial of various bizarre chemical wastes will have longer term effects on the nature of the Earth's surface; effects which already seem alarming. The writings of Plato reveal that some of these effects, such as deforestation and soil erosion, have long been a concern of intellectuals (Glacken, 1956). Presumably there was once a 'Golden Age,' though perhaps not as Plato imagined it, when human impact on the landscape was minimal and when many aspects of our character were imposed by an environment beyond our control. Evidence of that time and our long evolutionary career since, can be sought from paleosols, as well as from fossils of our early ancestors and the remains of their cultures.

A variety of extinct Miocene ape-like fossils have been considered ancestral to both humans and the living great apes. Ramapithecines especially have aroused interest, because some had short canines and thick-enameled molars arranged in divergent rows: intermediate in character between apes and humans. Recently discovered skulls of these creatures from Pakistan and China, on the other hand, indicate that Asiatic ramapithecines are more likely to have been ancestors of orangutans (Pilbeam, 1984). This interpretation is compatible with palaeoenvironments interpreted from paleosols associated with these fossils in India and Pakistan (Johnson, 1977; Johnson *et al.*, 1981). The author has made detailed studies of paleosols in the Late Miocene part (palaeomagnetically dated at 8.3 million years by Tauxe and Opdyke, 1982) of the Dhok Pathan Formation, Siwalik Group, near Khaur village in Northern Pakistan (Behrensmeyer and Tauxe, 1982; Retallack, 1985). Exceptionally well preserved ramapithecine fossils have been found here, including a face of *Sivapithecus* (Pilbeam *et al.*, 1980; Pilbeam, 1982). Former vegetation may be interpreted from six different kinds of paleosols (Retallack, 1985). Intimately associated with buff-coloured channel deposits of former streams draining Himalayan foothills, are paleosols with numerous large root traces and burrows, and obvious sedimentary relicts, such as bedding. These probably supported well drained, streamside, early successional woodlands. A second kind of paleosol associated with buff palaeochannels is brown to yellow and has a prominent zone of manganese staining below the surface (placic horizon of Soil Survey Staff, 1975). These may have supported early successional vegetation in poorly drained swales, associated with stream margins. A third kind of paleosol is thick, copiously bioturbated, clayey and red, with well differentiated subsurface (B) horizon and large root traces (Figure 12). It presumably formed under well drained streamside gallery forest. A fourth kind is yellow to brown and silty. The abundant fine root traces, rare large ones and weakly developed subsurface clayey (B) horizon are all features of savannah soils. These paleosols are interbedded with laterally impersistent red paleosols of the third kind. Thus, vegetation of dry floodplains may be more accurately spoken of as savannah groveland. The two remaining kinds of paleosols are associated with large, grey-green, sandstone palaeochannels, thought to represent the ancestral Indus River, which had a source, then as now, well within metamorphic rocks of the rising Himalayan Mountains. Both kinds of paleosols are grey in colour, contain large root traces, and presumably formed under different kinds of bottomland forest. Clayey, deeply cracked and partly brecciated, grey paleosols probably formed in seasonally dry swamps. Silty paleosols with abundant calcareous nodules presumably formed in levees and other slightly elevated parts of these alluvial bottomlands. This concept of the Late Miocene vegetative mosaic is well in accord with the nature of the abundant

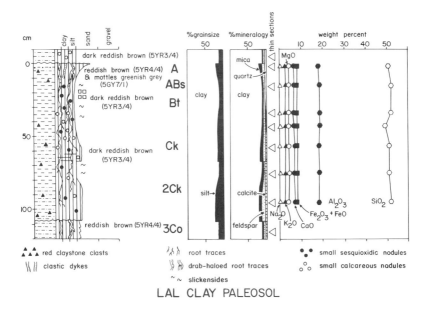

Figure 12. Columnar section, grainsize distribution and mineralogical and chemical composition of the Lal clay paleosol, an Oxic Haplustalf (Alfisol) in the Late Miocene, Dhok Pathan Formation, near Kaulial village, northern Pakistan (from Retallack 1985). Unspecified symbols are as for Figure 6.

vertebrate fauna known from these rocks, which includes crocodiles, elephants, antelopes, pigs and three-toed horses (Badgley and Behrensmeyer, 1980). Ramapithecines were very rare elements of these tropical faunas. They have only been found in palaeochannel deposits and paleosols of associated levees. These paleosols are of the first kind mentioned, and have been identified as Fluvents (of Soil Survey Staff, 1975). Presumably ramapithecines lived within these early successional woodlands and in adjacent gallery forests on the red (type 3 or Oxic Haplustalf) soils (Retallack, 1985). The life of these orangutan-like creatures would have had little impact on their environment, but these creatures were already a separate branch of the human evolutionary tree (Pilbeam, 1984). For evidence of the transition from ape to human, the more complete record of human evolution in East Africa is more promising.

During recent fieldwork in Kenya and Tanzania, the author was able to confirm reports of paleosols at a number of well known late Cenozoic vertebrate fossil localities. In Early Miocene (about 18 million years before present) localities for dryopithecine apes on Rusinga Island and near the villages of Songhor and Koru in southwestern Kenya (Andrews, 1981; Pickford and Andrews, 1981), abundant red paleosols are broadly similar to those thought to have formed under tropical forest in northern Pakistan. New kinds of soils formed in this region after uplift of the western margin of the East African Rift and extrusion of plateau phonolites by Middle Miocene time, some 14 million years ago. These are well illustrated by paleosols near the village of Fort Ternan, south-western Kenya (Shipman *et al.*, 1981). A paleosol presumably formed under woodland ('Braunlehm' of the classification of Kubiena, 1953) has been identified from here (Bishop and Whyte, 1962). This paleosol is poorly fossiliferous and situated stratigraphically below the most fossiliferous paleosols which are of a quite different type (Figure 13). These have simple (A-Ck) profiles, scattered large root traces and a dark granular surface layer like that of Mollisols. They probably formed under open grassy woodland or savannah, an interpretation amply supported by the nature of the fossil mammalian fauna from here (Evans *et al.*, 1981; Shipman *et al.*, 1981). The ramapithecine *Kenyapithecus* also has been found at Fort Ternan. Its ecology and habitat here, 14 million years ago, was probably very different from that of *Sivapithecus* in northern Pakistan, 8 million years ago. These preliminary findings underscore the need to obtain fossils of *Kenyapithecus* better than the few teeth and fragments of jaw now available. Perhaps it and savannah environments played some role in early human evolution? These ideas are neither new nor satisfactorily answered (Pilbeam, 1984). Evidence from paleosols may aid in guiding future efforts to answer them.

Figure 13. Middle Miocene (about 14 million years old) paleosols interbedded with tuffs in the main excavation at Fort Ternan National Monument, Kenya. The top of the upper paleosol is at the hammer head, and only the top of the lower paleosol is exposed at the base of the excavation. Both paleosols were probably Mollisols.

Paleosols now are recognized in association with fossil hominids from the East African Rift Valley from Ethiopia to Tanzania and ranging back as old as 4 million years (Hay, 1976; Leakey *et al.*, 1978; Aronson and Taieb, 1981; Burgraff *et al.*, 1981; White *et al.*, 1981; Cohen, 1982). In Plio-Pleistocene sediments of the East Turkana region of Kenya, and Olduvai Gorge, in Tanzania, studies of paleosols may be useful in interpreting the habitat preferences of sympatric species of hominids. Numerous root traces and paleosols have been recognized in the main sequence (Beds I to IV) of Olduvai Gorge, some 2.1 to 0.5 million years old (Figure 14). These sediments accumulated on alluvial uplands and the margins of alkaline lakes whose variation in extent and location with time has been beautifully documented by Hay (1976). During his own examination of these rocks, the author was especially struck by their palaeoenvironmental similarities with modern Amboseli National Park, Kenya (Williams, 1972; Sombroek *et al.*, 1982). Hay's lake margin sequences include zeolitic paleosols, which were presumably alkaline Entisols supporting scrubby, salt tolerant vegetation. Also in lake margin facies are dark granular paleosols with calcareous subsurface layers, which may have been Mollisols, formed under grassland and savannah. Hay's alluvial facies

Figure 14. The Plio-Pleistocene (2.1 to 0.5 million years old) sequence in Castle Rock (foreground) and Olduvai Gorge (background), northern Tanzania. Red paleosols form the thick dark layer (Bed III) in the upper part of Castle Rock. Underlying drab sediments include paleosols and lake deposits of Beds I and II.

is a prominent red unit (Bed III) composed of a number of well drained, calcareous paleosols, perhaps Inceptisols formed under open woodland. Calcrete-bearing paleosols have been found in the overlying Masek, Ndutu and Naisiusu Beds (Hay and Reeder, 1978), which appear to have formed in a more arid and less stable environment. Stone tools are associated with hominids in Olduvai Gorge and structures which may have been associated with permanent encampments have been found at low stratigraphic levels (Bed I; Hay, 1976). Chemical and petrographic studies of occupation floors could prove revealing. One of the later appearing hominid species (*Homo erectus*) found in Olduvai Gorge is thought to have discovered the use of fire by 0.7 million years ago in other parts of the world (Pilbeam, 1984). The use of fire, permanent camps and organized hunting are ways in which even primitive peoples may alter their environment. In some ways their effects may mimic climatic drying. This has been a persistent problem in untangling human and climatic effects in the origin of deserts, such as the Sahara (Huzayyin, 1956; Williams, 1979; Alimen, 1982). Perhaps the paleosols of Olduvai Gorge or other thick sequences of Pleistocene paleosols (Kukla, 1977; Firman, 1979) will some day yield evidence on the antiquity and extent of early human dominion over the landscape.

CONCLUSIONS

In this review of the diversification of soils through geological time, I have stressed not only what is known of the record of fossil soils, but also the record of fossils in them. Many non-biotic factors, such as climate, parent material, geographic setting and time, also play a role in the formation of soils, but it is organisms which give soils much of their distinctive character. The diversification of life on Earth is reflected in a diversification of soils and soil features with time (Figure 15). In some cases the link between organisms and environment was indirect and is difficult to assess, as in the likely oxygenation of Precambrian atmospheres by photosynthetic microbes and human desertification of our present environment. Some major events in the history of life on land such as the mid-Cretaceous dispersal and rise to dominance of angiosperms and the terminal Cretaceous extinction of dinosaurs do not yet appear to have had especially significant ramifications for the geological history of soils. Nevertheless, it is apparent that soils and ecosystems have developed in tandem and much can be learned about each by studying them together.

Paleosols are especially valuable in providing evidence of life in

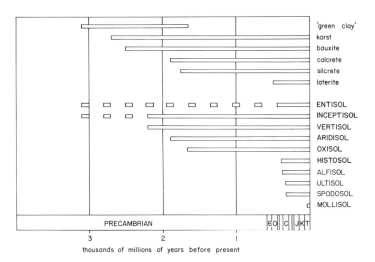

Figure 15. Geological range of soil features and USDA soil orders (of Soil Survey Staff 1975). Entisols and Inceptisols are assumed as precursors of other Precambrian paleosols, even though difficult to recognize in Precambrian rocks.

environments where fossils may not be preserved. Even though bones are not preserved in acidic paleosols, nor plants in oxidized paleosols (Retallack, 1984a), ecosystems of considerable biomass, such as tropical rainforest, commonly grow in such non-calcareous, red soils. Information from paleosols may provide clues to taphonomic biases in preserved fossil assemblages. Detailed evidence of behaviour and kinds of organisms may be obtained from the study of trace fossils in paleosols. In a sense paleosols themselves are trace fossils of ecosystems. Like other trace fossils they can be assumed to be in place and related to associated sedimentary environments. This is of value for assessing the original distribution of scattered remains of plants and animals in non-marine sediments. Fossil plants associated with paleosols may be more informative of past vegetation than mixed remains found in deposits of lakes and lagoons, although widely dispersed fossils may provide indications of vegetation not preserved in well drained paleosols (Retallack and Dilcher, 1981a). Similarly, bones associated with distinct kinds of paleosols are likely to reflect original vertebrate communities more accurately than transported remains found in fluvial palaeochannels (Bown and Kraus, 1981b; Behrensmeyer, 1982; Winkler, 1983).

Paleosols also provide evidence of former ecosystems independent of fossils. Interpretations based on paleosols can be used to check or enlarge reconstructions of the past based on fossil animals and plants. Many reconstructions using fossils alone have been based on the assumption that adaptive features of organisms were optimally suited to their environment (Gould and Lewontin, 1979). For example, open grassland environments have been inferred from assemblages of mammals with high crowned teeth and slender limbs. Using paleosols as independent evidence of environments, it is possible to recognize those evolutionarily interesting times when vegetation and animals were not well co-adapted (Retallack, 1983a,b). Degree of integration and coevolution within terrestrial ecosystems can thus be probed. The tentative early stages in the evolution of major new kinds of ecosystems are of special interest in allowing us to understand how presently complex, natural systems began.

Although research on fossil soils older than Quaternary is now accelerating, there is much to be done. The present inventory of described paleosols is especially inadequate to assess the earliest representative of various soil types and features. The author's tentative compilation (Figure 15) is based as far as possible on available data, but is also compatible with current concepts of soil forming processes and of changes in life and environments on land with time. The author is very conscious of the risk that such concepts may bias the way in which paleosols are interpreted. As more paleosols become known, they will be better assessed on their own terms. Perhaps then a history of soils on Earth may be constructed 'from the bottom up,' rather than 'from the top down.' At that time, studies of paleosols can be expected not only to illuminate specific events in the long history of life on land, but to contribute to general theories of soil formation and organic evolution.

ACKNOWLEDGEMENTS

Numerous colleagues have helped hone my understanding of the geological record of life and soils. I especially thank H. D. Holland (Harvard University, Boston, USA), A. J. Boucot (Oregon State University, Corvallis, USA), V. P. Wright (University of Bristol, England), W. G. Chaloner (Bedford College, England), P. S. Martin (University of Arizona, Tucson, USA), A. K. Behrensmeyer (Smithsonian Institution, Washington, D. C., USA) and P. Shipman (Johns Hopkins University, Baltimore, USA) for useful discussion of some of the issues treated in this article. I also owe a debt of gratitude to my paleopedology classes at the University of Oregon, for encouraging my evolving ideas on this subject. Fieldwork in Africa was supported by a Grant-in-Aid of Research from the Wenner-Gren Foundation. Other aspects of this research were funded by NSF Grant EAR8206183.

REFERENCES

Abbott, P.L. 1981. Cenozoic paleosols, San Diego area, California. *Catena* **8**, 223–37.
Abbott, P.L., Minch, J.A.K. and Peterson, G.L. 1976. Pre-Eocene paleosol south of Tijuana, Baja California, Mexico. *J. sedim. Petrol.* **46**, 355–61.
Abed, A.M. 1979. Lower Jurassic lateritic detritus from central Arabia. *Sedim. Geol.* **24**, 149–56.
Adams, A.E. 1980. Calcrete profiles in the Eyam Limestone (Carboniferous) of Derbyshire: petrology and regional significance. *Sedimentology* **27**, 651–60.
Adams, A.E. and Cossey, P.J. 1981. Calcrete development at the junction between the Martin Limestone and Red Hill Oolite (Lower Carboniferous), South Cumbria. *Proc. Yorks. geol. Soc.* **43**, 411–31.
Alimen, M.H. 1982. Le Sahara: grande zone désertique Nord-Africaine, in *The Geological Story of the World's Deserts*, T.L. Smiley, (ed.). *Striae*, **17**, 35–51.
Allen, J.R.L. 1973. Compressional structures (patterned ground) in Devonian pedogenic limestones. *Nature Phys. Sci.* **243**, 84–6.
Allen, J.R.L. 1974a. Sedimentology of the Old Red Sandstone (Siluro-Devonian) in the Clee Hill area, Shropshire, England. *Sedim. Geol.* **12**, 73–167.
Allen, J.R.L. 1974b. Studies in fluviatile sedimentation: implications of pedogenic carbonate units, Lower Old Red Sandstone, Anglo-Welsh outcrop. *Geol. J.* **9**, 181–208.
Allen, J.R.L. 1974c. Geomorphology of Siluro-Devonian alluvial plains. *Nature* **249**, 644–5.

Allen, P. 1947. Notes on Wealden fossil soil beds. *Proc. Geol. Ass.* **57**, 303–14.
Allen, P. 1959. The Wealden environment: Anglo-Paris Basin. *Phil. Trans. R. Soc.* **242B**,283–346.
Allen, P. 1976. Wealden of the Weald: a new model. *Proc. Geol. Ass.* **86**, 389–436.
Andreis, R.R. 1972. Paleosuelos de la Formación Musters (Eoceno medio), Laguna del Mate, Prov. de Chubut, Rep. Argentina. *Revta Asoc. Mineral. Petrol. Sediment. Argent.* **3**, 91–7.
Andrews, P. 1981. A short history of Miocene field palaeontology in western Kenya. *J. Human Evol.* **10**, 3–9.
Annovi, A., Fontana, D., Gelmini, R., Gorgoni, C. and Sighinolfe, G. 1980. Geochemistry of carbonate rocks and ferruginous horizons in the Verrucano in southern Tuscay—paleoenvironmental and paleogeographic implications. *Palaeogeogr. Palaeoclimatol. Palaeoecol.* **30**, 1–16.
Aronson, J.L. and Taieb, M. 1981. Geology and paleogeography of the Hadar hominid site, Ethiopia. In *Hominid Sites: Their Geologic Settings*, G. Rapp and C.F. Vondra, (eds.). Westview Press, Boulder, 165–95.
Badgley, C. and Behrensmeyer, A.K. 1980. Paleoecology of Middle Siwalik sediments and faunas, northern Pakistan. *Palaeogeogr. Palaeoclimatol. Palaeoecol.* **30**, 133–55.
Banks, H.P. 1975. The oldest vascular land plants: a note of caution. *Rev. Palaeobot. Palynol.* **20**, 13–25.
Banks, H.P. 1980. Floral assemblages in the Siluro-Devonian in *Biostratigraphy of Fossil Plants*, D.L. Dilcher and T.N. Taylor, (eds). Dowden, Hutchinson & Ross Inc., Stroudsburg, Pennsylvania. 1–24.
Barghoorn, E. 1977. *Eoastrion* and the *Metallogenium* problem, in *Chemical Evolution of the Early Precambrian*, C. Ponnanperuma, (ed). Academic Press, New York. 185–7.
Barghoorn, E.S. 1981. Aspects of Precambrian paleobiology: the early Precambrian, in *Paleobotany, Paleoecology and Evolution.* Vol. 1, K.J. Niklas (ed.). Praeger Publishers, New York. 1–16.
Barrell, J. 1913. The Upper Devonian delta of the Appalachian Geosyncline. *Am. J. Sci.* **36**, 429–72.
Basu, A. 1981. Weathering before the advent of land plants: evidence from detrital K feldspars in Cambrian-Ordovician arenites- reply. *Geology* **9**, 506–7.
Batten, D.J. 1973. Palynology of early Cretaceous soil beds and associated strata. *Palaeontology* **16**, 339–424.
Beck, C.B. 1964. Predominance of *Archaeopteris* in Upper Devonian flora of western Catskills and adjacent Pennsylvania. *Bot. Gaz.* **125**, 126–8.
Beck, C.B. 1971. Problems of generic delimitation in paleobotany. *Proc. 1st N. Am. Paleont. Conv.* **1**, 173–93.
Behrensmeyer, A.K. 1982. Time resolution in fluvial vertebrate assemblages. *Paleobiology* **8**, 211–27.
Behrensmeyer, A.K. and Tauxe, L. 1982. Isochronous fluvial systems in Miocene deposits of northern Pakistan. *Sedimentology* **29**, 331–52.
Bernoulli, D. and Wagner, L.W. 1971. Subaerial diagenesis and fossil caliche deposits in the Calcare Massico Formation (Lower Jurassic, Central Appenines, Italy). *N. Jb. Geol. Paläont. Abh.* **138**, 135–49.
Bertrand-Sarfati, J. and Moussine-Pouchkine, A. 1983. Pedogenetic and diagenetic fabrics in the upper Proterozoic Sarnyéré Formation (Gourma, Mali). *Precambrian Res.* **20**, 225–42.
Besly, B.M. and Turner, P. 1983. Origin of red beds in a moist tropical climate (Etruria Formation, Upper Carboniferous, U.K.), in *Residual Deposits: Surface Related Weathering Processes and Materials,* R.C.L. Wilson, (ed.). Blackwell Scientific Publications Oxford for the Geological Society of London. 131–47.

Bidwell, O.W. and Hole F.D. 1964. Man as a factor of soil formation. *Soil Sci.* **99**, 65–72.
Birkeland, P.W. 1984. *Soils and Geomorphology*. Oxford University Press, New York.
Bishop, W.W. and Whyte, F. 1962. Tertiary mammalian faunas and sediments in Karamoja and Kavirondo, East Africa. *Nature* **196**, 1283–7.
Blades, E.L. and Bickford, M.E. 1976. Volcanic ash-flows and volcaniclastic sedimentary rocks at Johnson Shut-ins, Reynolds Co., Missouri. *Abstr. Prog. Geol. Soc. Am.* **8**, 6.
Blank, H.R. 1978. Fossil laterite on bedrock in Brooklyn, New York. *Geology* **6**, 21–4.
Blaxland, A.B. 1973. Major element and Rb-Sr geochemistry of an ancient weathering zone in the Butler Hill Granite, Missouri. *EOS* **54**, 505.
Blodgett, R.H. 1984. Nonmarine depositional environments and paleosol development in the upper Triassic Dolores Formation, southwestern Colorado, in *Field Trip Guidebook*, D.C. Brew, (ed.). *Four Corners Geological Society*, Durango, Colorado. 46–92.
Blom, G.J. 1970. Buried palygorskite soils in the lower Triassic of the Moscow Syneclise. *Dokl. Acad. Sci. USSR Earth Sci.* **194**, 52–4.
Bohn, H.L., McNeal, B.L. and O'Connor, G.A. 1979. *Soil Chemistry*. Wiley-Interscience Publishers, New York.
Booth, W.E. 1941. Algae as pioneers in plant succession and their importance in erosion control. *Ecology* **22**, 38–46.
Bosellini, A. and Rossi, D.L. 1974. Triassic carbonate buildups of the Dolomites, northern Italy, in *Reefs in Time and Space*, L.F. Laporte, (ed.). *Spec. Publ. Soc. Econ. Paleont. Min.* **18**, Tulsa, Oklahoma. 209–33.
Boucot, A.J., Dewey, J.F., Dineley, D.L., Fletcher, R., Fyson, W.K., Griffin, J.G., Hickox, C.F., McKerrow, W.S. and Zeigler, A.M. 1974. The geology of the Arisaig area, Antigonish County, Nova Scotia. *Spec. Publ. Geol. Soc. Am.* **139**, 191 p.
Boucot, A.J. and Gray, J. 1982. Geologic correlates of early land plant evolution. *Proc. 3rd N. Am. Paleont. Conv.* **1**, 61–6.
Boucot., A.J., Gray, J., Fang, R.-S., Yang, X-C., Li, Z.-P. and Zang, N. 1982. Devonian calcrete from China: its significance as the first Devonian calcrete from Asia. *Can. J. Earth Sci.* **19**, 1532–4.
Bown, T.M. 1982. Ichnofossils and rhizoliths of the nearshore fluvial, Jebel Quatrani Formation (Oligocene), Fayum Province, Egypt. *Palaeogeogr. Palaeoclimatol. Palaeoecol.* **40**, 255–309.
Bown, T.M. and Kraus, M.J. 1981a. Lower Eocene alluvial palaeosols (Willwood Formation, northwest Wyoming, USA) and their significance for paleoecology, paleoclimatology and basin analysis. *Palaeogeogr. Palaeoclimatol. Palaeoecol.* **43**, 1–30.
Bown, T.M. and Kraus, M.J. 1983. Ichnofossils of the alluvial Willwood Formation Formation, northwest Wyoming, USA): implications for taphonomy, biostratigraphy and assemblage analysis. *Palaeogeogr. Palaeoclimatol. Palaeoecol.* **34**, 31–56
Bown, T.J. and Kraus, M.J. 1983. Ichnofossils of the alluvial Willwood Formation (lower Eocene), Bighorn Basin, northwest Wyoming, USA. *Palaeogeogr. Palaeoclimatol. Palaeoecol.* **43**, 95–128.
Braunagel, L.H. and Stanley, K.O. 1977. Origin of variegated red beds in the Cathedral Bluffs tongue of the Wasatch Formation (Eocene). *J. sedim. Petrol.* **47**, 1201–19.
Brewer, R. 1976. *Fabric and Mineral Analysis of Soils*. Krieger Publishers, New York.
Buchbinder, B., Margaritz, M. and Buchbinder, L.G. 1983. Turonian to Neogene palaeokarst in Israel. *Palaeogeogr. Palaeoclimatol. Palaeoecol.* **43**, 329–50.
Burggraf, D.R., White, H.J., Frank, H.J. and Vondra, C.F. 1981. Hominid habitats in the Rift Valley, part 2, in *Hominid Sites: Their Geologic Settings*, G. Rapp & Vondra, C.F., (eds). Westview Press, Boulder, Colorado. 115–47.
Button, A. 1979. Early Proterozoic weathering profile on the 2200 m. yr. old Hekpoort

Basalt, Pretoria Group, South Africa: preliminary results. *Inform. Circ. Econ. Geol. Res. Unit Univ. Witwatersrand* **133**, 20 pp.

Button, A. and Tyler, N. 1979. Precambrian palaeoweathering and erosion surfaces in southern Africa: a review of their character and economic significance. *Inform. Circ. Econ. Geol. Res. Unit Univ. Witwatersrand* **135**, 37 pp.

Buurman, P. 1980. Paleosols in the Reading Beds (Palaeocene) of Alum Bay, Isle of Wight, UK. *Sedimentology* **27**, 593–606.

Campbell, F.H.A. and Cecile, M.P. 1981. Evolution of the early Proterozoic Kilohigok Basin, Bathurst Inlet, Victoria Island, North West Teritories, Canada, in *Proterozoic basins of Canada*, F.H.A. Campbell, (ed.). *Pap. Geol. Surv. Canada* **81–100**, 103–31.

Campbell, S.E. 1979. Soil stabilization by a procaryotic desert crust: implications for a Precambrian land biota. *Origins of Life* **9**, 335–48.

Catt, J.A. and Weir, A.H. 1981. Soils, in *The Evoiving Earth*, L.R.M. Cocks, (ed.). British Museum (Natural History) and Cambridge University Press, London and Cambridge. 63–85.

Chaloner, W.G. and Sheerin, A. 1979. Devonian macrofloras, in *The Devonian System*, M.R. House, C.T. Scrutton and M.G. Bassett, (eds). *Spec. Pap. Palaeontology* **23**, 145–61.

Chalyshev, V.I. 1969. A discovery of fossil soils in the Permo-Triassic. *Dokl. Acad. Sci. USSR Earth Sci.* **182**, 53–6.

Chayka, V.M. 1970. Supergene leaching of detrital zircon during weathering of basaltoid rocks. *Dokl. Acad. Sci. USSR Earth Sci.* **192**, 173–6.

Chayka, V.M. and Zaviyaka, A.I. 1968. Precambrian basalt sheets on the western flank of the Anabar Shield. *Dokl. Akad. Sci. USSR Earth Sci.* **183**, 60–2.

Chernyakhovskii, A.G. and Khosbayar, P. 1973. The middle Oligocene weathering crust in west Mongolia. *Lithol. Mineral Resour.* **8(5)**, 618–24.

Chiarenzelli, J.R. 1983. *Mid-Proterozoic Chemical Weathering, Regolith and Silcrete in the Thelon Basin, North West Territories.* Unpub. MSc thesis, Carleton University, Ottawa.

Chiarenzelli, J.R., Donaldson, J.A. and Best, M. 1983a. Sedimentology and stratigraphy of the Thelon Formation and the sub-Thelon regolith. *Pap. Geol. Surv. Can.* **83–1A**, 443–5.

Chiarenzelli, J.R., Donaldson, J.A. and Miller, A.R. 1983b. Chemical weathering and diagenesis of the Proterozoic sub-Thelon regolith and Thelon Formation, North West Territories. *Abstr. Progm. Ann. Meet. Geol. Ass. Mineral. Ass. Can.* **8**, A11.

Chown, E.H. and Caty, J.L. 1983. Diagenesis of the Aphebian Mistassini Regolith, Quebec Canada. *Precambrian Res.* **19**, 285–99.

Clark, B.C. and van Hart, D.C. 1981. The salts of Mars. *Icarus* **45**, 370–8.

Clark, J., Beerbower, J.R. and Kietzke, K.K. 1967. Oligocene sedimentation in the Badlands of South Dakota. *Fieldiana Geol. Mem.* **5**, 158 p.

Clemmey, H. and Badham, N. 1982. Oxygen in the Precambrian atmosphere: an evaluation of the geological evidence. *Geology* **10**, 14–16.

Cloud, P.E. 1976. The beginnings of biosphere evolution and their biochemical consequences. *Paleobiology* **2**, 351–87.

Cohen, A.S. 1982. Paleoenvironments of root casts from the Koobi Fora Formation, Kenya. *J. sedim. Petrol.* **52**, 401–14.

Cotter, E. 1978. The evolution of fluvial style, with reference to the central Appalachian Paleozoic, in *Fluvial Sedimentology*, A.O. Miall, (ed.). *Mem. Can. Soc. Petrol. Geol.* **5**, 361–83.

Coultas, C.L. 1980. Soils of marshes in the Appalachicola, Florida, Estuary. *Proc. Soil Sci. Soc. Am.* **44**, 348–53.

Cummings, M.L. and Scrivner, J.V. 1980. The saprolite at the Precambrian-Cambrian

Contact, Irvine Park, Chippewa Falls, Wisconsin. *Trans. Wisc. Acad. Sci. Arts. Lett.* **68**, 22−9.

Daghlian, C.P. 1981. A review of the fossil record of monocotyledons. *Bot. Rev.* **47**, 517−55.

Danilov, I.S. 1968. The nature and geochemical conditions of formation of the variegated red Triassic and Permian rocks of the western Donetz Basin. *Lithol. Mineral Resour.* **3**, 612−18.

Dimroth, E. and Kimberley, M.M. 1976. Precambrian atmospheric oxygen: evidence of the sedimentary distributions of carbon, sulfur, uranium and iron. *Can. J. Earth Sci.* **13**, 1161−85.

Donaldson, J.A. and Ricketts, B. 1978. Silcretes and calcretes from Proterozoic successions in Canada. *Abstr. 10th Int. Congr. Sedimentology* **1**, 183.

Doner, H.E. and Lynn, W.C. 1977. Carbonate, halide, sulfate and sulfide minerals, in *Minerals in Soil Environments*, J.B. Dixon and S.B. Weed, (eds). Soil Science Society of America, Madison, Wisconsin. 75−98.

Dorn, R.I. and Oberlander, T.M. 1981. Microbial origin of desert varnish. *Science* **213**, 1245−7.

Duchafour, P. 1982. *Pedology* (translated by T.R. Paton). Allen & Unwin Publishers, Boston.

Dunham, R.J. 1969. Vadose pisolites in the Capitan Reef (Permian), New Mexico and Texas, in *Depositional Environments in Carbonate Rocks*. G.M. Friedman, (ed.). *Spec. Publ. Soc. Econ. Paleont. Mineral.* **14**, 182−91, Tulsa, Oklahoma.

Dunoyer de Segonzac, G. 1970. The transformation of clay minerals during diagenesis and low grade metamorphism: a review. *Sedimentology* **15**, 281−346.

Dury, G.H. and Habermann, G.M. 1978. Australian silcretes and northern hemisphere correlatives, in *Silcrete in Australia*, T. Langford-Smith, (ed.). University of New England, Armidale, N.S.W. 223−59.

Easton, R.M. 1981. Stratigraphy of the Akaitcho Group and the development of an early Proterozoic continental margin, Wopmay Orogen, North West Territories, in *Proterozoic basins of Canada*. F.H.A. Campbell, (ed.). *Pap. Geol. Surv. Can.* **81−10**, 79−95.

Edelman, M.J. Grandstaff, D.E. and Kimberley, M.M. 1983. Description and implications of two early Precambrian paleoweathering profiles from South Africa. *Abstr. Prog. geol. Soc. Am.* **15**, 565.

Eigen, M., Gardiner, W., Schuster, P. and Winkler-Oswatitsch, R. 1981. The origin of genetic information. *Sci. Am.* **244**, 88−118.

Elston, D.P. and Scott, G.R. 1976. Unconformity at the Cardenas-Nankoweap contact (Precambrian), Grand Canyon Supergroup, northern Arizona. *Bull. geol. Soc. Am.* **87**, 1763−72.

Emry, R.J. 1981. Additions to the mammalian fauna of the type Duchesnean, with comments on the status of the Duchesnean 'age.' *J. Paleont.* **55**, 563−70.

Epstein, J.B., Sevon, W.D. and Glaeser, J.D. 1974. Geology and mineral resources of the Lehighton and Palmerton Quadrangles, Carbon and Northampton Counties, Pennsylvania. *Atlas Geol. Surv. Penna.* **195**, 460 p.

Eskola, P. 1963. The Precambrian of Finland, in *The Precambrian*, K. Ranama, (ed.). Interscience Publishers, New York. 145−263.

Evans, E.M.N., Van Couvering, J.A.H. and Andrews, P. 1981. Palaeoecology of Miocene sites in western Kenya. *J. Human Evol.* **10**, 99−116.

Faugeres, L. and Robert, P. 1969. Precisions nouvelles sur les alterations contenues dans les Serie du Chainon de Viglia (Kozami, Macedoni, Grece). *C.R. Soc. Geol. Fr.* **3**, 97−8.

Feofilova, A.P. 1973. Characteristics of Paleozoic soils of the Donbas from different paleoclimatic environments. *Dokl. Acad. Sci. USSR Earth Sci.* **202**, 170−2.

Feofilova, A.P. 1977. Paleopedology and its significance in the reconstruction of ancient landscapes. *Lithol. Mineral. Resour.* **12**, 650–5.

Feofilova, A.P. and Rekshinskaya, L.G. 1973. Fossil soils in the unproductive deposits of the Namurian and Bashkirian stages in the western part of the Donets Basin. *Lithol. Mineral. Resour.* **8(2)**, 177–91.

Firman, J.B. 1979. Paleopedology applied to land use studies in southern Australia. *Geoderma* **22**, 105–17.

Fisher, G.C. and Yam, O.-L. 1984. Iron mobilization by heathland plant extracts. *Geoderma* **32**, 339–45.

Fitzpatrick, E.A. 1980. *Soils*. Longman, London.

Flannery, T. 1982. Hindlimb structure and evolution in the kangaroos (Marsupialia: Macropodoidea), in *The Fossil Vertebrate Record of Australasia*. P.V. Rich and E.M. Thompson, (eds). Monash University Press, Clayton, Victoria. 507–23.

Florensky, C.P., Basilevsky, A.T., Kryuchkov, V.P., Kusmin, R.O., Nikolaeva, O.V., Pronin, A.A., Chernaya, I.M., Tyuflin, Y.S., Selivanov, A.S., Naraeva, M.K. and Ronca, L.B. 1983. Venera 13 and Venera 14: sedimentary rocks on Venus? *Science* **221**, 57–9.

Folk, R.L. and McBride, E.F. 1976. Possible pedogenic origin of Ligurian ophicalcite: a Mesozoic calichified serpentinite. *Geology* **4**, 327–32.

Franks, P.C. 1980. Models of marine transgression-example from Lower Cretaceous fluvial and paralic deposits, north-central Kansas. *Geology* **8**, 56–61.

Frarey, M.J. & Roscoe, S.M. 1970. The Huronian Supergroup north of Lake Huron, in, *Symposium on basins and geosynclines of the Canadian Shield*. A.J. Baer, (ed.). *Pap. Geol. Surv. Can.* 70–40, 143–57.

Frenguelli, J. 1939. Nidos fósiles de insectos en el Terciario del Neuquén Rio Negro. *Notas Mus. La Plata 4, Paleont.* **18**, 379–402.

Freytet, P. 1973. Petrography and paleoenvironment of continental carbonate deposits, with particular reference to the upper Cretaceous and Lower Eocene of Languedoc (southern France). *Sedim. Geol.* **10**, 25–60.

Freytet, P. and Plaziat, J.-C. 1982. *Continental carbonate sedimentation and pedogenesis- Late Cretaceous and Early Tertiary of southern France*, Contributions in Sedimentology, 12, E. Schweizerbart'sche Verlagsbuchhandlung, Stuttgart. 213 pp.

Galloway, W.E. 1978. Uranium mineralization in a coastal plain fluvial aquifer system: Catahoula Formation, Texas. *Econ. Geol.* **73**, 1655–76.

Gariel, O., Charpal, O. and Bennacef, A. 1968. Sur la sedimentation des gres du Cambro-Ordovicien (Unite II) dans l'Ahnet et le Mouydir (Sahara central). *Bull. Serv. Geol. Algerie* **38**, 7–37.

Gay, A.L. and Grandstaff, D.E. 1980. Chemistry and mineralogy of Precambrian paleosols at Elliot Lake, Ontario, Canada. *Precambrian Res.* **12**, 349–73.

Gillespie, W.H., Rothwell, G.W. and Scheckler, S.E. 1981. The earliest seeds. *Nature* **293**, 462–4.

Glacken, C.J. 1956. Changing ideas of the habitable world, in *Man's role in changing the face of the earth*. W.L. Thomas, (ed.). University of Chicago Press, Chicago. 70–92.

Goldbery, R. 1979. Sedimentology of the Lower Jurassic flint clay bearing Mishhor Formation, Makhtesh Ramon, Israel. *Sedimentology* **26**, 229–51.

Goldbery, R. 1982a. Paleosols of the Lower Jurassic Mishhor and Ardon Formation ('laterite derivative facies'), Makhtesh Ramon, Israel. *Sedimentology* **29**, 669–90.

Goldbery, R. 1982b. Structural analysis of soil microrelief in palaeosols of the lower Jurassic 'laterite derivative facies' (Mishhor and Ardon Formations), Makhtesh Ramon, Israel. *Sedim. Geol.* **31**, 119–40.

Goldich, S.S. 1938. A study in rock weathering. *J. Geol.* **46**, 17–58.

Gould, S.J. and Lewontin, R.L. 1979. The spandrels of San Marco and the Panglossian

paradigm: a critique of the adaptationist program. *Proc. R. Soc. Lond.* **205**, 581–98.
Grabert, H. 1976. Alter und Geschichte der Roraima-Folge aus Guyama (SüdAmerika). *Münstersche Forsch. Geol. Paläont.* **39**, 29–45.
Graustein, W.C. and Velbel, M.A. 1981. Weathering before the advent of land plants: evidence from detrital K feldspars in Cambrian-Ordovician arenites- comment. *Geology* **9**, 505.
Gray, J. and Boucot, A.J. 1977. Early vascular land plants: proof and conjecture. *Lethaia* **10**, 145–74.
Gray, J., Massa, D. and Boucot, A.J. 1982. Caradocian land plant microfossils from Libya. *Geology* **10**, 197–201.
Grigor'ev, N.P. 1973. Petrographic indications of mineral soils in Jurassic deposits of the southwestern Siberian Platform. *Lithol. Mineral. Resour.* **8(6)**, 787–90.
Halffter, G. and Matthews, E.G. 1966. The natural history of dung beetles of the subfamily Scarabaeinae (Coleoptera, Scarabaeidae). *Folia Entomol. Mexicana* **12–14**, 312 p.
Hallbauer, D.K. and van Warmelo, K.T. 1974. Fossilized plants in thucolite from Precambrian rocks of the Witwatersrand, South Africa. *Precambrian Resour.*, **1**, 199–212.
Hay, R.L. 1976. *Geology of Olduvai Gorge.* University of California Press, Berkeley.
Hay, R.L. and Reeder, R.J. 1978. Calcretes of Olduvai Gorge and the Ndolanya Beds of northern Tanzania. *Sedimentology* **25**, 649–73.
Heling, D. 1974. Relationships between initial porosity of Tertiary argillaceous sediments and paleosalinity in the Rheintalgraben (S.W. Germany). *J. sedim. Petrol.* **39**, 246–54.
Herd, R.K., Chandler, F.W. and Ermanovics, I. 1976. Weathering of Archean granitoid rocks, Island Lake, Manitoba. *Prog. Abstr. Geol. Assoc. Mineral. Can.* **1**, 72.
Hlustik, A. 1974. The rootlet horizon in the Peruc Formation (Korinkove zony ve urstvach peruckych). *Nat. Muz. Cas. Oddil. Priroved* **141**, 153–4.
Hofmann, H.J. and Schopf, J.W. 1983. Early Proterozoic microfossils, in *Earth's Early Biosphere*, J. Schopf, (ed.). Princeton University Press, Princeton. 321–60.
Holland, H.D. 1984. *The Chemical Evolution of the Atmosphere and Oceans.* Princeton University Press, Princeton.
Hower, J., Eslinger, E.V., Hower, M.E. and Perry, E.A. 1976. Mechanisms of burial metamorphism of argillaceous sediments: I, Mineralogical and chemical evidence. *Bull. geol. Soc. Am.* **87**, 725–32.
Hubert, J.F. 1960. Petrology of the Fountain and Lyons Formations, Front Range, Colorado. *Qu. Colorado Sch. Mines* **55**, 1–242.
Hubert, J.F. 1977a. Paleosol caliche in the New Haven Arkose, Newark Group, Connecticut. *Palaeogeogr. Palaeoclimatol. Palaeoecol.* **24**, 151–68.
Hubert, J.F. 1977b. Paleosol caliche in the New Haven Arkose, Connecticut: record of semiaridity in Late Triassic-Early Jurassic time. *Geology* **5**, 302–4.
Huddle, J.W. and Patterson, S.H. 1961. Origin of Pennsylvanian underclays and related seat rocks. *Bull. geol. Soc. Am.* **72**, 1643–60.
Hunt, C.B. 1972. *Geology of Soils.* W.H. Freeman & Company, San Francisco.
Huzayyin, S. 1956. Changes in climate, vegetation and human adjustments in the Saharo-Arabian belt, with special reference to Africa, in *Man's Role in Changing the Face of the Earth.* W.L. Thomas, (ed.). University of Chicago Press, Chicago. 304–23.
Issac, K.P. 1983a. Tertiary lateritic weathering in Devon, England, and the Palaeogene continental environment of southwest England. *Proc. Geol. Ass.* **94**, 105–14.
Issac, K.P. 1983b. Silica diagenesis of Palaeogene residual deposits in Devon, England.

Proc. Geol. Ass. **94**, 181–6.
James, H.L., Clark, L.D., Lamey, C.A. and Pettijohn, F.J. 1961. Geology of central Dickinson County, Michigan. *Prof. Pap. U.S. Geol. Surv.* **310**, 176 p.
James, H.L., Dutton, C.E., Pettijohn, F.J. and Wier, K.L. 1968. Geology and ore deposits of the Iron River-Crystal Falls district, Iron County, Michigan. *Prof. Pap. U.S. Geol. Surv.* **570**, 134 p.
Johnson, G.D. 1977. Paleopedology of *Ramapithecus*-bearing sediments, north India. *Geol. Rdsch.* **66**, 192–216.
Johnson, G.D., Rey, P.H., Ardrey, R.H., Visser, C.F., Opdyke, N.D. and Tahirkheli, R.A.K. 1981. Paleoenvironments of the Siwalik Group, Pakistan & India, in *Hominid Sites: Their Geologic Settings*, G. Rapp and C.F. Vondra, (eds). Westview Press, Boulder Colorado. 197–254.
Jungerius, P.D. and Mücher, H.J. 1969. The micromorphology of fossil soils in the Cypress Hills, Alberta, Canada. *22 ez. Probl. Post. Nauk Rolnic*, Wroclaw **123**, 618–27.
Kabata-Pendias, A. and Ryka, W. 1968. Weathering of biotite gneiss in the Palaeozoic period. *Trans. 9th Int. Congr. Soil Sci. Adelaide* **4**, 381–90.
Kalliokosi, J. 1975. Chemistry and mineralogy of Precambrian paleosols in northern Michigan. *Bull. geol. Soc. Am.* **86**, 371–6.
Kalliokosi, J. 1977. Chemistry and mineralogy of Precambrian paleosols in northern Michigan—reply. *Bull. geol. Soc. Am.* **88**, 1376.
Keller, W.D., Wescott, J.F. and Bledsoe, A.O., 1954. The origin of Missouri fireclays, in *Clays and Clay Minerals*, A. Swineford and N. Plummer, (eds). *Publ. U.S. Nat. Acad. Sci.* **327**, 7–46.
Kennett, J.P. 1982. *Marine Geology*. Prentice-Hall, Englewood Cliffs, New Jersey.
Kent, L.E. 1980. Stratigraphy of South Africa. *Hdbk. Geol. Surv. South Africa* **8**, 690 p.
Kidston, R. and Lang, W.H. 1921. On Old Red Sandstone plants showing structure from the Rhynie Chert bed, Aberdeenshire. Part V. The thallophyta occurring in the peat bed; the succession of plants through a vertical section of the bed and the condition and preservation of the deposit. *Trans. R. Soc. Edinb.* **52**, 855–702.
Kimberley, M.M. Grandstaff, D.E. and Tanaka, R.T. 1984. Topographic control on Precambrian weathering in the Elliot Lake uranium district, Canada. *J. geol. Soc. Lond.* **141**, 229–33.
Kjellesvig-Waering, E.N. 1966. Silurian scorpions from New York. *J. Paleont.* **40**, 359–75.
Knight, I. and Morgan, W.C. 1981. The Aphebian Ramah Group, northern Labrador in, *Proterozoic basins of Canada*. F.H.A. Campbell, (ed.). *Pap. Geol. Surv. Can.* **81–10**, 313–30.
Koryakin, A.S. 1971. Results of a study of Proterozoic weathering crusts in Karelia *Int. Geol. Rev.* **13**, 973–80.
Koryakin, A.S. 1977. Precambrian metamorphosed weathered zones of Karelia (Dokembriiskie metamorfizovannie kori vivetrivaniya karelii), in *Contemporary problems of lithologies and useful sedimentary minerals* (*Problemii sovremennoi litologii i osadochnikh poleznikh iskopaemikh*). *Nauka, Sibiriskoe Otdelenie, Novosibirsk*, 72–80.
Kroonenberg, S.B. 1978. Precambrian paleosols at the base of the Roraima Formation in Surinam. *Geol. Mijnb.* **57**, 445–50.
Krumbein, W.C. and Giele, C. 1979. Calcification in a coccoid cyanobacterium associated with the formation of desert stromatolites. *Sedimentology* **26**, 593–604.
Kubiena, W.L. 1953. *The Soils of Europe*. T. Murby & Company, London.
Kukla, G.J. 1977. Pleistocene land-sea correlations. I, Europe. *Earth Sci. Rev.* **13**, 307–74.
Lang, W.H. 1937. On the plant remains from the Downtonian of England and Wales. *Phil. Trans. R. Soc. Lond.* **B227**, 245–91.

Leakey, R.E. Leakey, M.G. and Behrensmeyer, A.K. 1978. The hominid catalogue. in *Koobi Fora Research Project. Vol. 1, The Fossil Hominids and an Introduction to their Context 1968–1974*, M.G. Leakey and R.E. Leakey, (eds). Clarendon Press, Oxford. 82–182.

Lewan, M.D. 1977. Chemistry and mineralogy of Precambrian paleosols in northern Michigan: discussion. *Bull. geol. Soc. Am.* **88**, 1375–6.

Lindsay, J.F. 1976. *Lunar Stratigraphy and Sedimentology*. Elsevier Publishers, New York.

Loughnan, F.C. 1975. Laterites and flint clays in the Early Permian of the Sydney Basin, Australia, and their paleoclimatic implications. *J. sedim. Petrol.* **45**, 591–8.

Lucas, C. 1976. Vestiges de paleosols dans le Permien et le Trias inferieur, des Pyrenees et de l'Aquitaine. *C.R. Acad. Sci. Paris* **282**, 1419–22.

MacDonald, J.R. 1963. Miocene faunas from the Wounded Knee area of South Dakota. *Bull. Am. Mus. Nat. Hist.* **125**, 139–238.

MacDonald, J.R. 1970. Review of the Miocene Wounded Knee faunas of south-western South Dakota. *Bull. Los Angeles Co. Mus. Hist. Sci.* **8**, 82 p.

Maiklem, W.R. 1971. Evaporative drawdown- a mechanism for waterlevel lowering and diagenesis in the Elk Point Basin. *Bull. Can. Petrol. Geol.* **19**, 487–503.

Marbut, C.F. 1935. Atlas of American agriculture. Part III. *Soils of the United States*. U.S. Govt. Print. Office, Washington D.C.

Martin, L.D. and Bennett, D.K. 1977. The burrows of the Miocene beaver *Palaeocastor*, western Nebraska, U.S.A. *Palaeogeogr. Palaeoclimatol. Palaeoecol.* **22**, 173–93.

McBride, E.F. 1974. Significance of color in red, green, purple, olive, brown, and gray beds of the Difunta Group. *J. sedim. Petrol.* **44**, 760–73.

McBride, E.F., Lindemann, W.L. and Freeman, P.S. 1968. Lithology and petrology of the Gueydan (Catahoula) Formation in south Texas. *Rep. Invest. Bur. Econ. Geol. Univ. Texas* **63**, 122 p.

McFarlane, M.J. 1976. *Laterite and Landscape*. Academic Press, London.

McGowran, B. 1979. Comments on early Tertiary tectonism and lateritization. *Geol. Mag.* **116**, 227–30.

McPherson, J.G. 1979. Calcrete (caliche) paleosols in fluvial red beds of the Aztec Siltstone (Upper Devonian), southern Victoria Land, Antarctica. *Sedim. Geol.* **22**, 319–20.

Mehlich, A. and Drake, M. 1955. Soil chemistry and plant nutrition, in *Chemistry of the Soil*, F.E. Bear, (ed.). Reinhold Publishing Co, New York.

Meyer, R. 1976. Continental sedimentation, soil genesis and marine transgression in the basal beds of the Cretaceous in the east of the Paris Basin. *Sedimentology* **23**, 235–53.

Miller, A.R. 1983. A progress report: uranium-phosphorous association in the Helikian Thelon Formation and the Sub-Thelon Saprolite, central district of Keewatin. *Pap. Geol. Surv. Can.* **83–1A**, 449–55.

Milner, A.R. 1980. The tetrapod assemblage from Nyřany, Czechoslovakia, in, *The Terrestrial Environment and the Origin of Land Vertebrates*, A.L. Panchen, (ed.). Academic Press, London. 439–96.

Morand, F., Riveline-Bauer, J. and Trichet, J. 1968. Etudes sedimentologiques, paléogéographiques et géomorphologiques de la Butte Chaumont (Champlan, Essone). *Bull. Soc. geol. Fr.* **7**, 637–8.

Morey, G.B. (1972) Pre-Mt Simon regolith, in *Geology of Minnesota: a centennial volume*, P.K. Sims and G.B. Morey, (eds). Minnesota Geological Survey, St. Paul. 506–8.

Morris, S.C., Pickerill, R.K. and Harland, T.L. 1982. A possible annelid from the Trenton Limestone (Ordovician) of Quebec, with a review of fossil oligochaetes and other annulate worms. *Can. J. Earth Sci.* **19**, 2150–7.

Muller, J. 1981. Fossil pollen records of extant angiosperms. *Bot. Rev.* **47**, 1–142.

Nicholas, J. and Bildgen, P. 1979. Relations between the location of karst bauxites in the northern hemisphere, the global tectonics and the climatic variations during geological time. *Palaeogeogr. Palaeoclimatol. Palaeoecol.* **29**, 205−39.

Nikiforoff, C.C. 1943. Introduction to paleopedology. *Am. J. Sci.* **241**, 194−200.

Niklas, K.J. 1982. Chemical diversification and evolution of plants as inferred from paleobiochemical studies, in *Biochemical aspects of evolutionary biology*, M.H. Nitecki, (ed.). University of Chicago Press, Chicago. 29−91.

Niklas, K.J. and Smocovitis, V. 1983. Evidence for a conducting strand in early Silurian (Llandoverian) plants: implications for the evolution of land plants. *Paleobiology* **9**, 126−37.

Nilsen, T.H. 1978. Lower Tertiary laterite on the Iceland-Faeroe Ridge and the Thulean land bridge. *Nature* **274**, 786−8.

Nilsen, T.H. and Kerr, D.R. 1978. Palaeoclimatic and palaeogeographic implications of a lower Tertiary laterite (latosol) in the Iceland-Faeroe Ridge, North Atlantic region. *Geol. Mag.* **115**, 153−82.

Norton, S.A. 1969. Origin of the Chester Emery deposit, Chester, Massachusetts. *Abstr. Prog. geol. Soc. Am.* **2**, 44−5.

Nozette, S. and Lewis, J.S. 1982. Venus: chemical weathering of igneous rocks and buffering of atmospheric composition. *Science* **216**, 181−3.

Nussinov, M.D. and Vehkov, A.A. 1978. Formation of the early earth regolith. *Nature* **275**, 19−21.

Ortlam, D. 1971. Paleosols and their significance in stratigraphy and applied geology in the Permian and Triassic of southern Germany, in *Paleopedology: origin, nature and dating of paleosols*, (ed.). International Society for Soil Science and Israel University Press, Jerusalem. 321−7.

Ortlam, D. 1980. Erkennung und Bedeutung fossiler Bodenkomplexe im locker- und festgestein. *Geol. Rdsch.* **69**, 581−93.

Ortlam, D. and Zimmerle, W. 1982. Paläopedologische Ergebnisse der Bohrung Schwarzbachtal 1 (Givetium; Rheinisches Schiefergebirge). *Senckenberg. leth.* **63**, 293−313.

Parnell, J. 1983. Ancient duricrusts and related rocks in perspective: a contribution from the Old Red Sandstone, in *Residual Deposits: Surface Related Processes and Materials*, R.C.L. Wilson, (ed.). Blackwell Scientific Publications, Oxford for the Geological Society of London. 197−209.

Patel, I.M. 1977. Precambrian fossil soil horizon in southern New Brunswick and its stratigraphic significance. *Abstr. Prog. geol. Soc. Am.* **9**, 308.

Percival, C.J. 1983. A definition of the term ganister. *Geol. Mag.* **120**, 187−90.

Pettyjohn, W.A. 1966. Eocene paleosol in the northern Great Plains. *Prof. Pap. U.S. geol. Surv.* **550C**, 61−5.

Philobbos, E. and Hassan, K.E.-D.K. 1975. The contribution of paleosoil to Egyptian lithostratigraphy. *Nature* **253**, 33.

Pickford, M. and Andrews, P. 1981. The Tinderet Miocene sequence in Kenya. *J. Human Evol.* **10**, 11−33.

Pilbeam, D. 1982. New hominoid skull material from the Miocene of Pakistan. *Nature* **295**, 232−4.

Pilbeam, D. 1984. The descent of hominoids and hominids. *Sci. Am.* **250**, 84−96.

Pilbeam, D.R., Rose, M.D., Badgley, C. and Lipschutz, B. 1980. Miocene hominoids from Pakistan. *Postilla* **181**, 94 p.

Pomerol, C. 1964. Découverte de paléosols de type podzol au sommet de l'auversien (bartonien inférieur) de Moisselles (Seine-et-Oise). *C.R. Acad. Sci. Paris* **258**, 974−6.

Pomerol, C. 1982. *The Cenozoic Era, Tertiary and Quaternary* (translated by D.W. and E.E. Humphries). Ellis Horwood Publishers, Chichester.

Poty, E. 1980. Evolution and drowning of palaeokarst in Frasnian carbonates at Visé, Belgium. *Meded. Rijks Geol. Dienst.* **32**, 53–5.

Power, P.E. 1969. Clay mineralogy and palaeoclimatic signficance of some red regoliths and associated rocks in western Colorado. *J. sedim. Petrol.* **39**, 876–90.

Pratt, L.M., Phillips, T.L. and Dennison, J.M. 1978. Evidence of non-vascular plants from the Early Silurian (Llandoverian) Of Virginia, USA *Rev. Palaeobot. Palynol.* **25**, 121–49.

Ramaekers, P. 1981. Hudsonian and Helikian basins of the Athabaska region, northern Saskatchewan, in *Proterozoic Basins of Canada*. F.H.A. Campbell, (ed.). *Pap. geol. Surv. Can.*, **81–10**, 219–33.

Rankama, K. 1955. Geologic evidence of chemical composition of the Precambrian atmosphere. *Spec. Pap. geol. Soc. Am.* **62**, 651–64.

Ratcliffe, B.C. and Fagerstrom, J.A. 1980. Invertebrate lebensspuren of Holocene floodplains: their morphology, origin and paleoecological significance. *J. Paleont.* **54**, 614–30.

Reffay, A. and Ricq-Debouard, M. 1970. Contribution a l'étude des paléosols interbasaltiques a la Chauséé des Geants, comte d'Antrim, Irlande du Nord. *Rev. Geogr. Alpine* **58**, 301–38.

Retallack, G.J. 1976. Triassic palaeosols in the upper Narrabeen Group of New South Wales. Part I. Features of the palaeosols *J. geol. Soc. Aust.* **23**, 383–99.

Retallack, G.J. 1977. Triassic palaeosols in the upper Narrabeen Group of New South Wales. Part II. Classification and reconstruction *J. geol. Soc. Aust.* **24**, 19–35.

Retallack, G.J. 1979. Middle Triassic coastal outwash plain deposits in Tank Gully, Canterbury, New Zealand. *J. R. Soc. N.Z.* **9**, 397–414.

Retallack, G.J. 1980. Late Carboniferous to Middle Triassic megafossil floras from the Sydney Basin, in, *A Guide to the Sydney Basin*, C. Herbert and R.J. Helby, (eds). *Bull. geol. Surv. N.S.W.* **26**, 383–430.

Retallack, G.J. 1981a. Fossil soils: indicators of ancient terrestrial environments, in, *Paleobotany, Paleoecology and Evolution*. Vol. 1, K.J. Niklas, (ed.). Praeger Publishers, New York, 55–102.

Retallack, G.J. 1981b. Preliminary observations on fossil soils in the Clarno Formation (Eocene to Early Oligocene), near Clarno, Oregon. *Oregon Geology* **43**, 147–50.

Retallack, G.J. 1981c. Reinterpretation of the depositional environment of the Yellowstone fossil forests- comment. *Geology* **9**, 52–3.

Retallack, G.J. 1983a. A paleopedological approach to the interpretation of terrestrial sedimentary rocks: the mid-Tertiary paleosols of Badlands National Park, South Dakota. *Bull. geol. Soc. Am.* **94**, 823–40.

Retallack, G.J. 1983b. Late Eocene and Oligocene palaeosols from Badlands National Park, South Dakota. *Spec. Pap. geol. Soc. Am.* **193**, 82 p.

Retallack, G.J. 1983c. Middle Triassic estuarine deposits near Benmore Dam, southern Canterbury and northern Otago, New Zealand. *J. R. Soc. N.Z.* **13**, 107–27.

Retallack, G.J. 1984a. Completeness of the rock and fossil record: some estimates using fossil soils. *Paleobiology* **10**, 59–78.

Retallack, G.J. 1984b. Trace fossils of burrowing beetles and bees in an Oligocene paleosol, Badlands National Park, South Dakota. *J. Paleont.* **58**, 571–92.

Retallack, G.J. 1985. Fossil soils as grounds for interpreting the advent of large plants and animals on land. *Phil. Trans. R. Soc. Lond.* B. **309**, 105–42.

Retallack, G.J. and Dilcher, D.L. 1981a. A coastal hypothesis for the dispersal and rise to dominance of flowering plants, in, *Paleobotany, Paleoecology and Evolution*. Vol. 2, K.J. Niklas, (ed.). Praeger Publishers, New York. 27–77.

Retallack, G.J. and Dilcher, D.L. 1981b. Early angiosperm reproduction: *Prisca reynoldsii* gen. et sp. nov. from mid-Cretaceous coastal deposits in Kansas, USA. *Palaeontographica* **B179**, 103–37.

Retallack, G.J., Grandstaff, D. and Kimberley, M.M. 1984. The problems and promise of Precambrian paleosols. *Episodes* **7**, 8−12.

Ritzkowski, S. 1973. Böden des Tertiärs im nördlichen Hessen. *Mitt. Deutsch. Bödenkdl. Ges.* **17**, 119−22.

Roeschmann, G. 1971. Problems concerning investigations of paleosols in older sedimentary rocks, as demonstrated by the example of Wurzelböden of the Carboniferous System, in, *Paleopedology: Origin, Nature and Dating of Paleosols*, D.H. Yaalon, (ed.). International Society for Soil Science and Israel University Press, Jerusalem. 311−20.

Rolfe, W.D.I. 1980. Early invertebrate terrestrial faunas, in, *The Terrestrial Environment and the Origin of Land Vertebrates*, A.L. Panchen, (ed.). Academic Press, London. 117−57.

Roscoe, S.M. 1968. Huronian rocks and uraniferous conglomerates in the Canadian Shield. *Pap. geol. Surv. Canada* **68−40**, 205 p.

Rozen, O.M. 1967. Metamorphosed bauxite pebbles in conglomerate among the Precambrian schists of the Kokchetov Massif. *Dokl. Acad. Nauk. USSR* **51**, 1235−70.

Rubin, D.M. and Friedman, G.M. 1981. Origin of chert grains and a halite-silcrete bed in Cambrian and Ordovician Whitehall Formation of eastern New York state. *J. sedim. Petrol.* **51**, 69−72.

Russell, E.W. 1973. *Soil Conditions and Plant Growth*. Longman, London.

Salop, L.J. 1973. *Precambrian of the Northern Hemisphere*. Elsevier Publishing Company, Amsterdam.

Sanson, G.D. 1982. Evolution of feeding adaptations in fossil and recent macropodids, in, *The Fossil Vertebrate Record of Australasia*, P.V. Rich and E.M. Thompson, (eds). Monash University Press, Clayton, Victoria. 490−506.

Schau, M.K. and Henderson, J.B. 1983. Archaean weathering at three localities on the Canadian Shield. *Precambrian Res.* **20**, 189−202.

Schopf, J.M., Mencher, E., Boucot, A.B. and Andrews, H.N. 1966. Erect plants in the early Silurian of Maine. *Prof. Pap. U.S. Geol. Surv.* **550D**, 69−75.

Schopf, J.W., Hayes, J.M. and Walter, M.R. 1983. Evolution of earth's earliest ecosystems: recent progress and unsolved problems, in, *Earth's Earliest Biosphere*, J.W. Schopf, (ed.). Princeton University Press, Princeton. 361−84.

Schumm, S.A. 1968. Speculations concerning paleohydrologic controls of terrestrial sedimentation. *Bull. geol. Soc. Am.* **79**, 1575−88.

Schumm, S.A. 1977. *The Fluvial System*. Wiley-Interscience Publishers, New York.

Searle, A.B. 1930. *An Encyclopedia of the Ceramic Industries*. Vol. II. Ernest Benn Ltd, London.

Selleck, B.W., 1978. Syndepositional brecciation in the Potsdam Sandstone of northern New York. *J. sedim. Petrol.* **48**, 1177−84.

Sharp, R.P. 1940. The Ep-Archaean and Ep-Algonkian erosion surfaces, Grand Canyon, Arizona. *Bull. geol. Soc. Am.* **51**, 1235−69.

Shear, W.A., Bonamo, P.M., Grierson, J.D., Rolfe, W.D.I., Smith, E.L. and Norton, R.A. 1984. Early land animals in North America: evidence from Devonian age arthropods from Gilboa, New York. *Science* **224**, 492−4.

Shipman, P., Walker, A., Van Couvering, J.A., Hooker, P.J. and Miller, J.A. 1981. The Fort Ternan hominoid site, Kenya: geology age, taphonomy and paleoecology. *J. Human Evol.* **10**, 49−72.

Siegel, B.Z. 1977. Kakabekia, a review of its physiological and environmental features and their relation to its possible ancient affinities, in, *Chemical Evolution of the Early Precambrian*, C. Ponnamperuma, (ed.). Academic Press, New York. 143−54.

Singer, A. 1975. A Cretaceous laterite in the Negev Desert, southern Israel. *Geol. Mag.* **112**, 151−62.

Singer, M.J. and Nkedi-Kizza, P. 1980. Properties and history of an exhumed Tertiary Oxisol in California. *J. Soil Sci. Soc. Am.* **44**, 587–90.

Sloan, R.E. 1964. The Cretaceous System in Minnesota. *Rep. Invest. Univ. Minnesota* **5**, 64 p.

Sochava, A.V. 1979. Red-coloured formations of the Precambrian and Phanerozoic (Krasnoshvetnie formashii dokembriya i fanerozoya). *Akademia Nauk S.S.S.R. Leningrad.*

Sochava, A.V., Saveliev, A.A. and Schulenschko, I.K. 1975. Caliche in middle Proterozoic deposits of central Karelia (Kaliche v sredneproterozoiskikh otlozheniyah shentralinoi karelii). *Dokl. Akad. Nauk S.S.S.R.* **223**, 1451–4.

Soil Survey Staff 1975. Soil taxonomy. *Handb. U.S. Dep. Agric.* **436**, 754 p.

Sokolov, V.A. and Hieskanen, K.I. 1984. Developmental stages of Precambrian crusts of weathering. *Proc. 27th Int. Geol. Congr.* **5**, 73–94.

Sombroek, W.G. 1971. Ancient levels of plinthisation in N.W. Nigeria, in, *Paleopedology: Origin, Nature and Dating of Paleosols*, D.H. Yaalon, (ed.). International Society for Soil Science and Israel University Press, Jerusalem. 329–36.

Sombroek, W.G., Braun, H.M.H. and Van der Pouw, B.J.A. 1982. Exploratory soil map and agroclimatic zone map of Kenya 1980: scale 1:100,000. *Kenya Soil Survey, Nairobi.*

Spalleti, L.A. and Mazzoni, M.M. 1978. Sedimentologia del Grupo Sarmiento en el perfil ubicado al sudeste del Lago Colhue Huapi, Provincia de Chubut. *Obra Centen. Mus. La Plata* **4**, 261–83.

Staley, J.T., Palmer, F. and Adams, J.B. 1982. Microcolonial fungi: common inhabitants on desert rocks? *Science* **215**, 1093–5.

Stanworth, C.W. and Badham, J.P.N. 1984. Lower Proterozoic red beds, evaporites and secondary sedimentary uranium deposits from the East Arm, Great Slave Lake, Canada. *J. geol. Soc. Lond.* **141**, 235–42.

Stebbins, G.L. 1981. Coevolution of grasses and herbivores. *Ann. Mo. Bot. Gard.* **68**, 75–86.

Stebbins, G.L. and Hill, G.J.C. 1980. Did multicellular plants invade the land? *Am. Nat.* **165**, 342–53.

Steel, R.J. 1974. Cornstone (fossil caliche)—its origin, stratigraphic and sedimentological importance in the New Red Sandstone, Scotland. *J. Geol.* **82**, 351–69.

Stevenson, F.J. 1969. Pedohumus: accumulation and diagenesis during the Quaternary. *Soil Sci.* **107**, 470–9.

Størmer, L. 1955. Merostomata, in, *Treatise on Invertebrate Paleontology. Part P. Arthropoda.* Vol. 2, R.C. Moore (ed.). Geological Society of America & University of Kansas Press, Boulder & Lawrence.

Størmer, L. 1977. Arthropod invasion of land during Late Silurian and Devonian times, *Science* **197**, 1362–64.

Strickland, E.L. 1979. Martian soil stratigraphy and rock coatings observed on color enhanced Viking Lander Images. *Proc. 10th Lunar Planet. Sci. Conf.* **3**, 3055–77.

Strother, P.K. and Traverse, A. 1979. Plant microfossils from Llandoverian and Wenlockian rocks of Pennsylvania. *Palynology* **3**, 1–21.

Sturt, B.A., Dalland, A.B. and Mitchell, J.L. 1979. The age of the sub-mid-Jurassic tropical weathering profile of Andoya, northern Norway, and its implications for late Palaeozoic palaeogeography. *Geol. Rdsch.* **68**, 523–42.

Summerfield, M.H. 1983. Petrography and diagenesis of silcrete from the Kalahari Basin and Cape Coastal zone, southern Africa. *J. sedim. Petrol.* **53**, 895–909.

Swain, T. and Cooper-Driver, G. 1981. Biochemical evolution in early land plants, in, *Paleobotany, Paleoecology and Evolution.* K.J. Niklas, (ed.). Vol. 1, Praeger Publishers, New York. 103–34.

Tauxe, L. and Opdyke, N.D. 1982. A time framework based on magnetostratigraphy for the Siwalik sediments of the Khaur area, northern Pakistan. *Palaeogeogr. Palaeoclimatol. Palaeoecol.* **37**, 43–61.

Thomassen, J.R. 1979. Late Cenozoic grasses and other angiosperms from Kansas, Nebraska and Colorado: biostratigraphy and relationships to living taxa. *Bull. geol. Surv. Kansas* **218**, 68 p.

Thompson, G.R., Fields, R.W. and Alt, D. 1982. Land-based evidence for Tertiary climatic variations: northern Rockies. *Geology* **10**, 413–7.

Thompson, I. & Jones, D.S. 1980. A possible onychophoran from the Middle Pennsylvanian Mazon Creek beds of northern Illinoia. *J. Paleont.* **54**, 588–96.

Tiffney, B.H. 1981. Diversity and major events in the evolution of land plants, in, *Paleobotany, Paleoecology and Evolution*. Vol. 2, K.J. Niklas, (ed.). Praeger Publishers, New York. 193–230.

Tremblay, L.P. 1983. Some chemical aspects of the regolithic and hydrothermal alterations associated with uranium mineralization in the Athabaska Basin, Saskatchewan. *Pap. geol. Surv. Can.* **83–1A**, 1–14.

Valeton, I. 1972. *Bauxites*. Elsevier Publishers, Amsterdam.

Van Couvering, J.A.H. 1980. Community evolution in East Africa during the Late Cenozoic, in, *Fossils in the Making*. A.K. Behrensmeyer and A.P. Hill, (eds). University of Chicago Press, Chicago. 272–98.

Van Moort, J.C. 1971. A comparative study of the diagenetic alteration of clay minerals in Mesozoic shales from Papua, New Guinea, and in Tertiary shales from Louisiana, USA. *Clays Clay Mineral.* **19**, 1–20.

Vogel, D.E. 1975. Precambrian weathering in acid metavolcanic rocks from the Superior Province, Villebon Township, South Central Quebec. *Can. J. Earth Sci.* **12**, 2080–5.

Vogl, R.J. 1974. Effects of fire on grasslands, in, *Fire and Ecosystems*, T.T. Kozlowski and C.E. Ahlgren, (eds). Academic Press, New York. 139–94.

Walkden, G.M. 1974. Palaeokarstic surfaces in upper Visean (Carboniferous) limestone of the Derbyshire Block, England. *J. sedim. Petrol.* **44**, 1242–47.

Walker, B.H., Ludwig, D., Hollig, C.S. and Peterman, R.M. 1981. Stability of semiarid savanna grazing systems. *J. Ecol.* **69**, 473–98.

Walker, J.C.G., Klein, C., Schidlowski, M., Schopf, J.W., Stevenson, D.J. and Walter, M.R. 1983. Environmental evolution of the Archaean-Early Proterozoic earth, in, *Earth's Earliest Biosphere*, J.W. Schopf, (ed.). Princeton University Press, Princeton. 260–90.

Walker, R.G. and Harms, J.C. 1971. The 'Catskill Delta': a prograding muddy shoreline in central Pennsylvania. *J. Geol.* **79**, 381–99.

Walls, R.A., Harris, W.B. and Nunan, W.E. 1975. Calcareous crust (caliche) profiles and early subaerial exposure of Carboniferous carbonates, northeastern Kentucky. *Sedimentology* **22**, 417–40.

Wardlaw, N.C. and Reinson, G.E. 1971. Carbonate and evaporite deposition and diagenesis, Middle Devonian Winnepegosis and Prairie evaporite formations of south central Saskatchewan. *Bull. Am. Ass. Petrol. Geol.* **55**, 1759–87.

Watts, N.L. 1976. Paleopedogenic palygorskite from the basal Permo-Triassic of northwest Scotland. *Am. Mineral.* **61**, 299–302.

Watts, N.L. 1978. Displacive calcite: evidence from recent and ancient calcretes. *Geology* **6**, 699–703.

Weaver, C.E. 1969. Potassium, illite and the ocean. *Geochim. Cosmochim. Acta* **31**, 2181–96.

Webb, S.D. 1977. A history of savanna vertebrates in the New World. Part I. North America. *Ann. Rev. Ecol. Syst.* **8**, 355–80.

Webb, S.D. 1978. A history of savanna vertebrates in the New World. Part II. South America and the Great Interchange. *Ann. Rev. Ecol. Syst.* **9**, 393–426.

West, I.M. 1975. Evaporites and associated sediments of the basal Purbeck Formation (Upper Jurassic) of Dorset. *Proc. Geol. Ass.* **86**, 205–25.
West. I.M. 1979. Review of evaporite diagenesis in the Purbeck Formation of southern England. *Spec. Publ. Ass. Sedim. Fr.* **1**, 407–16.
White, H.T., Burggraf, D.R., Bainbridge, R.B. and Vondra, C.F. 1981. Hominid habitats in the Rift Valley: Part I, in, *Hominid Sites: their Geologic Settings*, G. Rapp and C.F. Vondra, (eds). Westview Press, Boulder, Colorado. 57–113.
Willard, B. 1938. Evidence of Silurian land plants in Pennsylvania. *Proc. Penna Acad. Sci.* **12**, 121–4.
Williams, G.E. 1968. Torridonian weathering and its bearing on Torridonian palaeoclimate and source. *Scott. J. Geol.* **4**, 164–84.
Williams, M.A.J. 1979. Droughts and long-term climatic change: recent French research in and North Africa. *Geogr. Bull.* 11, 82–96.
Williams, L.A.J. 1972. Geology of the Amboseli area. *Rep. geol. Surv. Kenya* **90**, 86 p.
Winkler, D.A. 1983. Paleoecology of the Early Eocene mammalian fauna from paleosols in the Clarks Fork Basin, northwestern Wyoming (U.S.A.). *Palaeogeogr. Palaeoclimatol. Palaeoecol.* **43**, 261–98.
Winkler, H.G.F. 1979. *Petrogenesis of Metamorphic Rocks* (5th edition). Springer Verlag, New York.
Wolfe, J.A. 1978. A paleobotanical interpretation of Tertiary climates in the northern hemisphere. *Am. Sci.* **66**, 694–703.
Wopfner, H. 1978. Silcretes of northern South Australia and adjacent regions, in, *Silcrete in Australia*, T. Langford-Smith, (ed.). University of New England, Armidale, N.S W 93–141.
Wright, V.P. 1982. The recognition and interpretation of palaeokarsts: two examples from the Lower Carboniferous of South Wales. *J. sedim. Petrol.* **52**, 83–94.
Wright, V.P. 1982. Calcrete palaeosols from the Lower Carboniferous, Llanelly Formation, South Wales. *Sedim. Geol.* **33**, 1–33.
Wright, V.P. 1983. A rendzina from the Lower Carboniferous of South Wales. *Sedimentology* **30**, 159–79.
Yaalon, D.H. 1971. Soil forming processes in time and space, in *Palepedology: Origin, Nature and Dating of Paleosols*, D.H. Yaalon, (ed.). International Society for Soil Science and Israel University Press, Jerusalem. 29–39.
Zeigler, A.M. and McKerrow, W.S. 1975. Silurian marine red beds. *Am. J. Sci.* **275**, 31–56.

Chapter 2

PEDOGENIC CALCRETES IN THE OLD RED SANDSTONE FACIES (LATE SILURIAN—EARLY CARBONIFEROUS) OF THE ANGLO-WELSH AREA, SOUTHERN BRITAIN

J. R. L. ALLEN
Department of Geology, Reading University, Great Britain

INTRODUCTION

The records of simple forms of life and of an oxygenated atmosphere extend so far back in geological time that sedimentologists have been forced to accept that all but the very oldest continental red-beds may be expected to include materials either modified or created by pedological processes. Such materials, however, arranged vertically in horizons to form paleosols, have so far been generally recognized in only the most obvious cases. This is partly because many of the criteria for a pedogenic origin are seemingly destroyed or modified beyond recognition during burial and lithification, but also because of the relative neglect by neopedologists, from whom comparative data must in the first instance be sought, of the evolution of soil environments on a geological time-scale. Thus paleosols have so far been claimed from only two of the three great continental red-bed successions present in the British Isles. None are reported from within the thick and extensive Torridonian (late Precambrian) rocks of north-west Scotland, although this formation rests locally on a paleosol (Williams, 1968). In the Old Red Sandstone (late Silurian-early Carboniferous) facies, however, paleosols have proved to be widespread and quite varied, although the soil materials so far recognized are restricted to the most resistant to diagenesis. Calcretes and some silcretes are known from the rocks in Scotland and the Scottish Border country (Burgess, 1961; Leeder, 1976; Parnell, 1983a, 1983b). Both the Lower and the Upper Old Red Sandstone in the Anglo-Welsh area abound in calcretes (Pick, 1964; Allen, 1965, 1974a), some of which are associated with a little barytes and others with concentrations of iron and manganese minerals. Calcretes are also present in the Old Red Sandstone facies to the south in North Devon (Tunbridge, 1981a). It is only in the exceptionally thick and rapidly accumulated Upper Old Red Sandstone (mid to late Devonian) of southern Ireland that calcretes have proved to be infrequent and poorly developed (e.g. MacCarthy et al., 1978; Gardiner and Horne, 1981). The Middle Devonian fluvial rocks of the Belgian Ardennes are also known to include carbonate-bearing paleosols (Molenaar, 1984). Well-developed calcretes are locally important in the New Red Sandstone (Permo-Triassic), the youngest major red-bed succession in the British Isles (Steel, 1974). Fossil calcretes and carbonate-bearing soils also have a wide distribution outside the British Isles, as shown by the recent work of Hubert (1978), McPherson (1979), Bown and Kraus (1981) and Goldbery (1982).

The purpose of this chapter is to give a general review and discussion of the character and implications of the calcretes developed so plentifully within the Lower and the Upper Old Red Sandstone facies of the Anglo-Welsh area in southern Britain. The claim (Allen, 1974a, 1974b) that these accumulations of carbonate minerals are primarily pedogenic is made essentially on the basis of (1) their repeated occurrence in vertical profile according to a simple general pattern, and (2) the repetition from profile to profile of the same (or largely the same) set of features, some microscopic and other macroscopic, which have been closely matched in calcretes forming either today or in the recent past (Netterberg, 1967, 1971; Goudie, 1973, 1983; Reeves, 1976; Milnes and Hutton, 1983). The lack of reports from the Anglo-Welsh area of catenas involving calcretes may cause some neopedologists either to suspend judgement on or even to dismiss these broadly 360–410 Ma old soils, but the fact that the other two criteria are satisfied is to the palaeopedologist a *prima facie* ground for the interpretation offered. On the basis of their presently observable characteristics—a set limited by intraformational erosion and by protracted diagenesis—these Old Red Sandstone paleosols may tentatively be classified largely amongst the Aridisols and Vertisols. Further work on them is necessary, particularly in Scotland and Ireland, and it is probable that other kinds of soil material will as a consequence be detected. However, it seems unlikely that the broad interpretation so far advanced will require other than minor modification.

STRATIGRAPHICAL CONTEXT OF THE CALCRETES

The stratigraphy of the Old Red Sandstone facies at outcrop in the Anglo-Welsh area is outlined in Figure 1, based on Allen's (1977) review and later contributions by Allen and Williams (1978, 1981),

Squirrell and White (1978), Allen *et al.* (1981, 1982), and Williams *et al.* (1982). Details of thickness, lithology and evidence for correlation should be sought in these sources.

The Lower Old Red Sandstone, ranging from the Pridoli (late Silurian) to possibly the Emsian (late Lower Devonian), is an upward-coarsening red-bed succession some 2–4 km thick. Except in south-west Wales, where a thin fluvial sequence fills a palaeovalley, the lowermost beds are thin shallow-marine sandstones and, except locally (mid Wales), subordinate mudstones. The overlying Pridoli and early Gedinnian rocks are mudstones with very subordinate sandstones. A restricted marine fauna typifies the earlier of these beds, which on the whole appear to represent mixed marine and fluvial influences and extensive coastal mudflats. The overlying beds are wholly fluvial, consisting of intraformational conglomerates, sandstones and mudstones in various kinds of erosively-based and commonly upward-fining sequence (e.g. Allen, 1974b, 1983; Tunbridge, 1981b). Upward from the base of these fluvial measures the sandstones become progressively coarser grained and thicker relative to the mudstones until, in the upper part of the Woodbank Group, Brownstone Group, Brownstones, and Cosheston Group, mudstones disappear except as a ghost facies (intraformational debris) and exotic pebbles manifest themselves abundantly.

The Upper Old Red Sandstone (Frasnian–earliest Tournasian) is an order of magnitude thinner than the Lower division and at most localities passes upward without a break into marine Carboniferous rocks. In South Wales, and possibly in the Clee Hills, a mid Famennian disconformity is present within the beds. The rocks below the disconformity in central South Wales are shallow-marine, and the Gupton Formation in south-west Wales may possibly be of this origin.

The Old Red Sandstone in Anglesey is of uncertain age (mid Silurian to early Carboniferous) but, on strong lithological grounds, may be correlated to the Gedinnian formations of the Welsh Borders and South Wales.

Calcretes occur in all but the most marine portions of the Lower Old Red Sandstone succession, typically within mudstones, but in places either within or as a capping to sandstones. They are first seen in the Temeside Shale Formation, in a close and repeated association with tidal-flat sandstones and mudstones bearing a restricted marine invertebrate fauna. Calcrete profiles, many of a thick and advanced type, occur throughout the Pridoli and early Gedinnian mud-dominated part of the succession (Allen and Williams, 1979), culminating in the great concentration of thick calcretes known as the '*Psammosteus*' Limestones. King (1934) recognized that the '*Psammosteus*' Limestones

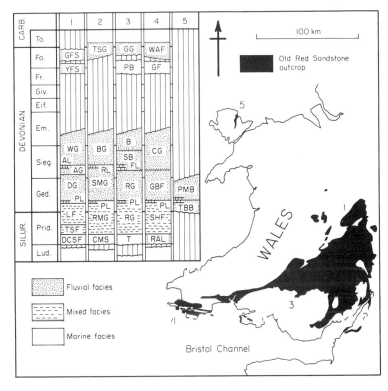

Figure 1. Stratigraphical outline of the Old Red Sandstone facies at outcrop in the Anglo-Welsh area (sources listed in text). Key to stratigraphic columns: 1—Clee Hills; 2—Forest of Dean and area south-east of the R. Severn; 3—Central South Wales; 4—South-west Wales; 5—Anglesey. Key to lithostratigraphic units: AG—Abdon Group; AL—Abdon Limestone Formation; B—Brownstones; BG—Brownstone Group; CG—Cosheston Group; CMS—Clifford's Mesne Sandstone; DCSF—Downton Castle Sandstone Formation; DG—Ditton Group; GBS—Gelliswick Bay Formation; GG—Grey Grits; LF—Ledbury Formation; PB—Plateau Beds; PL—'*Psammosteus*' Limestones; PMB—Porth y Mor and Traeth Lligwy beds; RG—Red Marl Group; RL—Ruperra Limestone; RMG—Raglan Marl Group; SB—Senni Beds; SHF—Sandy Haven Formation ('*Psammosteus*' Limestones inserted from a correlative to the south); SMG—St. Maughan's Group; T—Tilestones; TBB—Traeth Bach and Bodafon beds; TSF—Temeside Shale Formation; TSG—Tintern Sandstone Group (with Quartz Conglomerate); WAF—West Angle Formation; WG—Woodbank Group; YFS—Yellow Farlow Sandstone.

could be traced as a facies throughout the Welsh Borders and South Wales; they appear to be represented also in Anglesey (Allen, 1965). In the more southerly parts of this tract, for example, in south-west Wales (Dixon, 1921; Allen, 1974a; Allen and Williams, 1979) and the Forest of Dean (Allen, 1974a; Allen and Dineley, 1976), the facies is 40–50 m thick and consists predominantly of mudstones (with calcretes) but with few sandstones. Gradually northward the facies roughly doubles in thickness, as fluvial sandstones in upward-fining sequences become more numerous and appear at progressively lower horizons (Allen, 1974b; Allen and Williams, 1979). White's (1950a,b) contention that the '*Psammosteus*' Limestones facies is diachronous is difficult to accept, since it is in these fluvial sandstones that the vertebrates used for dating are preserved. Calcretes are no less frequent above than below the '*Psammosteus*' Limestones, except in the sand-dominated late Siegenian and Emsian measures, but are generally speaking thinner and less well-developed. An important exception, representing a stratigraphically higher concentration of calcretes, occurs in the mid Siegenian beds. Massive calcretes are developed near the top of the St. Maughan's Group in the Forest of Dean (Allen and Dineley, 1976), and approximate equivalents appear to be the Ruperra Limestone and other named beds near Monmouth and Newport (Welch and Trotter, 1961; Squirrell and Downing, 1969), what have been called the Ffynnon Limestones of the Black Mountains (Ball and Dineley, 1961), and the Abdon Limestone Formation of the Clee Hills (Allen, 1974b).

Calcretes are less prevalent and not as well developed in the Upper Old Red Sandstone, in which they are restricted to the fluvial measures. They are best seen in the West Angle Formation (Williams *et al.*, 1982) of south-west Wales but are also present in the Tintern Sandstone Group and its correlatives (Pick, 1964) of the Forest of Dean area and the Grey Farlow Sandstone of the Welsh Borders.

Mesozoic and younger strata conceal an extensive development of the Old Red Sandstone facies in central and eastern England. Nodular mudstones, possibly including calcretes, are described from an Emsian fluvial facies cored at Canvey Island in the Thames Estuary (Smart *et al.*, 1964). Unquestionable calcretes are present in some abundance in the Famennian fluvial rocks penetrated at Merevale No. 2 east of Birmingham (Taylor and Rushton, 1972).

FIELD CHARACTERISTICS OF THE CALCRETES

Profiles

In terms of vertical profile, most of the calcretes can be assigned to one of three intergrading types (Figure 2). Type A profiles, with affinities to Gile's stage II (Gile *et al.*, 1966) and Goudie's (1983) nodular calcrete, range from a few decimetres to between 2 and 3 m in thickness, and consist of scarce to common, diffuse and irregular to crudely cylindrical, rounded or discoidal calcite glaebules (Brewer, 1964) set in

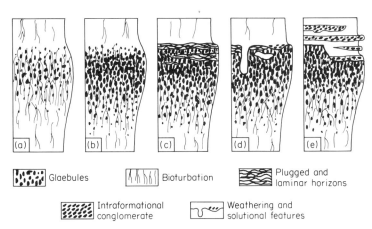

Figure 2. Schematic calcrete profiles (complete and truncated). (a) Type A. (b) Type B. (c) Type C. (d, e) Modified profiles.

the host sediment (Figures 3, 4). Equant to bedding-normal glaebules tend to be restricted to mudstone hosts (Figures 3, 4(a)), whereas discoidal ones, aligned parallel with the lamination, are normally found only in the sandstones (Figures 4(c), (d)). The glaebules vary from a few millimetres to many centimetres across or long, and may

Figure 3. Truncated type A profile bordering on type B, Freshwater West Formation, Greenala Point (British National Grid Reference SS 008 967). Geological hammer for scale.

locally fuse with each other. Untruncated profiles (comparatively rare) show the glaebules to have their highest density about mid-way up or toward the top of the calcrete. Closely resembling Gile's type III profiles, and Goudie's (1983) honeycombe calcretes, are profiles of type B. These locally attain 15 m in thickness (Figure 5(a)) but are generally between 1 and 5 m thick (Figure 5(b)). They are characterized by a gradual upward increase in glaebule size and density, from small and scarce near the base, to large and partly fused at intermediate levels, to closely packed and extensively joined at the top (Figures 5, 6). Particularly where the host sediment is silty or sandy (Figure 6(a)), a three-dimensional meshwork of irregular calcite veins may occur in the middle and lower parts of the profile. Glaebules near the top are in many cases so large and close-packed as to be crudely prismatic and partly separated by discontinuous mudstone screens (Figure 7). Type C profiles compare well with the stage IV calcretes of Gile et al. (1966) and Goudie's (1983) hardpan and laminar calcretes. They are similar to profiles of type B except that, at and near the top, there is evidence

Figure 4. Variation amongst glaebules in calcrete profiles. (a) Equant to bedding-normal glaebules, Red Marl Group, Llanstephan (SN 350 099). (b) Very irregular glaebules with some subhorizontal calcite vein fillings and carbonate-filled tubes, West Angle Formation, West Angle Bay (SM 850 038). (c) Bedding-parallel glaebules in very fine sandstone, Moor Cliffs Formation, south of Angle (SM 867 008). (d) Equant to bedding-parallel glaebules, silty sandstone, West Angle Formation, West Angle Bay (SM 850 038). Geological hammer and pocket measuring tape for scale.

OLD RED SANDSTONE CALCRETES 65

Figure 5. Type B calcretes. (a) Truncated profile (younging to right) overlain by thin patches of intraformational conglomerate, Moor Cliffs Formation, Presipe (SS 069 969). Note wave-like pseudoanticlines and fanning of glaebules (rotate photograph). (b) Calcrete profiles (mainly type B) in the '*Psammosteus*' Limestones, Moor Cliffs Formation, Chapel Point, Caldey Island (SS 141 958). Beds young to right and thickest profile attains about 6 m. Note pseudoanticlines in the two youngest profiles.

Figure 6. Type B calcretes. (a) Red Marl Group, Llanstephan (SN 349 099). Geological hammer for scale is at horizon of maximum glaebule density. Note system of combined horizontal and vertical calcite veins in lower half of profile. (b) Advanced type B profile with local type C features, West Angle Formation, West Angle Bay (SM 867 008). Geological hammer for scale.

for the repeated fracture and rebinding by crystallaria (Brewer, 1964) of extensively fused nodules, and the development of substantial horizontal sheets of irregularly laminated to mammilar crystallaria (Figure 8).

Stratigraphically, the three types of profile are unequally distributed.

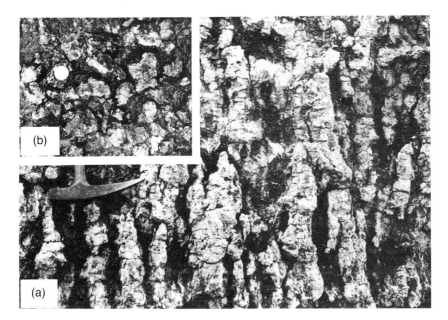

Figure 7. Details of subprismatic glaebules in a type B calcrete (11 m thick), '*Psammosteus*' Limestones, Raglan Marl Group, Lydney (SO 654 021) in (a) vertical profile (note local fusion of glaebules and thin mudstone screens; geological hammer for scale) and (b) bedding-parallel view (coin for scale 0.028 m across).

In the Lower Old Red Sandstone, type A profiles occur throughout the succession and are perhaps most representative of the fluvial measures above the '*Psammosteus*' Limestones. Type B profiles are typical of the Pridoli and early Gedinnian calcretes, including those of the '*Psammosteus*' Limestones. Profiles of type C are comparatively rare, being restricted to the Pridoli and early Gedinnian mud-dominated part of the succession, and to the mid Siegenian calcretes (e.g. Abdon Limestone, Ffynnon Limestones) developed in sandstone hosts. The highest substantial calcrete is developed in the Brownstones above the Pebbly Beds (Squirrell and White, 1978) in Central South Wales. Profiles of all three kinds are present in the Upper Old Red Sandstone, and particularly in equivalents of the Tintern Sandstone Group (Pick, 1964) and the West Angle Formation. The calcretes at this level tend to be much thinner than their counterparts in the Lower Old Red Sandstone and are developed in sandier hosts.

Lateral Extent

There can now be little doubt that individual calcretes in the Old Red Sandstone are laterally very extensive, although the direct evidence for this is not surprisingly meagre, given the nature of the exposures

Figure 8. Vertical polished surfaces in type C calcretes. Scale bars 0.01 m long. (a) '*Psammosteus*' Limestones, Ditton Group, near Ludlow (SO 511 786). Fused glaebules are separated by screens of calcite-cemented quartz sand, and the mass is extensively cut by both horizontal and vertical crystallaria. (b) Abdon Limestone Formation, Abdon Group, near Ludlow (SO 587 868). A complex of very subordinate glaebules with predominant internal, coating, and cross-cutting crystallaria from a laminar horizon. The most extensive crystallaria are sub-horizontal.

available. Only at a few places can calcretes be walked out for any distance. One of these is on the cliffs east of Freshwater East (southwest Wales), where thick profiles of type B can be traced laterally for distances of up to 800 m (Williams *et al.*, 1982). A calcrete 11 m thick within the '*Psammosteus*' Limestones can be walked out a similar distance at Lydney in the Forest of Dean (Welch and Trotter, 1961). Using resistivity methods, Dineley and Gossage (1959) were able to trace calcretes for up to 700 m laterally at a similar stratigraphic level in the Clee Hills. The Abdon Limestone of the same area has been mapped over an extent of roughly 100 km^2 without a significant alteration in character (Allen, 1974b), and so must originally have been part of a much more extensive sheet (embracing the Forest of Dean and south-east Wales?).

Convincing evidence that individual calcretes are as extensive as the preceding observations imply comes from the Pridoli rocks of south-west Wales, where Allen and Williams (1982) were able to correlate with precision numerous thin airfall tuffs just a few metres apart stratigraphically over an area measuring approximately 12 by 35 km. The mudstones between the tuffs include type A and B calcretes, some of which could be recognized over the whole area, except where cut out by an erosively-based sandstone complex.

Figure 9. Upright calcite-filled tubes and small glaebules in type A calcrete, Red Marl Group, Llanstephan (SN 349 099). Coin for scale 0.028 m across.

BIOTURBATION

The mudstones of the Old Red Sandstone are characteristically unlaminated, massive and blocky-fracturing. These features, during Pridoli and early Gedinnian times, resulted from organic destratification contemporaneous with deposition, to judge from the faeces-strewn mudstone surfaces repeatedly smothered by the airfall tuffs (Allen and Williams, 1981, 1982), and the same explanation may largely apply throughout the Old Red Sandstone. Conventional trace fossils in an as yet poorly assessed variety are also plentiful in the Old Red Sandstone facies, and two apparently distinct forms are abundantly present in association with the calcretes, especially those older than mid Siegenian.

One kind, typically surrounded by a diffuse halo of blue mudstone or sandstone, consists of an irregular, downward-branching (once or twice) tube, several to many decimetres long and a few millimetres across with uneven, longitudinal corrugations on the wall. The fill may be either calcite or mudstone. Whether these structures record plant or animal activities remains uncertain (Allen and Williams, 1982). The other kind of trace fossil (Figure 9) is a cylindrical, straight to slightly winding, smooth-walled, subvertical tube up to 8 mm wide and many centimetres long. The fill varies from mud or silt, with evidence for back-packing, to calcite frequently serving as the nucleus of a subsequent glaebule. These structures almost certainly have an invertebrate origin. The striking abundance of these two expressions of bioturbation in the calcretes points to the marked biological activity of the Old Red Sandstone soils, if not throughout the development of each profile then certainly in the earlier stages.

FRACTURE SYSTEMS

Allen (1973, 1974a,b) described patterns of fold-like fractures from certain Lower Old Red Sandstone calcretes, and compared them in both scale and origin to the *gilgai* widely reported from certain clay-rich alluvial soils in present-day warm dry regions (Hallsworth and Beckmann, 1969). Further work has shown that a rather wider range of fracture geometries is represented in Lower Old Red Sandstone calcretes and that perhaps the majority of profiles display one or more of these patterns, which appear schematically in Figure 10. Fracture patterns are rare in the Upper Old Red Sandstone calcretes, developed in generally sandier sediments than in the Lower division. Similar structures have subsequently been found to occur in other ancient fluvial formations with paleosols (e.g. Goldbery, 1982; Wright, 1982).

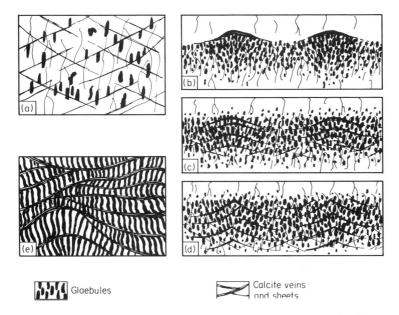

Figure 10. Schematic representation of fracture systems in Old Red Sandstone calcretes. (a) Conjugate planar fractures. (b) Pseudoanticlines expressed by an undulose profile top and internal glaebule fans. (c) Pseudoanticlines in the form of rounded calcite-filled fractures with glaebule fans. (d) Pseudoanticlines expressed by glaebule fans and calcite-filled fractures in the form of pointed folds. (e) Pseudoanticlines with thick calcite-mud fracture-fills and distorted glaebules.

Many profiles of type A display in vertical profile an open meshwork of extensive, conjugate, slickensided fractures (Figure 10(a)), which may be either unfilled or sealed with millimetre-scale calcite veins (Figure 11(a)). The fractures vary from planar (the majority) to weakly concave up, and lie at 20–25° from the general bedding. There is no spatial variation in the distribution of dip-angles on the fractures.

A very few profiles of type B (advanced) and type C reveal a regularly undulose top on a scale of several metres beneath which the glaebules show a systematic variation in both size and density as well as orientation (Figure 10(b)). This type of structure, for which the term pseudoanticline (Watts, 1977) is appropriate, is strongly suggestive of *gilgai* and particularly recalls the variety involving puffs (mounds) and shelves (Allen, 1974a).

Much more common in profiles of both types A and B are the fold-like (wavelength 2–10 m) distributions of glaebules and calcite-filled slickensided fractures shown schematically in Figure 10(c), (d). The first variety displays, in two-dimensional profile, fan-like arrangements of glaebules (cf. cleavage fans) and extensive calcite-filled fractures in

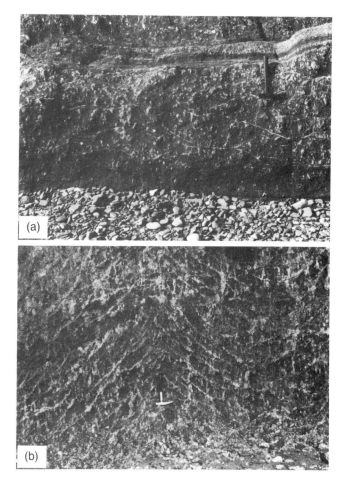

Figure 11. Fracture systems in calcrete profiles. Geological hammer for scale. (a) Type A calcrete with truncated top and conjugate planar to weakly concave calcite-filled fractures, Red Marl Group, Llanstephan (SN 349 099). (b) Pseudoanticlines (see Fig. 10d) in a type B calcrete 11 m thick, '*Psammosteus*' Limestones, Raglan Marl Group, Lydney (SO 654 021).

the form of rounded 'synclines' and less rounded 'anticlines'. The second variety differs from the first in showing sharply pointed anticlines and a more obvious severance and/or displacement of glaebules across the fractures (Figure 11(b)). The three-dimensional geometry of these fracture systems has not yet been fully explored, but their general form can be readily appreciated from certain favourable cliff exposures, east of Freshwater East and at Greenala Point near Stackpole, both in south-west Wales. Most of the fractures prove to be roughly bowl-like, with the slickensides radiating upward away from the axis of the concave-up bowl. These fractures become less strongly developed downward in the calcrete profiles and may be accompanied,

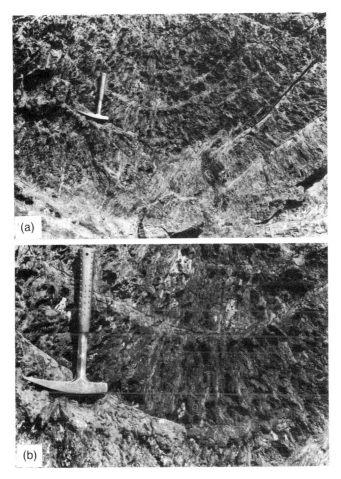

Figure 12. General view (a) and close-up (b) of pseudoanticlines defined by thick calcite-mud fracture fillings (note open cracks) and distorted glaebules (partly weathered out), Moor Cliffs Formation, south of Angle, (SM 867 009). Geological hammer for scale.

at intermediate and low levels, by systems of less conspicuous planar cracks (Figure 10(a)).

The thick types B and C profiles developed in the Pridoli and early Gedinnian mudstones, especially of south-west Wales, display a fifth type of pseudoanticlinal structure (Figure 10(e)). The calcrete is divided up into large concavo-convex to corrugated lens-like masses by curved centimetre-thick sheets composed of irregularly interlaminated calcite and terrigenous mud (Figures 5(a), 12). The glaebules, closely packed between mudstone screens and almost invariably highly elongate, in many places bend toward the sheets in a manner reminiscent of strata dragged against a fault (Figure 12(b)). An open crack may be present at the centre of each sheet.

Figure 13. Features of the eroded tops of calcrete profiles. Geological hammer and camera lens-cap for scale. (a) Calcrete profile overlain by several centimetres of intraformational pebble conglomerate (arrowed) and a (calcretized) mudstone, Freshwater West Formation, Greenala Point (SS 008 967). (b) Interbedded mudstones and calcrete conglomerates overlying the uneven, eroded top of a calcrete in the '*Psammosteus*' Limestones, Moor Cliffs Formation, Caldey Island (SS 143 958). (c) Part of a thick complex of interbedded mudstones and intraformational conglomerates overlying a calcrete, Red Marl Group, Llanstephan (SN 350 099). (d) Pebbly mudstone infilling channeled top of calcrete, '*Psammosteus*' Limestones, Moor Cliffs Formation, Caldey Island (SS 143 958).

PROFILE TRUNCATION

A substantial proportion of calcrete profiles, especially the thicker ones, are sharply truncated upward by a surface of either erosion (Allen and Williams, 1979) or, much less commonly, weathering.

Weathered tops, as can be seen at Lydney in the Forest of Dean, are slightly uneven surfaces, with the hollows infilled by a mud-bound rubble of slightly disturbed angular calcrete fragments ranging downward to millimetre-scale crumbs (Figure 2(d)). The bosses between, instead of being irregular, are in some places smoothly curved, as if rounded by solution. Features closely resembling vertical solution pipes and cavities have been seen to range downward for a few decimetres from the sharp top of one profile.

Eroded tops to calcrete profiles are much more common (Figure 2(e)). The thick profile shown in Figure 5(a) has a sharp, even top overlain by patches up to a few centimetres thick of an intraformational conglomerate formed of sorted and rounded calcrete debris (see also Figure 13(a)). In many instances this kind of overlay is a mere single clast thick, and is easily overlooked in the field, particularly where access is difficult. A fairly common type of overlay is a complex as thick as 2 m of interbedded mudstone and intraformational conglomerate units resting on an even to slightly irregular surface (Figure 13(b), (c)). Locally these conglomerates are internally channelled. Some conglomerate units are well sorted and internally bedded, with a sparry calcite cement, whereas others are ill-sorted and poorly if at all stratified internally, mud serving to bind the clasts. Quartz sand grains are invariably lacking from the conglomerates and beds of quartzose sandstone do not figure amongst the interbedded units of conglomerate and mudstone. Another feature which distinguishes intraformational conglomerates in this setting from those associated with upward-fining quartzose sandstone complexes is the absence of vertebrate remains (Allen and Williams, 1979). Locally the eroded tops of calcretes reveal broad, shallow flat-bottomed channels (Figure 13(d)).

MICROSCOPIC FEATURES OF THE CALCRETES

Compositionally, the calcrete glaebules and massive limestones consist of carbonate minerals together with a variable quantity of quartz, feldspar, rock fragment and clay-mineral particles representative of the host sediment (Allen, 1965, 1974a,b). Except in the folded sequences of south-west Wales and Anglesey, where secondary ferroan dolomite and locally quartz are present in most calcretes, the profiles are dominated by low-magnesian calcite. Locally in the Welsh Borders a little barytes in the form of small rosettes is present, and in one calcrete profile in south-west Wales developed on a tuff (Allen and Williams, 1978), manganiferous and ferruginous nodules are recorded.

Thin sections, peels and polished surfaces reveal in the unrecrystallized calcretes from the Old Red Sandstone a wealth of textures and fabrics which have been described in detail (Allen, 1965, 1974a,b). Commonest is Brewer's (1964) undifferentiated crystic plasmic fabric, represented mainly by a lightly mottled mosaic of microcrystalline calcite accompanied by floating skeleton grains of quartz, feldspar, rock fragments, mica and clay minerals (Figure 14(a), (b)). Many of the skeleton quartz grains reveal corroded margins and some of the feldspars (rare from the start) have been entirely replaced by calcite. Exfoliation tends to be shown by cleaved or foliated skeleton grains

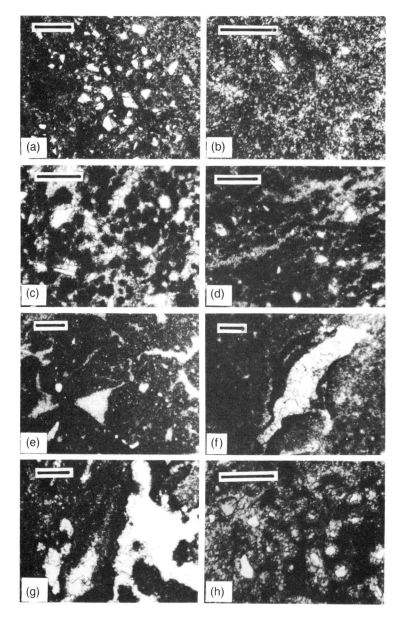

Figure 14. Microscopical features of calcretes in vertical section (stratigraphical top upward). (a) Undifferentiated crystic plasmic fabric. (b) Undifferentiated crystic plasmic fabric with fewer skeleton grains than (a). (c) Agglomeratic fabric. (d) Subhorizontal cracks. (e) Irregular to concavo-triangular crystallaria. (f) Laminated crystallaria lining an irregular void. (g) Complex crystallaria from a large void filling. (h) Skeleton grains with coarse-textured calcite halos. All samples from the Clee Hills. (a, h) Abdon Limestone (SO 587 868). (b, g) *Psammosteus* Limestones (SO 511 786). (c, d, e) Abdon Limestone (SO 581 848). (f) *Psammosteus* Limestones (SO 576 888). Scale bar 250 μm long.

(feldspar, mica, schistose and phyllitic rock fragments). Brewer's (1964) agglomeratic fabric is well represented in the glaebules and glaebular masses (Figure 14(c)). The voids preserved in these vary from subhorizontal systems of irregular hair-like cracks (Figure 14(d)), to irregular to concavo-triangular spaces between disrupted and partly displaced masses with a crystic plasmic fabric (Figure 14(e)), to septaria-like patterns (Allen, 1965, Pl. 13, Fig. 2), and finally large planar to irregular voids lined by laminated calcite crystallaria (Figure 14(f)). The largest voids of all were large open cracks, in many cases partly plugged by irregularly laminated to mammilar crystallaria before being broken open again (Figures 8, 14(g)). A final common fabric, a variant of Brewer's (1964) agglomeratic fabric and attributed to corrosion by some workers (e.g. Goudie, 1983), shows skeleton grains surrounded by a diffuse halo of microcrystalline calcite coarser textured than the bulk of the plasma (Figure 14(h)). Some smaller glaebules in Old Red Sandstone calcretes reveal a crudely concentric structure, but nothing comparable to the oolitic and pisolitic structures found in some Quaternary calcretes has so far been recorded.

THE PEDOGENIC MODEL

The arguments for the pedological origin of the (Lower) Old Red Sandstone calcretes are set out fully and in detail elsewhere (Allen, 1973, 1974a,b), on the basis of an extensive direct and literature comparison with Quaternary calcretes and carbonate-bearing soils from many parts of the world. Only the main points need therefore be mentioned here, and details of the setting, character and age of these Quaternary examples used for comparison should be sought in the papers mentioned.

Essentially, the pedogenic model is adopted because the calcretes are closely comparable in scale, profile, macrostructure, composition and microfabric to many Quaternary pedogenic calcretes and carbonate-bearing soils, as reviewed by Netterberg (1967, 1971) and Goudie (1973), and more recently by Read (1976), Reeves (1976), Goudie (1983), and Milnes and Hutton (1983). The evidence points to the combined replacive and, more important, displacive introduction of calcite into deep alluvial sediments, particularly overbank muds but in many instances the upper levels of abandoned channel sand-bodies. The rocks are distinctly different from groundwater calcretes (Mann and Horwitz, 1979), principally in field development, although limestones and other calcareous sediments were exposed in the Old Red Sandstone source areas (Allen, 1974b); the amounts of plagioclase feldspar which reached the depositional site are far too small to have afforded, by their decay, the quantity of carbonate minerals locked up in the calcretes. Hence the *per descensum* scheme of calcrete formation (Goudie, 1973, 1983) is preferred, with the carbonate entering the Anglo-Welsh area as windblown dust.

The widely and abundantly developed fracture systems are also consistent with a pedological origin for the calcretes. Bearing a marked resemblance to the larger cracks analysed by Knight (1980) from an Australian gilgaied soil, their shape, attitude and slickensided nature, combined with the evidence of displaced and curved glaebules, strongly suggests that they arose during the formation of the calcrete and in response to disruptive horizontal stresses (compressive) that substantially exceeded the vertical ones. As Allen (1973) pointed out, such a system of forces permitting upward relief could arise because of either or both (1) the displacive introduction of calcite to form glaebules, and (2) the swelling and shrinking of the clay-rich host in response to climatically controlled moisture variations. Probably both causes operated, although changing in relative importance as the calcrete profile matured. Moisture-related changes must be invoked where the glaebule density is low (e.g. Figure 11(a)), but the continued introduction of calcite is sufficient to explain the relative movement of large masses of soil-material suggested by the fractured carbonate-rich calcretes (Figures 5(a), 12). Significant for the former process is the fact that smectite clays, perhaps of penecontemporaneous volcanic origin (Parker *et al.*, 1983), occur in significant quantities in most of the Pridoli and earliest Gedinnian mudstones (e.g. Allen, 1974b). Watts (1977) in his discussion of pseudoanticlines has emphasized the considerable crystallization pressure exerted by calcite, but the displacive fibrous calcites he occasionally found have not so far been detected in the Old Red Sandstone calcretes.

SIGNIFICANCE OF THE CALCRETES

TIME-SCALES

Various lines of evidence (Gile, 1970, 1977; Baker, 1973; Goudie, 1973; Hay and Reeder, 1978; Shlemon, 1978), including radiocarbon dating (Gile, 1970; Williams and Polach, 1971; Williams, 1973), now known to have unreliable features (Callen *et al.*, 1983), have been used to age Quaternary calcretes absolutely and to estimate durations of pedogenesis. Visible carbonate accumulations can arise in as short a time as 10^2 years. Well-developed profiles can be formed within as short a period as of the order of 10^3 years or may necessitate pedogenosis over as long as 10^5 or even 10^6 years. The rate of calcrete formation thus appears to be highly variable and is probably dependent on many factors, of which the rainfall regime (Gile, 1977) and the source and availability of carbonate may be the most important. Nothing directly is known of the durations of pedogenesis represented by Old Red Sandstone calcretes, but some constraints can be indicated by reference to stratigraphical evidence and our general understanding of the fluvial environment, the context of Old Red Sandstone pedogenesis. This provides a basis for looking at the implications of other time-scales.

As has been noted, there is within the Lower Old Red Sandstone some stratigraphical variation in the frequency, maturity, and mode of preservation of the calcretes. So far as the younger beds are concerned, the continued generation of calcrete soils within the depositional basin can be inferred only from the prevalence of calcrete debris in the intraformational conglomerates which floor the fluvial channel sandstones. Nonetheless, a fair average for the stratigraphical frequency of the calcretes would be one profile for each 5 m of the succession upward from the base of the Temeside Shale Formation and its correlatives (Figure 1). Assigning a mean total thickness of 3 km to the Lower Old Red Sandstone, we arrive at a total of about 600 calcretes, or pedogenic events, in a vertical section such as a borehole core through the beds. Arguably the Lower Old Red Sandstone was accumulated over a period of about 20 Ma (Harland et al., 1982), whence soil-forming conditions at a site could have occurred on the average roughly once every 3.33×10^4 years. The average duration of profile generation therefore cannot have exceeded this figure which, if it had actually been operative, would have left no time for the deposition of the host mudstones and sandstones. Of the mean total thickness of 3 km for the Lower Old Red Sandstone, about 1.54 km comprise non-calcretized floodplain mudstones. On the basis of our understanding of fluvial environments they are likely to have accumulated much more slowly than the partly laterally-deposited channel sandstones. If Leeder (1975) is correct in identifying a floodplain deposition rate of order 2×10^{-3} m a^{-1} as the limit to calcrete generation, the mudstones in the Lower Old Red Sandstone cannot in total represent more than about 1.15 Ma (consolidation by 33% to present thickness assumed). This result implies a maximum average duration of pedogenesis per calcrete profile of roughly 3.14×10^4 years, a figure not significantly different from the previous value based on total formation thickness. Thus the Lower Old Red Sandstone calcretes may not have horizonated at rates greater on the average than the order of 10^4 years per profile, and it is possible that only a small fraction of the 20 Ma represented by the Lower Old Red Sandstone may, at any one site, have been spent in the normal accumulation of sediment.

PALAEOGEOMORPHOLOGY

The formation of a soil requires the exposure of a parent material to the atmosphere over a sufficient period of time, so that the relevant physical and chemical processes and biological communities can become established, together with a state of geomorphic stability, so that soil horizons can be initiated and matured. In the context of an extensive region of alluviation, such as the Anglo-Welsh area during Old Red Sandstone times, the requirement of geomorphic stability at a site implies a pause in sedimentation, or at the very least a reduction in the sediment supply to a level at which soil-horizonation became possible. The abundance in the calcretes of features due to bioturbation is a clear and independent proof of the required reduction.

The required pauses must have been at least sub-regional in spatial extent, to judge from the evidence summarized above on the lateral range of individual calcretes. Such evidence, together with the truncated tops to many profiles, suggests a simple geomorphological model for the Anglo-Welsh area in Old Red Sandstone times (Allen and Williams, 1979). If one had been able to fly over the region at any time during the Pridoli, one might have seen extensive calcreted plains separating narrower but shallow valleys containing estuarine channels. Similarly, at any time during the Gedinnian and at least the bulk of the Siegenian, one might have witnessed similar plains divided by broad shallow valleys watered by rivers. The valleys in both cases reached back into the source-lands of the Old Red Sandstone, but the intervening plains bore their own smaller drainage systems, fed by local rain. To judge from the geomorphological expression of Quaternary calcretes (Goudie, 1973, 1983; Reeves, 1976), the Old Red Sandstone profiles in the later stages of their development could have formed caprocks underlying, and escarpments bordering, these level plains. The Abdon Limestone and its possible correlatives is a particularly mature and perhaps very extensive calcrete. Geomorphologically, it invites comparison with the plateau-forming calcretes of the Reynosa Cuesta, Edwards Plateau, and High Plains between the Kansas River and the Rio Grande in the south-western USA (Price, 1958; Swineford et al., 1958; Goudie, 1973), and could record some especially significant event in the development of the Old Red Sandstone succession. There is certainly a noticeable change in fluvial facies across the Abdon Limestone in the Clee Hills (Allen, 1974b). The '*Psammosteus*' Limestones lower down in the succession undeniably record a substantial (0.25−0.5 Ma?) and regionally effective period of marked sediment starvation in the depositional basin. As there is a significant change of both depositional environment and sediment provenance vertically across this complex, it is possible that these calcretes owe their existence to a geodynamic event which, by creating new uplifts, caused the eventual redirection of existing regional drainage and the establishment of new river systems (Allen 1974b).

PALAEOCLIMATE

Calcretes have been reported from some polar situations but, speaking generally, they are today most prevalent in and characteristic of a warm to hot climate (mean annual temperature 16–20°C) marked by a low but seasonal rainfall (100–500 mm) (Goudie, 1973, 1983). The Old Red Sandstone may therefore have formed under similar conditions, a conclusion consistent with the moderately low latitudinal position assigned on various independent grounds to the southern British Isles in the Devonian (e.g. Heckel and Witzke, 1979). The above suggestion can of course refer only to the depositional site, and must not be taken to imply that the Old Red Sandstone rivers were necessarily ephemeral or even flashy, as a significantly different climate could have prevailed in the source lands where their discharges originated (e.g. Colorado River of California).

ALLUVIAL ARCHITECTURE

One of the major current problems of fluvial sedimentology concerns the way in which the bodies of channel-related coarse sediment on the one hand and of overbank fines on the other fit together three-dimensionally to make up a fluvial formation. Another way of putting this is to ask how the essential palaeogeomorphology (Allen and Williams, 1979) sketched above may be projected through time to give us an assembly of sedimentary surfaces and enclosed deposits. The spatial scale on which this fit or assembly must be considered— kilometres to tens of kilometres horizontally and metres to tens of metres vertically—means that the question cannot generally be resolved from outcrop evidence but must be approached in other ways. Although this is not the place in which to review this now substantial field as a whole, it is worth noting that the pedogenic interpretation of the Old Red Sandstone calcretes is consistent to a degree with a number of models of fluvial architecture, depending on the factor(s) considered to be in control (Allen, 1974a).

Allen's first model envisaged pedogenesis as controlled by climatically-dependent proximal-distal shifts in the range of rivers entering the depositional basin. This model could be discounted for the Old Red Sandstone, primarily because it implied a pattern and occurrence of facies not recorded from the succession.

In three models an autocyclic cause is envisaged, pedogenesis being controlled by lateral movements of the rivers, either continuously, in small steps, or in large avulsive steps. The last of these accords best with the behaviour of modern rivers and leads to quite realistic facies distributions.

A model in which the rivers shift only vertically in response to a purely allocyclic control (climatic change, base-level change, tectonism) is unacceptable, because channel sands, overbank muds and palaeosols could not then occur together in the one vertical profile. A combination of allocyclic and autocyclic factors, however, can create realistic sequences and the interfingering in three dimensions of channel with overbank sediment bodies. Remembering that the Anglo-Welsh area in late Silurian and Devonian times bordered a sea to the south, base-level change driving cyclic valley cutting and filling may have been the most significant factor controlling the geomorphology and sedimentology of the region. Evidence that base-level change may indeed have been a major allocyclic control comes from the Temeside Shale Formation and its correlatives, in which, on a scale of metres, sandstone complexes with a restricted marine fauna are interbedded vertically with mudstones (some also with lingulids) and calcretes (e.g. Dixon, 1921; Allen, 1974b; Allen and Williams, 1978). A lowering of base-level by no more than a few metres would have been all that was necessary to have brought a coastal mudflat into the vadose zone and the realm of pedogenesis.

CONCLUSIONS

(1) As with fluvial formations in many parts of the stratigraphic record, calcretes are abundant and in many instances well-developed in the Old Red Sandstone facies (late Silurian–early Carboniferous) of the Anglo-Welsh area in the southern British Isles.

(2) The calcrete profiles vary widely in vertical extent, but are typically a few metres thick, revealing an upward increase in the size and density of calcite glaebules contained in a mudstone host and, in the case of the better developed ones, a massive to laminated uppermost part. Individual calcretes appear to have a lateral range measuring kilometres to tens of kilometres.

(3) Features due to bioturbation abound in the calcrete profiles and provide a clear and independent proof of an intermittent and drastic reduction in the local rate of sediment deposition.

(4) Systems of organized and slickensided fractures, some of which may be classified as pseudoanticlines and compared to *gilgai*, are common in the calcrete profiles and suggest superficial compressive effects related to seasonal clay swelling and drying and/or the displacive introduction of carbonate minerals.

(5) Many of the calcrete profiles are truncated upward by a surface of either weathering or erosion, overlain in the first instance by calcrete rubble and displaying in some cases solution phenomena, and in the second by an intraformational conglomerate or conglomerate complex.
(6) In terms of microscopic fabrics and textures, as well as field characteristics, the Old Red Sandstone calcretes compare closely with Quaternary calcretes and carbonate-bearing soils and may tentatively be referred largely to the Aridsol and Vertisol classes.
(7) The calcretes point to a warm, dry, seasonal climate in the Old Red Sandstone depositional basin, and to a basin landscape (known on other grounds to be fluvial) in which geomorphological change was drastic and frequent and, on the larger scale, under the control of allocyclic factors (mainly base-level oscillations?). The calcretes have a recurrence interval of about 3×10^4 years. In many cases a calcrete came to underlie a low plateau characterized by local drainage systems which fed laterally into one or more of the larger rivers crossing the alluvial plain from the distant source-lands of the Old Red Sandstone. Major geodynamic events affecting the accumulation of the Old Red Sandstone had a significant influence on calcrete pedogenesis within the depositional basin.

REFERENCES

Allen, J.R.L. 1965. Sedimentation and palaeogeography of the Old Red Sandstone of Anglesey, North Wales. *Proc. Yorks geol. Soc.* **35**, 139–85.

Allen, J.R.L. 1973. Compressional structures (patterned ground) in Devonian pedogenic limestones. *Nature, Phys. Sci.* **243**, 84–6.

Allen, J.R.L. 1974a. Studies in fluviatile sedimentation: implications of pedogenic carbonate units, Lower Old Red Sandstone, Anglo-Welsh outcrop. *Geol. J.* **9**, 181–208.

Allen, J.R.L. 1974b. Sedimentology of the Old Red Sandstone (Siluro-Devonian) in the Clee Hills area, Shropshire, England. *Sedim. Geol.* **12**, 73–167.

Allen, J.R.L. 1977. Wales and the Welsh Borders, in *A Correlation of the Devonian Rocks in the British Isles*, M.R. House, (ed.). *Spec. Rep. geol. Soc. Lond. No. 8.* 40–54.

Allen, J.R.L. 1983. Studies in fluviatile sedimentation: bars, bar-complexes, and sheet sandstones (low-sinuosity braided streams) in the Brownstones (L. Devonian), Welsh Borders. *Sedim. Geol.* **33**, 237–93.

Allen, J.R.L. and Dineley, D.L. 1976. The succession of the Lower Old Red Sandstone (Siluro-Devonian) along the Ross-Tewkesbury Spur Motorway (M.50), Hereford and Worcester. *Geol. J.* **11**, 1–14.

Allen, J.R.L. and Williams, B.P.J. 1978. The sequence of the earlier Lower Old Red Sandstone (Siluro-Devonian), north of Milford Haven, southwest Dyfed (Wales). *Geol. J.* **13**, 113–36.

Allen, J.R.L. and Williams, B.P.J. 1979. Interfluvial drainage on Siluro-Devonian alluvial plains in Wales and the Welsh Borders. *J. geol. Soc. Lond.* **136**, 361–6.

Allen, J.R.L. and Williams, B.P.J. 1981. Sedimentology and stratigraphy of the Townsend Tuff Bed (Lower Old Red Sandstone) in South Wales and the Welsh Borders, *J. geol. Soc. Lond.* **138**, 15–29.

Allen, J.R.L. and Williams, B.P.J. 1982. The architecture of an alluvial suite: rocks between the Townsend Tuff and Pickard Bay Tuff beds (early Devonian), southwest Wales. *Phil. Trans. R. Soc.* **B297**, 51–89.

Allen, J.R.L., Thomas, R.G. and Williams, B.P.J. 1981. Field Meeting: The facies of the Lower Old Red Sandstone, north of Milford Haven, southwest Dyfed, Wales. *Proc. Geol. Ass.* **92**, 251–67.

Allen, J.R.L., Thomas, R.G. and Williams, B.P.J. 1982. The Old Red Sandstone north of Milford Haven, in *Geological Excursions in Dyfed, South-West Wales*, M.G. Bassett (ed.). National Museum of Wales and University of Wales Press, Cardiff. 123–49.

Baker, V.R. 1973. Paleosol development in Quaternary alluvium near Colorado. *Mtn. Geol.* **10**, 127–33.

Ball, H.W. and Dineley, D.L. 1961. The Old Red Sandstone of Brown Clee Hill and the adjacent area. I. Stratigraphy, *Bull. Brit. Mus. Nat. Hist. (Geology)* **5**, 177–242.

Bown, T.M. and Kraus, M.J. 1981. Lower Eocene alluvial paleosols (Willwood Formation, northwest Wyoming, U.S.A.) and their significance for paleoecology, paleoclimatology, and basin analysis. *Palaeogeogr. Palaeoclimatol. Palaeoecol.* **34**, 1–30.

Brewer, R. 1964. *Fabric and Mineral Analysis of Soils*. Wiley, New York. 470 pp.

Burgess, I.C. 1961. Fossil soils in the Upper Old Red Sandstone of south Ayrshire. *Trans. geol. Soc. Glasgow* **24**, 138–53.

Callen, R.A., Wasson, R.J. and Gillespie, R. 1983. Reliability of radiocarbon dating of pedogenic carbonate in the Australian arid zone. *Sedim. Geol.* **35**, 1–14.

Dineley, D.L. and Gossage, D.W. 1959. The Old Red Sandstone of the Cleobury Mortimer area, Shropshire. *Proc. Geol. Ass.* **70**, 221–38.

Dixon, E.E.L. 1921. *Geology of the South Wales Coalfield. Part XIII. The Country around Pembroke and Tenby*. Mem. geol. Surv. UK. 220 pp.

Gardiner, P.R.R. and Horne, R.R. 1981. The stratigraphy of the Upper Devonian and Lower Carboniferous clastic sequence in southwest County Wexford. *Bull. geol. Surv. Ireland* **3**, 51–77.

Gile, L.H. 1970. Soils of the Rio Grande Valley border in southern New Mexico. *Proc. Soil Sci. Soc. Am.* **34**, 465–72.

Gile, L.H. 1977. Holocene soils and soil-geomorphic relations in a semi-arid region of southern New Mexico. *Quat. Res.* **7**, 112–32.

Gile, L.H., Peterson, F.F. and Grossman, R.B. 1966. Morphological and genetic sequences of carbonate accumulation in desert soils. *Soil Sci.* **101**, 347–60.

Goldbery, R. 1982. Palaeosols of the Lower Jurassic Mishhor and Ardan Formation ('Laterite Derivative Facies'), Maktesh Ramon, Israel. *Sedimentology* **29**, 669–90.

Goudie, A. 1973. *Duricrusts in Tropical and Subtropical Landscapes*. Clarendon Press, Oxford. 174 pp.

Goudie, A.S. 1983. Calcrete, in *Chemical Sediments and Geomorphology*, A.S. Goudie and K. Pye (eds). Academic Press, London. 93–131.

Hallsworth, E.G. and Beckmann, G.G. 1969. Gilgai in the Quaternary. *Soil. Sci.* **107**, 409–20.

Harland, W.B., Cox, A.V., Llewellyn, P.G., Pickton, C.A.G., Smith, A.G. and Walters, R. 1982. *A Geologic Time Scale*. Cambridge University Press, Cambridge. 131 pp.

Hay, R.L. and Reeder, R.J. 1978. Calcretes of the Olduvai Gorge and the Ndolanga Beds of northern Tanzania. *Sedimentology* **25**, 649–73.

Heckel, P.H. and Witzke, B.J. 1979. Devonian world palaeogeography determined from distribution of carbonates and related lithic palaeoclimatic indicators. *Spec. Pap. Palaeont.* No. **23**, 99–123.

Hubert, J.R. 1978. Paleosol caliche in the New Haven Arkose, Newark Group, Connecticut. *Palaeogeogr. Palaeoclimatol. Palaeoecol.* **24**, 151–68.

King, W.W. 1934. The Downtonian and Dittonian strata of Great Britain and north-western Europe. *Q. J. geol. Soc. Lond.* **90**, 526–70.
Knight, M.J. 1980. Structural analysis and mechanical origins of gilgai at Boorook, Victoria, Australia. *Geoderma* **23**, 245–83.
Leeder, M.R. 1975. Pedogenic carbonates and flood sediment accretion rates: a quantitative model for alluvial arid-zone lithofacies. *Geol. Mag.* **112**, 257–70.
Leeder, M.R. 1976. Palaeogeographic significance of pedogenic carbonates in the topmost Upper Old Red Sandstone of the Scottish Border Basin. *Geol. J.* **11**, 21–8.
MacCarthy, I.A.J., Gardiner, P.R.R. and Horne, R.R. 1978. The lithostratigraphy of the Devonian-early Carboniferous succession in parts of Counties Cork and Waterford, Ireland. *Bull. geol. Surv. Ireland* **2**, 265–305.
Mann, A.W. and Horwitz, R.C. 1979. Groundwater calcrete deposits in Australia: some observations from Western Australia. *J. geol. Soc. Aust.* **26**, 293–303.
Milnes, A.P. and Hutton, J.T. 1983. Calcrete in Australia, in *Soils: an Australian Viewpoint*, Division of Soils CSIRO, CSIRO/Academic Press, London. 119–162.
Molenaar, N. 1984. Palaeopedogenic features and their palaeoclimatological significance for the Nevremont Formation (Lower Givetian), the northern Ardennes, Belgium. *Palaeogeogr. Palaeoclimatol. Palaeoecol.* **46**, 325–44.
McPherson, J.G. 1979. Calcrete (caliche) paleosols in fluvial redbeds of the Aztec Siltstone (Upper Devonian), southern Victoria Land, Antarctica. *Sedim. Geol.* **22**, 267–85.
Netterberg, F. 1967. Some roadmaking properties of South African calcretes. *Proc. Fourth Conf. for Africa on Soil Mechanics and Foundation Engng.* Cape Town, South Africa. 77–81.
Netterberg, F. 1971. Calcrete in road construction. *Nat. Inst. Road Res. Bull. (CSIR Pretoria)* **10**, 1–73.
Parker, A., Allen, J.R.L. and Williams, B.P.J. 1983. Clay mineral assemblages of the Townsend Tuff Bed (Lower Old Red Sandstone), South Wales and the Welsh Borders. *J. geol. Soc. Lond.* **140**, 769–79.
Parnell, J. 1983a. The Cothall Limestone. *Scott. J. Geol.* **19**, 215–18.
Parnell, J. 1983b. Ancient duricrusts and related rocks in perspective: a contribution from the Old Red Sandstone, in *Residual Deposits: Surface Related Weathering Processes and Materials*, R.C.L. Wilson, (ed.). Published by Blackwell Scientific Publications, Oxford for The Geological Society of London. 197–209.
Pick, M.C. 1964. The stratigraphy and sedimentary features of the Old Red Sandstone, Portishead section, north-east Somerset. *Proc. Geol. Ass.* **75**, 199–221.
Price, W.A. 1958. Sedimentology and Quaternary geomorphology of south Texas. *Trans. Gulf Cst. Ass. geol. Socs.* **8**, 41–75.
Read, J.F. 1976. Calcretes and their distinction from stromatolites, in *Stromatolites*, M.R. Walter (ed.). Elsevier, Amsterdam. 55–71.
Reeves, C.C. 1976. *Caliche*. Estecado Books, Lubbock, Texas. 233 pp.
Shlemon, F.J. 1978. Quaternary soil-geomorphic relationships, southeastern Mojave Desert, California and Arizona, in *Quaternary Soils*, W.C. Mahaney (ed.). Geoabstracts, Norwich. 187–207.
Smart, J.G.O., Sabine, P.A. and Bullerwell, W. 1964. The Geological Survey boring at Canvey Island, Essex. *Bull. geol. Surv. Gt. Br.* No. **21**, 1–36.
Squirrell, H.C. and Downing, R.A. 1969. *Geology of the South Wales Coalfield. Part I. The country around Newport (Mon.)*. Mem. geol. Surv. UK. 333 pp.
Squirrell, H.C. and White, D.E. 1978. Stratigraphy of the Silurian and Old Red Sandstone of the Cennen Valley and adjacent areas, south-east Dyfed, Wales. *Rep. Inst. geol. Sci. Lond.* No. **78/6**, 1–45.
Steel, R.J. 1974. Cornstone (fossil caliche)—its origin, stratigraphic, and sedimentologic importance in the New Red Sandstone, western Scotland. *J. Geol.* **83**, 351–69.
Swineford, A., Leonard, A.B. and Frye, J.C. 1958. Petrology of the Pliocene pisolitic

limestones in the Great Plains. *Bull. geol. Surv. Kansas*, No. **130**, 97–116.

Taylor, K. and Rushton, A.W.A. 1972. The pre-Westphalian geology of the Warwickshire Coalfield. *Bull. geol. Surv. Gt. Br.* No. **35**, 1–152.

Tunbridge, I.P. 1981a. The Yes Tor Member of the Hangman Sandstone Group (North Devon). *Proc. Ussher Soc.* **5**, 7–12.

Tunbridge, I.P. 1981b. Old Red Sandstone sedimentation—an example from the Brownstones (highest Lower Old Red Sandstone) of south central Wales. *Geol. J.* **16**, 111–24.

Watts, N.L. 1977. Pseudo-anticlines and other structures in some calcretes of Botswana and South Africa. *Earth Surf. Proc.* **2**, 63–74.

Welch, F.B.A. and Trotter, F.M. 1961. *Geology of the Country around Monmouth and Chepstow*. Mem. geol. Surv. UK. 164 pp.

White, E.I. 1950a. The vertebrate faunas of the Lower Old Red Sandstone of the Welsh Borders. *Bull. Br. Mus. Nat. Hist. (Geology)* **1**, 51–67.

White, E.I. 1950b. *Pteraspis leathensis* White, a Dittonian zone-fossil. *Bull. Brit. Mus. Nat. Hist. (Geology)* **1**, 69–89.

Williams, B.P.J., Allen J.R.L. and Marshall, J.D. 1982. Old Red Sandstone facies of the Pembroke Peninsula, S.W. Dyfed—south of the Ritec Fault, in *Geological Excursions in Dyfed, South-West Wales*, M.G. Bassett (ed.). National Museum of Wales and University of Wales Press, Cardiff. 151–74.

Williams, G.E. 1968. Torridonian weathering and its bearing on Torridonian palaeoclimate and source. *Scott. J. Geol.* **4**, 164–84.

Williams, G.E. 1973. Late Quaternary piedmont sedimentation, soil formation and palaeoclimates in arid South Australia. *Z. Geomorph.* **17**, 102–25.

Williams, G.E. and Polach, H.A. 1971. Radiocarbon dating of arid-zone calcareous palaeosols. *Bull. geol. Soc. Am.* **82**, 3069–86.

Wright, V.P. 1982. Calcrete palaeosols from the Lower Carboniferous Llanelly Formation, South Wales. *Sedim. Geol.* **33**, 1–33.

Chapter 3

PALEOSOLS CONTAINING AN ALBIC HORIZON: EXAMPLES FROM THE UPPER CARBONIFEROUS OF NORTHERN ENGLAND

C. J. PERCIVAL
BP Alaska Exploration Inc., San Francisco, USA

INTRODUCTION

The Upper Carboniferous sequence of northern England consists of a clastic dominated succession deposited in a predominantly deltaic environment. This sequence contains a wide variety of paleosols, many of which formed in delta plain environments which were established as a result of progradation of fluvial dominated deltas. A small but significant number of these paleosols contain a conspicuous white to light grey, quartz arenitic sandstone horizon underlying any originally surficial organic or organo-mineral horizons and overlying a more clay rich horizon. Such paleosols form the subject of this paper. The mineralogically mature quartz arenite is interpreted as the albic (eluvial) horizon of the paleosol profile.

In some cases it can be demonstrated that the quartz arenite developed from a less mature parent material due to weathering and leaching during pedogenesis. However, in many instances the possibility that the horizon developed from an original distinct sedimentary quartz arenite layer cannot be discounted.

Due to their high quartz content, many of these quartz arenites were extracted for use as the raw material in the manufacture of siliceous refractory bricks which were used by the local iron and steel industry. Horizons used in this manner often attained the name ganister (Percival, 1981, 1983a).

THE UPPER CARBONIFEROUS ENVIRONMENT OF NORTHERN ENGLAND: FACTORS AFFECTING SOIL FORMATION

Soils develop due to the interplay of a variety of factors, the most important of which are climate, parent material, relief, organisms and time. Some general comments can be made regarding the nature of the first three factors (and consequently their influence on soil formation) during deposition of the Upper Carboniferous sequence of northern England. These are outlined below:

(1) Climate. Palaeomagnetic data suggest that during Upper Carboniferous times Britain lay close to the equator (Faller and Briden, 1978). Together with the abundant plant fossils, common presence of *in situ* coals and the lack of features indicative of aridity within the Namurian and Westphalian strata, this suggests that during these periods a humid tropical climate generally existed. There is no evidence within much of the Upper Carboniferous sequence to suggest any marked seasonality.

Towards the end of the Upper Carboniferous, however, this humid tropical climate gradually gave way to a more arid setting which eventually became fully established during the succeeding Permian period.

(2) Parent materials. The Upper Carboniferous sequence in northern England is composed principally of claystones, siltstones and sandstones. Deposition of this sequence took place primarily in environments ranging from shallow-marine to delta plain and possibly fluvial plain; locally, e.g. in the Central Pennine Basin, deeper water sediments were deposited. Periods of emergence and soil development were generally associated with the establishment of a delta plain environment produced as a result of delta progradation. Consequently, the parent materials for many Upper Carboniferous soils of northern England were unconsolidated deltaic and associated fluvial sands, silts and clays.

(3) Relief. Due to the broadly delta plain type of environment that existed during most emergent periods, relief is likely to have been extremely limited. Local small topographic highs may have existed, e.g. levees, beach ridges, but most emergent areas were probably flat. The occurrence of *in situ* coals and paucity of desiccation cracks within the Upper Carboniferous sequence suggests that most areas were continually waterlogged, and that freely drained conditions were rare. Consequently most Upper Carboniferous soils developed in a hydromorphic environment.

MODERN SOILS CONTAINING AN ALBIC HORIZON

According to the Food and Agriculture Organization (FAO-UNESCO, 1974) an albic (E/A_2) horizon is one from which clay and free iron oxides have been removed, or in which the oxides have been segregated to the extent that the colour of the horizon is determined by the colour of the primary sand and silt particles rather than by coatings on these particles. Typically, therefore, such horizons are white to ash grey in colour and composed of sand or silt. They occur in a variety of soil types, including podzols, podzoluvisols, albic luvisols, planosols, etc. To aid comparison of the paleosols to be described with modern soils containing an albic horizon some of these modern soils are briefly described below:

(1) Podzols. The term podzol is derived from the Russian words *pod* = under, and *zola* = ash, and was originally applied to sandy soils containing a bleached, ash grey (albic) horizon underlying a thin organic–rich horizon (Muir, 1961). Generally, the bleached (albic) horizon has lost importance as a diagnostic feature of podzols and at present the term podzol is commonly applied to all soils containing a spodic B horizon, e.g. FAO-UNESCO (1974).

Typically, a spodic horizon consists of an illuvial accumulation of some combination of organic carbon, iron and aluminium (Soil Survey Staff, 1960). Although variations exist, many podzols consist of surficial organic horizons underlain by a bleached sandy albic horizon which lies on a more strongly coloured (typically red, brown, or black) illuvial horizon.

Podzols typically form on freely drained sandy parent materials in areas where precipitation exceeds evapotranspiration. Percolating meteoric water causes leaching and the development of a bleached eluvial (albic) horizon. Deposition of some of the leached constituents (typically sesquioxides and humus) takes place lower within the soil profile, resulting in a horizon of accumulation. An acidophyllous vegetation and a poorly humified (mor–type) surface organic horizon aids the podzolization process. Typically, therefore, podzols form on freely drained sandy parent materials, colonized by coniferous or heath vegetation under a humid, cool, temperate climate. Although podzols are most widespread in the latter climatic zone, examples occur elsewhere, e.g. lowland tropical regions (Klinge, 1965; Andriesse, 1969) where suitable conditions exist.

(2) Luvisols. The term luvisol is derived from the Latin word *luo* = to wash. Luvisols contain an argillic B horizon due to the illuvial accumulation of clay, but except in the albic luvisols a bleached albic horizon is generally absent.

Luvisols are a widely distributed soil type which typically form under a humid climate with a marked dry season. Free drainage of water through the upper and middle parts of the profile is generally necessary for clay translocation to take place. Deposition of clay often results in the formation of clay coatings (cutans; Brewer, 1964) on any available surfaces, e.g. surface of peds, grains, rootlet channels and also clay pore-fillings. In sandy or loamy parent materials, clay deposition may result in the formation of an argillic horizon consisting of a series of clay rich laminae spaced at intervals from a few centimetres to several tens of centimetres. The parent material of luvisols is commonly unconsolidated sediments of medium to fine texture, and the vegetation is variable although deciduous forest is most common (Fitzpatrick, 1983).

(3) Podzoluvisols. Podzoluvisols span the gap between the podzols and the luvisols and contain both a bleached albic (E/A_2) horizon and an argillic B-horizon. The junction between these two horizons is generally irregular or broken due to tonguing of the A_2 into the B, or from the formation of discrete nodules with iron enriched rims (FAO-UNESCO, 1974). The latter form due to water accumulating within the profile during winter, spring and early summer. Iron is reduced to the ferrous state and much becomes redistributed to form the nodules (Fitzpatrick, 1983). Translocation of clay takes place during periods of free drainage and results in the development of the argillic horizon. Deposition of clay may result in the formation of cutans. Podzoluvisols typically form on parent materials of fine to medium texture in cool, moist, continental areas of flat to gently sloping topography where a deciduous forest cover is present.

(4) Planosols. The term planosol originates from the Latin *planus* = flat, level. Such soils develop in level or depressed topography with poor drainage. They are characterized by the presence of an albic horizon which exhibits hydromorphic properties, at least in part, overlying a slowly permeable horizon, e.g. a clay pan or cemented hard pan within 125 cm of the surface. There is commonly an abrupt textural change between these two horizons (FAO-UNESCO, 1974).

Planosols are generally waterlogged for at least part of the year. Clay destruction and translocation from the upper horizons leads to development of the albic horizon and enhancement of any textural differences with the lower part of the solum. Weathering *in situ* in the B-horizon aids the latter. Planosols are widely distributed, but are most commonly developed in the transition zone between semi-arid and humid climates (Dudal, 1973). Stratified parent materials may aid planosol formation, but are not necessary for strong differentiation to develop.

PALEOSOLS CONTAINING AN ALBIC HORIZON FROM THE UPPER CARBONIFEROUS OF NORTHERN ENGLAND

In northern England, Upper Carboniferous paleosols containing an albic horizon occur primarily in strata of Namurian to lower Westphalian A age. Within this sequence they are not uniformly distributed, suggesting that only at certain periods, e.g. lower Westphalian A times, were conditions conducive for their formation and preservation. Three paleosols are described, as together they show many of the features typical of the varied suite of paleosols containing an albic horizon present within the Upper Carboniferous succession.

THE FIRESTONE SILL PALEOSOL; ROUND HILL QUARRY, COUNTY DURHAM

Introduction and description
The Firestone Sill is a sandstone of Pendleian, (E_1) Namurian age (Hull, 1968) which outcrops on the Alston Block in the northern Pennines (Figures 1 and 2). It forms the top of the White Band cyclothem (Figure 2) which was deposited during a period of delta progradation following an initial marine transgression which resulted in deposition of the fossiliferous White Band (Percival, 1981).

Within the confines of the Alston Block the Firestone Sill is broadly divisible into two facies, a fine to very coarse grained distributary channel sandstone facies and a fine to medium grained sandstone facies which was deposited in a wave influenced interdistributary environment. Paleosols are developed at the top of both facies, but a particularly good example occurs at the top of the latter facies at Round Hill Quarry (British National Grid Reference NZ01203835) (Figure 3).

At this locality the Firestone Sill is composed of interbedded fine-to medium-grained subarkosic sandstones, with sporadic thin shale laminae. This is overlain by a profile which, when unmodified by subsequent erosion, is generally composed of the following horizons from the top downwards:

(1) Several centimetres of impure coal with thin shale laminae. This is overlain by calcareous fossiliferous shales and limestones (Crag Limestone) which were deposited in a shallow-marine environment produced during a transgressive phase subsequent to paleosol development.

(2) A black carbonaceous-rich sandstone up to a few tens of centimetres thick. This contains large *Stigmaria sp.* roots (the rhizophore of the arborescent lycopods/Lepidodendrales), abundant associated

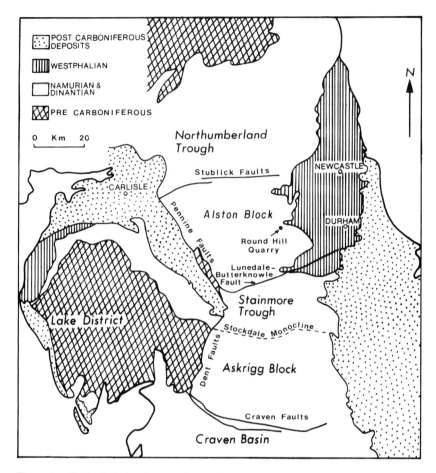

Figure 1. Geological sketch map of part of northern England showing the location of Round Hill Quarry.

carbonaceous rootlet impressions, admixed very fine-grained and somewhat irregularly disseminated carbonaceous material and sericite/illite; burrows are also present in places. Mineralogically, the sandstone consists primarily of quartz with some muscovite, sericite/illite, and feldspar being present. The latter is generally altered to sericite/illite, at least in part.

The admixed carbonaceous material forms coatings on some sand grains and together with the abundant rootlets gives the rock its dark colouration. Towards the base of this unit there is a decrease in the content of carbonaceous material and hence colouration in places. The base of this unit is generally sharp and may be slightly undulating. Calcite cement occurs within this horizon in places, and from its distribution and textural relationships can be seen to be a post pedogenic

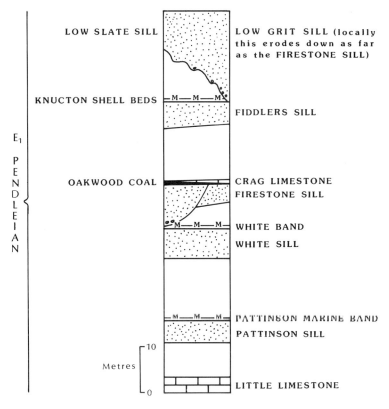

Figure 2. Section through part of the E_1/Pendleian sequence of the Alston Block showing the stratigraphic position of the Firestone Sill. The White Band cyclothem includes all strata from the White Band up to the base of the Crag Limestone. The base of the Namurian occurs approximately 45 m below the base of the section. -M—M—M- = Marine Band

diagenetic phase derived from the closely overlying carbonate rich sequence (Crag Limestone).

(3) A conspicuous white quartz arenite, varying from several tens of centimetres to approximately 1.1 m thick (Figure 3). This horizon is structureless except for the presence of outlines of rootlets and some large stigmarian roots, and consists of a moderately well sorted, fine- to medium-grained sandstone containing approximately 94–97% quartz. Other constituents include altered feldspar, muscovite, sericite/illite, ?kaolinite, heavy minerals including zircon, tourmaline and opaques, and rare chert and siltstone rock fragments.

The base of this unit is sharp and irregular, typically varying from wavy to lobate with downward projections of the quartz arenite into the underlying horizon up to 36 cm deep. Rootlet channels which pass through this contact are commonly infilled with the white quartz

Figure 3. Top of the Firestone Sill, Round Hill Quarry, showing white irregularly based quartz arenite (1) overlying subarkosic sandstone containing crescentic structures (2). Hammer (arrowed) is 33 cm long.

arenite in the underlying horizon, which imparts a tongued-like nature to the junction.

(4) A brownish, fine-to medium-grained sandstone horizon, typically several tens of centimetres to perhaps a metre or more in thickness (Figure 3). This horizon displays some sedimentary structures including horizontal lamination and low angle cross–bedding and contains more clay (principally sericite/illite), feldspar, mica and carbonaceous material than the overlying quartz arenite. In places, clays form tangentially orientated coatings to framework grains (Figure 4) and well developed laminated (often meniscate) void fillings (Figure 5). These grain coatings and void fillings are particularly common in the crescentic structures which occur within this horizon. The crescentic structures are very common in the top few tens of centimetres and are thin, low amplitude (generally several millimetres, but up to a few centimetres high), long wavelength (up to several tens of centimetres long) features which commonly have convex upward top surfaces and convex upward to flat bottom surfaces (Figures 6 and 7). They are up to a few centimetres thick and consist of clay and to a lesser extent carbonaceous rich sandstone which may be finer grained than the enclosing sediment. In rare instances they occur above low permeability or impermeable layers, e.g. large wood fragments. Many are parallel to bedding but some are developed along low angle cross bedding

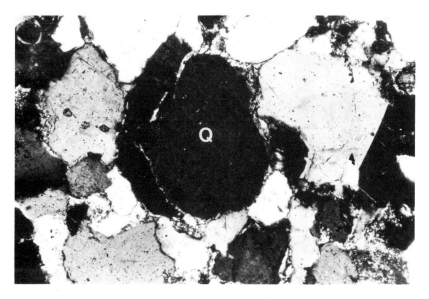

Figure 4. Quartz grain (Q) displaying tangentially orientated clay coating. Subarkosic sandstone beneath quartz arenite, Firestone Sill, Round Hill Quarry. Length of photomicrograph 1.3 mm (crossed polarized light).

Figure 5. Laminated clay void-filling from crescentic structure in subarkosic sandstone beneath quartz arenite, Firestone Sill, Round Hill Quarry. Length of photomicrograph 0.33 mm (plane polarized light).

96 CHAPTER 3

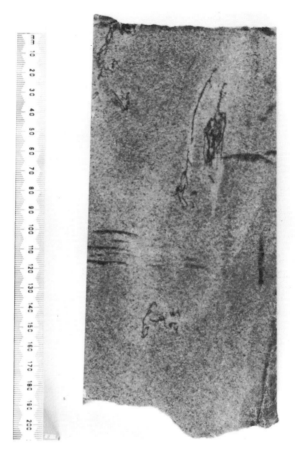

Figure 6. Sawn block of subarkosic sandstone from beneath the quartz arenite, showing subvertical carbonaceous rootlet traces and subhorizontal small crescentic structures, Firestone Sill, Round Hill Quarry. Scale is in millimetres.

foresets and others cut across bedding at angles up to 20°. Examples that bifurcate or amalgamate are also present. Sedimentary lamination passes through the structures in places, but is commonly discordant with their external form, (Figure 8). Rootlets and burrows often distort and truncate these features, but in a few instances the structures are essentially unaffected by rootlets which pass through them.

These crescentic structures decrease in abundance downwards and this horizon passes transitionally into the underlying subarkosic sandstones of the Firestone Sill which contain a lower proportion of clay and carbonaceous material (Figure 9). Rootlets decrease in abundance in the same direction, whereas sedimentary structures become more evident.

Figure 7. Crescentic structures in subarkosic sandstone from beneath the quartz arenite, Firestone Sill, Round Hill Quarry. Left hand margin of rule is in centimetres and is 30 cm long.

Feldspar is generally present throughout the described profile. Although it is fairly common and moderately fresh in the sandstones beneath the quartz arenite, it generally decreases in abundance and becomes more altered upwards.

Interpretation

Many of the features present in the topmost portion of the Firestone Sill at Round Hill Quarry suggest that pedogenesis has taken place and led to the development of a leached paleosol profile. These include:

(1) The presence of large stigmarian roots and associated rootlets provides positive evidence of emergence and colonization by plants.

(2) The occurrence of a thin coal at the top of the profile suggests surficial accumulation of plant debris similar to that which takes place to form the surface organic horizons of some modern soils. However, the entire thickness of coal is unlikely to represent a surface organic horizon formed during the development of a leached soil profile, and it is quite likely that most of this horizon accumulated during a subsequent phase of waterlogging.

Establishment of such hydromorphic conditions may have been associated with the relative rise in base level which eventually resulted in marine transgression and deposition of the overlying Crag Limestone. Similar relationship between coals and marine bands have been des-

Figure 8. Dark clay and carbonaceous-rich crescentic structure showing a discordant relationship between internal lamination and external form. Internal lamination is concordant with lamination perserved in the enclosing sediment. Sample is 14 cm wide at centre.

cribed from slightly older Namurian strata on the Alston Block by Elliott (1974, 1975).

(3) The underlying black sandstone which contains abundant rootlets and admixed carbonaceous material and sericite/illite. The disseminated and irregular distribution of the admixed material is unlikely to have been produced by physical sedimentary processes, and is more likely to be a product of biogenic mixing. The occurrence of roots, rootlets and some burrows supports this and suggests that admixing probably took place in a terrestrial environment.

Similar horizons (A_1) to those described above are common in some modern soil profiles where soil organisms, roots, rootlets and other processes incorporate organic matter into the topmost mineral fraction of the soil to produce a dark organic stained mineral horizon (Soil Survey Staff, 1960; Hunt, 1972; Birkeland, 1974).

(4) The quartz arenite horizon lacks any sedimentary structures and lies with an irregular contact on the underlying horizon. The nature of the contact is not erosive, and is generally too irregular to

UPPER CARBONIFEROUS PALEOSOLS FROM NORTHERN ENGLAND   99

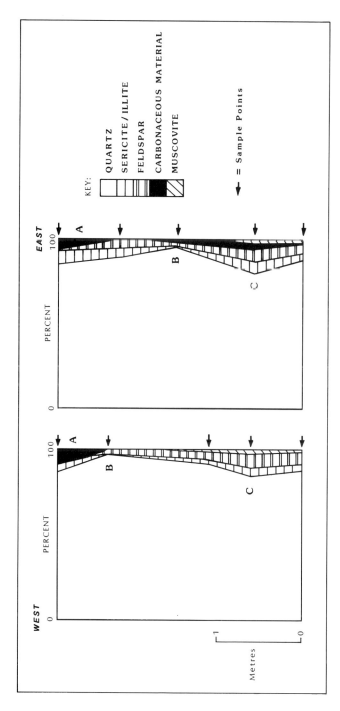

Figure 9. Two vertical profiles through the topmost portion of the Firestone Sill, Round Hil. Quarry. Note: A = Carbonaceous top; B = Zone of quartz enrichment; C = Zone of enrichment of sericite, Feldspar carbonaceous material and muscovite. Spacing between profiles approximately 20 m. Percentages obtained by point count analysis of thin sections.

have resulted from loading. Similar white quartz-rich horizons with an irregular basal contact are common in many modern soils, e.g. podzoluvisols etc. where differential leaching of the albic (E/A_2) horizon has taken place (Soil Survey Staff, 1960; Daniels, *et al.*, 1967; FAO-UNESCO, 1974).

(5) The occurrence of crescentic structures, and associated tangentially orientated clay coatings on framework grains and laminated void fillings in the horizon beneath the quartz arenite. The crescentic structures often display a discordant relationship between their external morphology and lamination or bedding. This suggests a diagenetic rather than a sedimentary origin for these features. Burrows and rootlets which truncate and distort the structures indicate a very early diagenetic origin. In some instances rootlets pass through the structures with little effect, and it appears that much of the structure formed subsequent to rootlet penetration. The most likely explanation for these various relationships is that rootlet penetration and structure formation were essentially contemporaneous.

Somewhat similar clay and, to a lesser extent, carbonaceous rich-laminae do occur in modern soils. They are particularly common in soils which have undergone clay translocation (lessivage), e.g. luvisols, where they may form the argillic horizon (Soil Survey Staff, 1960; FAO-UNESCO, 1974).

Tangentially orientated clay coatings to grains and laminated void fillings can form in a variety of environments (Brewer, 1964, 1972; Bullock and Mackney, 1970; Wilson and Pittman, 1977). However, those present within the Firestone Sill generally lack the crystallinity and morphology of truly diagenetic clay coatings and fillings and are similar to argillans (clay cutans). Many, particularly the meniscate clay pore fills, are similar to illuviation argillans. The latter form by deposition of clays transported by percolating water, and are particularly common in argillic soil horizons.

(6) The general upward decrease in abundance and increase in alteration of feldspars through the profile suggests that felspars in the upper horizons probably underwent weathering concomitant with plant colonization of the sand body. In view of the postulated humid tropical climate, the presence of up to a few percent of feldspar in these upper horizons suggests that this period of weathering was probably fairly short.

(7) The horizon beneath the quartz arenite commonly shows an increase in the percentage of clay and carbonaceous material relative to the enclosing strata (Figure 9). This increase is due primarily to the presence of grain coatings and void fillings which led to the formation of the crescentic structures. Somewhat similar horizons enriched in

clay and carbonaceous material do occur in present day soils where translocation of these components has taken place, e.g. luvisols, podzoluvisols.

Taken together, these features suggest that the top of the Firestone Sill at Round Hill Quarry was emergent for a sufficient time for a paleosol profile to develop (Percival, 1983b). The nature of the horizons formed, their contacts, and the evidence of clay translocation suggests comparison with modern podzoluvisols and albic luvisols. These both contain an albic horizon overlying an argillic horizon. A tonguing contact between these two horizons is common in the podzoluvisols but absent in the albic luvisols (FAO-UNESCO, 1974). Consequently, where the paleosol exhibits a highly irregular tongued-like basal contact to the interpreted albic horizon it is similar to a podzoluvisol, where the contact is less irregular it is more akin to an albic luvisol.

Discussion
The profile developed suggests that during paleosol formation the accumulation of some organic matter at the surface, and incorporation of part of this organic matter into the topmost mineral fraction of the soil took place. This resulted in the development of the A_0 and A_1 horizon respectively. This was accompanied by the breakdown of unstable minerals in the upper part of the solum and downward translocation of their alteration products, along with clays and some carbonaceous material. Removal of these constituents resulted in the development of the quartz-rich albic (E/A_2) horizon. Variation in the thickness and homogeneity of the albic horizon may have resulted from differential leaching due to the presence of root holes or burrows, protection by overlying large roots, or the localized effect of particular plants which colonized the soil surface (Butuzova, 1962, Buol, et al., 1973; Buurman, 1980). Slight variations in parent material or depth to the water table probably also exerted some influence.

Deposition of some of the constituents removed from the upper horizons (particularly clays, together with some organic matter and iron) took place in the lower part of solum. This led to development of cutans, crescentic structures and formation of an illuvial (B ?argillic) horizon. Pedogenically produced cutans do not form below a permanent water table (Buurman, 1980). This suggests that the permanent water table was probably at least 2.4 m below the soil surface during pedogenesis. Deposition in the B-horizon appears to have commonly taken place preferentially in finer grained or slightly less permeable zones or above local permeability barriers, e.g. large wood fragments, resulting in the crescentic structures.

During soil formation, bioturbation by plant rootlets and other soil organisms, together with other pedogenic processes, led to obliteration of most sedimentary structures originally present in the upper horizons. The parent material on which the soil developed appears to have been a fine-to medium-grained subarkosic sand similar to the underlying unmodified Firestone Sill. Under the prevailing humid tropical climate, leaching and development of the paleosol profile would have progressed rapidly, providing the water table remained fairly low. The presence of feldspar in the albic horizon may suggest that the period of pedogenesis was relatively brief.

Broadly similar paleosol profiles containing an albic horizon occur at the top of the wave influenced interdistributary facies of the Firestone Sill at a number of localities in the central northern part of the Alston Block (Figure 1). This suggests that a widespread period of emergence, leaching and paleosol development followed deposition of the Firestone Sill in this area. Subsequent to paleosol development, a widespread marine transgression took place leading to deposition of the overlying Crag Limestone. This transgressive phase resulted in modification or removal of the topmost parts of the paleosol profile in some areas.

THE POT CLAY GANISTER PALEOSOL, PORTER BROOK, SOUTH YORKSHIRE

Introduction and description
The Pot Clay is a kaolinitic fireclay lying at the very top of the Namurian sequence in South Yorkshire and adjacent areas. It is of Yeadonian (G_1) age (Ramsbottom et. al., 1978) occurring above the Rough Rock and below the thin Pot Clay Coal. The latter is overlain by the Pot Clay (*Gastrioceras subcrenatum*) Marine Band (Eden et. al., 1957) which marks the base of the Westphalian (Figure 10). Deposition of this topmost Namurian strata is thought to have taken place during a period of delta progradation, with the Pot Clay and Pot Clay Coal being deposited in a delta plain environment. Subsequent transgression over the delta plain at the onset of Westphalian times led to deposition of the overlying marine band.

At a few localities a thin hard quartz arenite occurs at the horizon of the Pot Clay, and has been termed the Pot Clay Ganister. An exposure of this quartz arenite occurs in Porter Brook, Sheffield (British National Grid Reference SK 31758535) (Figure 11).

The relevant part of the section which is visible at this locality is shown in Figure 12. A small gap occurs in the section directly above

UPPER CARBONIFEROUS PALEOSOLS FROM NORTHERN ENGLAND 103

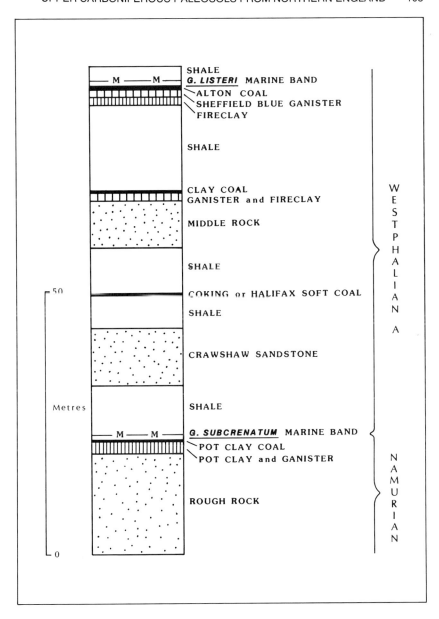

Figure 10. Section through the uppermost Namurian and lowermost Westphalian of the Sheffield region showing the stratigraphic position of the Pot Clay and Sheffield Blue Ganisters. Modified from Strahan (1920).

the quartz arenite. The Pot Clay Coal is thought to occur within this gap in the section.

At this locality the quartz arenite is approximately 35 cm thick, but varies in thickness due to a highly irregular base which contains undulations up to 20 cm in amplitude. The quartz arenite is light grey, fine-grained, moderately well sorted and consists of approximately 98% quartz with minor mixed layer clay, muscovite, zircon, tourmaline and opaque heavy minerals and diagenetic pyrite. In the top 7–8 cm the clay content often increases to several percent forming a darker coloured horizon which may break away from the rest of the rock. The quartz arenite is structureless except for the presence of large stigmarian roots and associated rootlets, some of which pass down into the underlying light brown sandstone. The latter is approximately 90 cm thick and is generally fairly poorly cemented. Sedimentary structures are absent from much of this unit, but towards the base, as the proportion of rootlets decreases, horizontal lamination becomes evident and this horizon passes down transitionally into the underlying laminated fine grained sandstones (Figure 12).

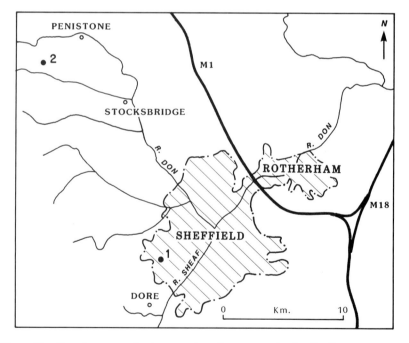

Figure 11. Location map of relevant paleosol exposures in the Sheffield region. 1 = Pot Clay Ganister Paleosol, Porter Brook (British National Grid Reference SK 31758535); 2 = Sheffield Blue Ganister Paleosol, Langsett (British National Grid Reference SE 21300105).

The proportion of clay in the sandstone below the quartz arenite is approximately 20% and includes both kaolinite and mixed layer illite-smectite. Quartz forms much of the remaining 80%, with only a small proportion of feldspar, muscovite, zircon and tourmaline being present.

Kaolinite occurs as well developed diagenetic pore fills, whereas the mixed layer clay often forms tangentially orientated coatings to framework grains (Figure 13). These coatings resemble free grain cutans, particularly illuviation cutans (Brewer 1964).

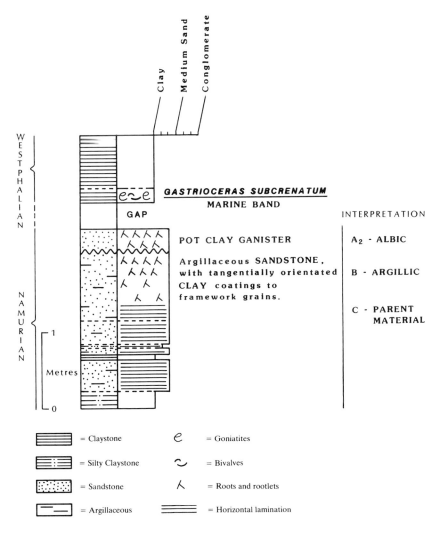

Figure 12. Vertical section through the Pot Clay Ganister and associated strata, Porter Brook, South Yorkshire.

106 CHAPTER 3

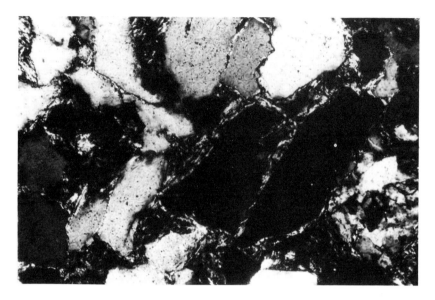

Figure 13. Photomicrograph of the sandstone beneath the Pot Clay Ganister in Porter Brook showing some tangentially orientated mixed layer clay coatings on framework grains. Length of photomicrograph 0.52 mm (crossed polarized light).

Interpretation

The presence of fossilized roots, rootlets and the occurrence of a thin, irregularly based, mature quartz arenite overlying a sandstone enriched in clay (and containing probable illuviation cutans) suggests that these horizons represent part of a leached paleosol profile. The parent material from which the paleosol developed was probably similar to the underlying argillaceous, micaceous fine-grained sandstones (Figure 12). The texture, mineralogy and micromorphology of the horizons present suggest that the main processes operating during paleosol development were probably the breakdown of unstable grains in the upper part of the solum and downward translocation of the breakdown products and clays under the influence of percolating water. This resulted in the development of a bleached, quartz rich, albic (E/A_2) horizon (Figure 12). Deposition of some of the leached components (particularly mixed layer clay) resulted in a B (argillic) horizon of accumulation containing probable cutans. If the latter are true illuviation cutans they suggest that during pedogenesis the permanent water table was probably over a metre below the soil surface.

The thin, clay enriched zone at the top of the ganister probably formed by clays being incorporated into the top of the coarse mineral fraction of the soil by plant roots and rootlets and the soil fauna. However, the possibility that this horizon formed by later reworking of the albic horizon cannot be excluded. The presence of an albic horizon overlying a probable argillic horizon suggests that the paleosol profile, like the Firestone Sill paleosol may represent a podzoluvisol or albic luvisol.

SHEFFIELD BLUE GANISTER PALEOSOL, LANGSETT, SOUTH YORKSHIRE

Introduction and description
The Sheffield Blue Ganister is a quartz arenite occurring below the Halifax Hard Mine/Alton Coal of Westphalian A age (Ramsbottom *et al.*, 1978) (Figure 10). It outcrops extensively in South Yorkshire, particularly in the area between Dore and Penistone (Figure 11), but although it was once extensively worked in this area for the raw material for refractory bricks, present day exposures are poor. One of the best exposures occurs in some old shallow quarries just north of Langsett (British National Grid Reference SE 21300105). At this locality the base and top of the ganister are visible along with the overlying Alton Coal and Marine Band (Figure 14). Deposition of the

Figure 14. Vertical section through the Sheffield Blue Ganister and associated Strata, Langsett, South Yorkshire.

sequence below the Alton Marine Band probably took place in a delta plain environment. This was followed by a transgression which established shallow-marine conditions in which the Alton Marine Band was deposited. At the Langsett locality the Sheffield Blue Ganister varies in thickness from 35 to 84 cm due to an undulatory top surface and highly irregular base (Figure 15). The latter is conspicuous and consists of abundant lobate downward projections of the ganister into the underlying kaolinitic claystone. The latter is a few tens of centimetres thick, contains fossilized rootlets and abruptly overlies a 14 cm thick argillaceous fine-grained sandstone with more rootlets in places.

The ganister comprises a light grey, very fine-grained, moderately well sorted, quartz arenite, containing approximately 99% quartz and some very fine-grained kaolinite. Internally, the quartz arenite lacks any sedimentary structures, but contains large stigmarian roots and associated rootlets, some of which pass down into the underlying strata. Grain size does not vary vertically within the quartz arenite, but near the base there is commonly a slight increase in the small propor-

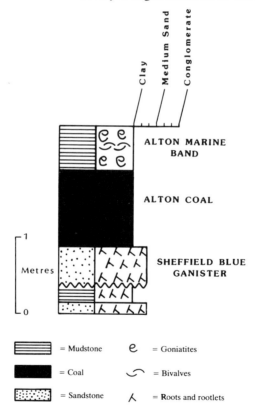

Figure 15. Exposure of the Sheffield Blue Ganister at Langsett showing the undulating top surface and the highly irregular base. Hammer (arrowed) is 33 cm long.

tion of kaolinite present, causing the rock to be slightly darker grey. The top few centimetres of the quartz arenite although similar petrographically, contains a small amount of diagenetic pyrite and breaks away from the rest of the rock. This is overlain by the Alton Coal which is approximately 1 m thick.

Interpretation
The presence of fossilized roots, rootlets, and the highly irregular base to the mature quartz arenite suggests that it represents the albic horizon of a paleosol profile. The abrupt textural change from the quartz arenite into the underlying kaolinite claystone indicates that the paleosol developed on a bisequent parent material composed of sand overlying clay. Pedogenesis seems to have accentuated this original lithology contrast. The mineralogy of the sandstone from which the albic horizon developed is uncertain, but in view of its very fine grain size and by analogy with other closely associated Westphalian A sandstones of similar grain size, the parent material was probably less mineralogically mature, and is likely to have been an argillaceous, very fine-grained sand containing < 92% quartz. This suggests that quartz enrichment, by breakdown of unstable grains and removal of their breakdown products, together with clays, probably took place. Leaching and translocation of these components probably occurred by percolating water. Vertical drainage within the soil profile must have been impeded by the underlying clays, causing some deposition of clays close to this interface, and resulting in at least some lateral drainage taking place within the albic horizon.

The thick peat which developed on top of the quartz arenite and eventually formed the overlying Alton Coal is unlikely to have been produced during this period of pedogenesis. It is suggested that much of this peat probably accumulated during a subsequent phase of complete waterlogging as such conditions are generally necessary for thick *in situ* organic horizons to develop. The development of waterlogged conditions may have been associated with the relative rise in base level which eventually resulted in marine transgression and deposition of the overlying Alton (*G. listeri*) Marine Band.

The paleosol profile, formed prior to the interpreted period of complete waterlogging, is similar to some types of planosol in exhibiting an abrupt textural change between an upper albic horizon and a lower claypan horizon.

CONCLUSIONS

The Upper Carboniferous succession of northern England contains a wide variety of paleosols. The majority developed in delta plain environments from fluvial and deltaic sand, silt and clay parent materials under a humid tropical climate. Due to the predominantly low relief of the delta plain, and the generally high water table most paleosols developed in a waterlogged environment and show hydromorphic affinities.

A small, but significant, proportion of the paleosols contain a conspicuous white to light grey quartz arenite horizon underlying any original surficial organic or organo-mineral horizons and overlying a more clay rich horizon. Such quartz arenites contain fossilized roots and rootlets and have a characteristic highly irregular basal contact with underlying strata. They are interpreted as representing the albic horizon of the paleosol profile. In many instances the albic horizon seems to have developed from a less mature sand parent material during pedogenesis, by the breakdown of unstable mineral grains and removal of most of the breakdown products and any clays present. Deposition of some of these components (particularly clays) often took place lower in the profile, resulting in a horizon of accumulation. Where the underlying horizon was sand, illuvial clay and carbonaceous laminae, and tangentially orientated clay coatings and laminated void fillings were often developed.

The paleosols described exhibit many similarities with modern soils containing albic horizons, in particular the podzoluvisols, albic luvisols, and planosols. With the possible exception of planosols, such soils require free drainage through the upper parts of solum for at least part of the year for their development. This suggests that at times, relatively well drained sites did exist within the generally waterlogged delta plain environments of the Upper Carboniferous of northern England. The presence of paleosols similar to modern albic luvisols, may also indicate that a somewhat seasonal climate existed at least periodically in northern England during the Upper Carboniferous, as modern luvisols generally require a distinct dry season for their formation.

ACKNOWLEDGEMENTS

This work forms part of the author's Ph.D. research which was funded by a NERC studentship. Thanks go to Dr A.P. Heward and Dr G.A.L. Johnson, who supervised this research, and to the staff of BP Alaska Exploration, Inc., in particular, Debbie Linton, Glen Schumacher, Pamela Demory, and Joyce Meyer, who were of invaluable assistance in preparation of this manuscript.

REFERENCES

Andriesse, J.P. 1969. A study of the environment and characteristics of tropical podzols in Sarawak (East-Malaysia). *Geoderma* **2**, 201–27.

Birkeland, P.W. 1974. *Pedology, Weathering and Geomorphological Research*. Oxford University Press, New York. 285 pp.

Brewer, R. 1964. *Fabric and Mineral Analysis of Soils*. Wiley, New York. 470 pp.

Brewer, R. 1972. The basis of interpretation of soil micromorphological data. *Geoderma* **8**, 81–94.

Bullock, P. and Mackney, D. 1970. Micromorphology of strata in the Boyn Hill Terrace deposits, Buckinghamshire, in *Micromorphological Techniques and Applications*, D.A. Osmond and P. Bullock (eds). Agric. Res. Council Soil Survey, Tech. Monograph No. 2. 97–105.

Buol, S.W., Hole, F.D. and McCracken, P.J. 1973. *Soil Genesis and Classification*. Iowa University Press, Ames. 360 pp.

Butuzova, O.V. 1962. Role of the root system of trees in the formation of micro-relief. *Soviet Soil Sci.* **4**, 364–72.

Buurman, P. 1980. Palaeosols in the Reading Beds (Paleocene) of Alum Bay, Isle of Wight, U.K. *Sedimentology* **27**, 593–606.

Daniels, R.B., Gamble, E.E. and Nelson, L.A. 1967. Relation between horizon characteristics and drainage in some fine loamy ultisols. *Soil Sci.* **104**, 364–9.

Dudal, R. 1973. Planosols, in *Pseudogley and Gley Genesis and Use of Hydromorphic Soils*, E. Schlichting and U. Schwertmann (eds). Yerlag Chemie, Weinheim/Bergstr. 275–85.

Eden, R.A., Stevenson, I.P. and Edwards, W. 1957. Geology of the country around Sheffield. *Mem. geol. Surv. UK.*, 238 pp.

Elliott, T. 1974. Abandonment facies of high constructive lobate deltas, with an example from the Yoredale Series. *Proc. Geol. Ass.* **85**, 359–65.

Elliott, T. 1975. The sedimentary history of a delta lobe from a Yoredale (Carboniferous) cyclothem. *Proc. Yorks. geol. Soc.* **40**, 505–36.

Faller, A.M. and Briden, J.C. 1978. Palaeomagnetism of Lake District rocks, in *The Geology of the Lake District*, F. Moseley (ed.). Yorks geol. Soc. Occ. Pub. 3. 17–24.

FAO-UNESCO 1974. FAO-Unesco soil map of the world, 1:5,000,000. Vol. 1, Legend (Legend sheet and memoir). UNESCO, Paris. 59 pp.

Fitzpatrick, E.A. 1983. *Soils: Their Formation, Classification and Distribution*. Longman, London. 353 pp.

Hull, J.H. 1968. The Namurian stages of northeast England. *Proc. Yorks. geol. Soc.* **36**, 297–308.

Hunt, C.B. 1972. *Geology of Soils*. Freeman, San Francisco. 344 pp.

Klinge, H. 1965. Podzol soils in the Amazon basin. *J. Soil. Sci.* **16**, 95–103.

Muir, A. 1961. The podzol and podzolic soils. *Adv. Agron.* **13**, 1–56.

Percival, C.J. 1981. *Carboniferous Quartz Arenites and Ganisters of the Northern Pennines*. Unpubl. Ph.D. thesis, University of Durham. 353 pp.

Percival, C.J. 1983a. A definition of the term ganister. *Geol. Mag.* **120**, 187–90.

Percival, C.J. 1983b. The Firestone Sill ganister, Namurian, northern England—the A_2 horizon of a podzol or podzolic paleosol. *Sedim. Geol.* **36**, 41–9.

Ramsbottom, W.H.C., Calver, M.A., Eager, R.M.C., Hodson, F., Holliday, D.W., Stubblefield, C.J. and Wilson, R.B. 1978. A correlation of Silesian rocks in the British Isles. *Spec. Rep. geol. Soc. Lond.* **10**, 81 pp.

Soil Survey Staff 1960. *Soil Classification a Comprehensive System*. 7th Approximation, US Department of Agriculture, Washington. 265 pp.

Strahan, A. 1920. Refractory materials: ganister and silica—rock—sand for open-hearth steel furnaces—dolomite: resources and geology. *Mem. geol. Surv. spec. Rep. Miner. Resour. Gt. Br.*, 6 (2nd edition). 241 pp.

Wilson, M.D. and Pittman, E.D. 1977. Authigenic clays in sandstones: recognition and influence on reservoir properties and palaeoenvironmental analysis. *J. sedim. Petrol.* **47**, 3–31.

Chapter 4

THE CALCAREOUS PALEOSOLS OF THE BASAL PURBECK FORMATION (UPPER JURASSIC), SOUTHERN ENGLAND

JANE E. FRANCIS*
Department of Geology, University of Southampton, Southampton, Great Britain
*now at, Department of Geology and Geophysics,
University of Adelaide, Box 498 GPO,
Adelaide, S. Australia 5001.

INTRODUCTION

A wide variety of criteria are used for the recognition of fossil soils, e.g. soil microfabric (Meyer, 1976), ganisters (Retallack, 1976), lateritic horizons (Goldbery, 1982), carbonate accumulations (McPherson, 1979; Wright, 1983) and gley phenomena (Buurman, 1980). Of these, the most diagnostic is the presence of fossil roots preserved in their original growth positions in the soil (Retallack, 1981). Plant material usually decays rapidly in well-drained, aerobic soils and the roots are often preserved only as sedimentary infills or coalified traces. However, if the plant cells are infiltrated by a permineralizing solution before cellular decay occurs, plants may be retained in the fossil soil as structurally preserved petrifactions. They are not only important factors in the identification of a paleosol but can also yield additional information concerning the nature of the plant/soil interactions.

The presence of petrified plant remains facilitated the recognition of paleosols within the Upper Jurassic Purbeck Formation of southern England (Arkell, 1947). Large silicified stumps of coniferous trees can be found with their roots spreading through the dark, pebbly, carbonaceous horizons. The morphology, classification and origin of the paleosols, and the nature of the soil-forest environment is discussed in this paper.

GEOLOGICAL SETTING

During the Late Jurassic an extensive shallow basin (the Purbeck basin) extended across southern England, in which fresh water, brackish and hypersaline sediments of the Purbeck Formation were deposited

(Howitt, 1964). On the western margins of the basin (the Dorset region) (Figure 1), the basal part of the Purbeck Formation consisted of intertidal and supratidal deposits formed during minor regressive and transgressive phases (West, 1975). The sediments consist mainly of algal stromatolitic limestones, called the 'Caps', (Figures 2 and 3), formed as mounds of algal-bound sediment on hypersaline tidal-flats (Brown, 1963). Pelletoid and intraclastic limestones surround the stromatolites and crypt-algal laminites with evaporite pseudomorphs also occur. Interbedded with these limestones are dark carbonaceous marls, the Dirt Beds, former forest soils in which silicified tree stumps are preserved (Arkell, 1947).

The Great Dirt Bed, between the Hard Cap and the Soft Cap (Figure 2), is the best developed paleosol with a dark carbonaceous horizon containing conspicuous black and white limestone pebbles. It was recognized as a fossil soil in early reports, e.g. Webster (1826); Fitton (1835, 1836); Buckland and De La Beche (1836); Gray (1861). Multer (in Damon, 1884) even tested its fertility with mustard and cress seed and compared it to a Russian Tchornozem (sic) due to its origin as 'decomposed debris of an ancient forest'. West (1979) suggested it was more like a rendzina.

The Great Dirt Bed horizon has been recorded throughout most of the Purbeck outcrop in Dorset (Figure 1) (Arkell, 1947; West, 1975) but it is only present as a carbonaceous marl with tree stumps (i.e. a paleosol)

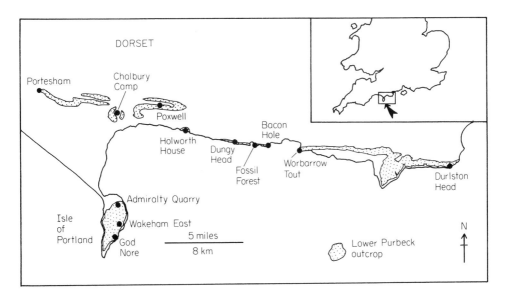

Figure 1. Sketch map showing location of some exposures of the basal Purbeck Formation mentioned in this paper.

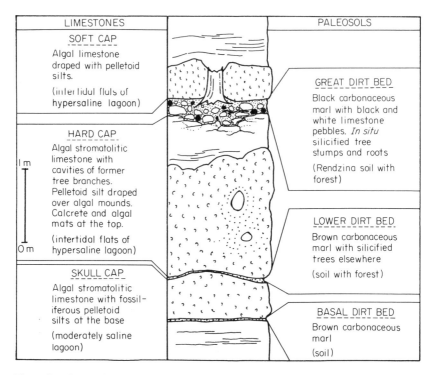

Figure 2. Generalized section of the basal Purbeck Beds on Portland showing terminology and environmental interpretation.

over a limited area in the Lulworth region (Dungy Head to Bacon Hole) and on the Isle of Portland. The development of the soil over this area appears to have been influenced by the presence of an elevated barrier (West, 1975; Francis, 1983a). The paleosol appears to grade laterally into lagoonal clays to the west and evaporitic sediments to the east (West, 1975).

Two thinner Dirt Beds occur in the basal Purbeck sequence, the Lower Dirt Bed between the Hard Cap and Skull Cap and the Basal Dirt Bed below the Skull Cap (Figures 2 and 3). Both are only a few centimetres thick and often impersistent, although large silicified conifer stumps are preserved in places within the Lower Dirt Bed. The marly layers of these Dirt Beds can be traced from Portesham to Swanage (Figure 1), (though the Basal Dirt Bed is locally absent at Dungy Head and the Fossil Forest), but east of Worbarrow Tout the sections are not easily correlated (Strahan, 1898; Arkell, 1947; West, 1975). At Bacon Hole and Worbarrow Tout the Lower Dirt Bed is represented by a laminated black and white shale containing ostracods, fish remains and conchostracan branchiopod crustacea (Francis, 1984).

Figure 3. The basal beds of the Purbeck Formation in Wakeham East Quarry, Portland (British National Grid Reference SY 699 714). SC: Soft Cap, GDB: Great Dirt Bed, HC: Hard Cap, LDB: Lower Dirt Bed, SkC: Skull Cap, BDB: Basal Dirt Bed. The Basal and Lower Dirt Beds are only 3−4 cm thick here.

These shales are interpreted as seasonally-influenced marginal lagoonal deposits. Similar Dirt Beds, some with fossil trees, have been recorded from inland exposures of the Purbeck Formation, e.g. the Vale of Wardour and Swindon (Arkell, 1947).

The marl matrices of the Dirt Beds were not lithified so for thin-section study samples were collected in metal tubes hammered horizontally into the matrix. Samples were then impregnated with resin and thin-sectioned. The carbonate and organic carbon content were analysed using an infra-red gas analyser.

PALAEOBOTANY OF THE PALEOSOLS

The large conifers which once grew in the Great and Lower Dirt Bed paleosols are now preserved as silicified logs lying on the surface of the palcosols or as large tree stumps in their original growth positions with their roots in the paleosols (Figure 4). The upright stumps are encased in circular burrs or domes of algal stromatolitic limestone which originated as algal-bound sediment that covered the tree bases when the forest was inundated with rising lagoon water (Francis, 1983b, P1.38, Figures 1–5).

The fossil plant assemblage from the basal Purbeck sediments is dominated by one type of wood named *Protocupressinoxylon purbeckensis*, conifer foliage with small, scale-like leaves called *Cupressinocladus valdensis* and small male cones yielding *Classopollis* pollen. These parts have all been attributed to the same type of conifer (see the reconstruction in Francis, 1983b) which appears to have dominated the Purbeck forests. These conifers were members of the Mesozoic family Cheirolepidiaceae, which had a widespread distribution at this time and were often associated with sediments indicating a semi-arid climate (Vakhrameev, 1970; Alvin, 1982). The silicified remains indicate that the Purbeck conifers were monopodial trees with low branches and shallow roots which spread laterally through the soils (Francis, 1983b. text-fig. 3) but did not ever penetrate the underlying limestone.

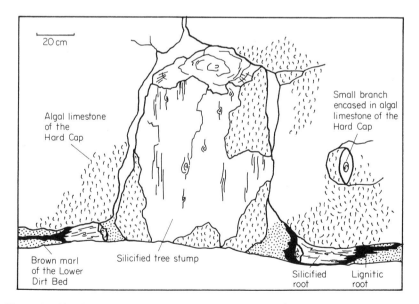

Figure 4. Sketch of an *in situ* fossil tree stump within the Lower Dirt Bed, Chalbury Camp (to scale).

A few araucarian conifers were also present in the forests (indicated by fossil wood) and Bennettitalean cycadophytes, represented by silicified stumps. Miospores and megaspores from the Great Dirt Bed (Francis, 1983a) and the Purbeck limestones (Norris, 1969) suggest that ferns and lycopods also grew in the forest. The density of the Great Dirt Bed forest, estimated from the spacing of the burrs, was approximately one tree in 15 m^2 (Francis, 1984).

Fossil wood is also present in all three paleosols as fusain or fossil charcoal, usually as black, shiny millimetre-sized, rectangular blocks with a sooty texture (Francis, 1983b, Pl. 39, Figures 1–3). Fossil charcoal is generally considered to be the product of past forest fires (Cope and Chaloner, 1980; Cope, 1981; Harris, 1981), thus this suggests that the fires also occurred in the Purbeck forests though the small quantity of fusain and the absence of charred tree stumps suggests that fires also occurred in the Purbeck forests though the on the forest floor, as observed in modern forests by Hill (1982).

The narrow and variable growth rings in the trees indicate that they grew very slowly and irregularly in a very seasonal climate with warm, wet periods followed by drought in which the poor supply of water was the most important limiting factor for their growth (Francis 1984). This semi-arid, Mediterranean-type climate was also proposed by West (1975) for the evaporites, Arkell (1947) for the macrofaunas and by Anderson (1971) for the ostracods.

DESCRIPTIONS OF THE PALEOSOLS

The Great Dirt Bed

This paleosol consists of a dark brown-black carbonaceous marl approximately 10–25 cm thick, containing black and white limestone pebbles. The base of the bed grades down into the Hard Cap below, which consists of micritized pelletoid sediment (Figure 5(b)) overlying mounds of algal stromatolitic limestone. The top of the Hard Cap is generally nodular, the nodules gradually becoming dispersed within the marl above as pebbles (Figure 3). At some localities (e.g. the Fossil Forest, Figure 1) the junction between the marl and the underlying limestone is abrupt rather than gradational due to the lithification of the pebbles at the base of the marl by secondary carbonate.

The carbonaceous marl has a waxy lustre and a poorly developed, granular structure which when dried disintegrates into small granules rather than large peds, although layers of marl and small clasts firmly adhere to surfaces of large pebbles. Within the marl there is a continuous size range of limestone clasts from those of boulder size (Wentworth scale: > 25 cm) down to small clasts incorporated in the

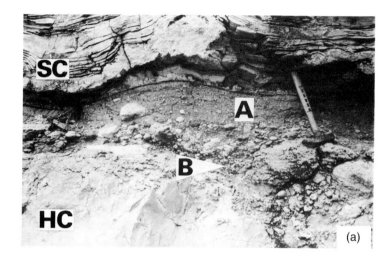
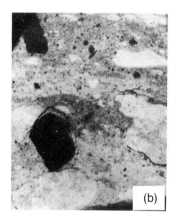
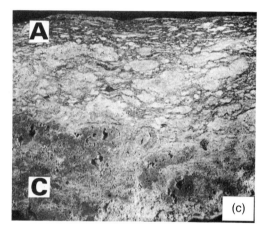

Figure 5. (a) The Great Dirt Bed at God Nore, Portland. The paleosol here consists of small black and white limestone pebbles in a micritic matrix (A) with laminae of fossil plant debris. This overlies black carbonaceous marl containing larger black and white limestone pebbles (B), infilling hollows in the Hard Cap (HC) limestone below. The Great Dirt Bed is overlain by hummocky algal stromatolitic limestone with thinly-bedded pelletoid limestone draped over it, constituting the Soft Cap (SC) (the Soft Burr and Aish of Arkell (1974)).

soil matrix (the S-matrix of Brewer, 1964a). An arbitrary grain size of 2 mm was taken to separate the clasts considered to be skeleton grains in the S-matrix (< 2 mm) from the large black and white limestone clasts, conspicuous in the field, which here are referred to generally as 'pebbles'.

The skeleton grains consist of small pelletoid clasts of micrite which are rounded, elliptical or irregular in shape, ranging in size from 2 mm to 9 µm and are poorly sorted (Figure 6(a)). The micrite is predominantly light brown in colour and without internal structure and resembles the micritic sediment in the Hard Cap below. Some of the peloids have external coatings of clay minerals or orange/brown organic matter, up to 30 µm thick. Some skeleton grains are composed of oolitically-coated clasts which are similar to those present in intraclastic sediment at the top of the Hard Cap, and some consist of micrite enclosing oolitically-coated calcitized pseudomorphs after lenticular gypsum. There are no smaller or well-sorted pellets which indicate a faecal origin, such as those described by Wright (1983) in a similar calcareous paleosol. Small pieces of bone and black wood (lignite) are present. Silt and sand-sized, sub-angular quartz grains (7–120 µm) constitute 3–5% of the S-matrix, although there are no clasts of chert or evidence of secondary silicification of the soil matrix itself. No vertical zonation is apparent within the S-matrix although a vague lamination is detectable in places where it appears to 'flow' around large pebbles. This is, however, probably the result of compaction.

Empty burrows and root moulds are, in general, not apparent due to the irregular fabric of the S-matrix. Since there is no secondary calcite cement in the matrix, all types of fenestrae (voids in the sense of Brewer, 1964a), root moulds or burrows in the marl were not infilled and preserved. However one burrow in the marl was observed in thin-section (Figure 6(e)). It is indicated by a 'swirl texture' composed of concentric layers of fine-grained orange organic matter, 98 µm wide (similar to that figured by Buurman, 1980, p. 600, Figure 4). No trails of faecal pellets indicative of faunal activity were observed.

Plant roots are present within the matrix, not as calcified rhizoliths

(b) The micritized sediment containing blackened pebbles from the top of the Hard Cap at God Nore, Portland. Many of the pebbles have thin laminar calcrete rinds. Width of field of view 4 cm. (c) The base of the Lower Dirt Bed, Holworth House. The lower part of the carbonaceous A-horizon grades down into the algal stromatolitic limestone of the Skull Cap (C) via a zone of mottling. Width of field of view 10 cm. (d) Black, carbonized conifer root (R) within the Great Dirt Bed at the Fossil Forest, Lulworth. The core of the root is silicified (S) and the cellular structure preserved, enabling identification of this root wood as *Araucarioxylon*. The cellular details are not present in the outer 'lignitic' wood. Lens cap 7 cm diameter.

but as brown, waxy, structureless lignite (compressed wood). The roots are present as isolated horizontal, sometimes branching, streaks 2–6 cm thick, and up to 30 cm long or can be seen to extend horizontally from *in situ* silicified tree stumps (Figure 5(d)).

The skeleton grains are incorporated in a plasma consisting mainly of finely disseminated organic material on a base of clay minerals and microcrystalline calcite (Figure 6(a)). The organic matter consists predominantly of small particles of orange material, 2–4 µm in size, which may be fossil resin. Similar material is present in the wood cells. Long strips of orange plant cuticle (Figure 6(d)), black and brown wood fibres and dark amorphous masses are also common, giving the marl its characteristic dark colour. Wood is also present as small, millimetre blocks of black, sooty fusain (fossil charcoal). The plasma has an asepic fabric (*sensu* Brewer, 1964b) and shows a flecked extinction pattern, indicating that the plasma constituents are randomly orientated. In patches a random crystic fabric is apparent, where calcite crystals dominate the plasma.

The top of the Great Dirt Bed is level and often marked by a 1–2 cm black band composed dominantly of organic material. This is overlain by thinly-bedded micrite and spongiostromata-type stromatolitic limestone of the Soft Cap (Brown, 1963; Pugh, 1968). Very little of the organic matter from the top of the Great Dirt Bed is incorporated in the sediments immediately above.

The lithology of the Great Dirt Bed changes in south-east Portland (e.g. Breston Quarry, British National Grid Reference SY 6889 6960). The S-matrix of the paleosol here is devoid of organic debris, consisting instead of pelmicrite and microspar which encloses small, black and white limestone pebbles (2–8 cm only) and forms a well-lithified sediment (Figure 5(a)). The amorphous carbonaceous material is absent from the matrix although some well-preserved conifer cuticle, cones and pollen are preserved on organic-rich laminae (Francis, 1983b).

Figure 6. (a) The S-matrix of the Great Dirt Bed, Admiralty Quarry, Portland. Skeleton grains (S) of micrite incorporated in a plasma of clay, microcrystalline carbonate and dark flecks of organic matter. Resin-impregnated thin-section. Plane polarized light. Width of field of view 1.9 mm. (b) The brown A-horizon of the Lower Dirt Bed, Poxwell, consisting of flecks and streaks of organic matter on a base of clay and microcrystalline carbonate. Thin-section. Plane polarized light. Width of field of view 1.4 mm. (c) The A-horizon of the Basal Dirt Bed, Chalbury Camp, showing same texture and composition as the Lower Dirt Bed in Figure 5(b). Thin-section. Plane polarized light. Width of field of view 1.2 mm. (d) The S-matrix of the Great Dirt Bed, Admiralty Quarry, Portland. The dark colour of the plasma is due to abundance of black and brown fragments of amorphous organic matter (M). Poorly preserved plant cuticle (C) and quartz silt (Q) are also present. Resin impregnated thin-section. Plane polarized light. Width of field of view

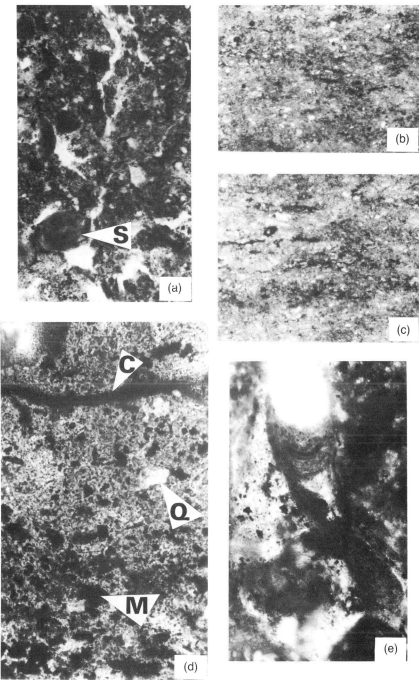

0.5 mm. (e) Burrow of a soil organism in the Great Dirt Bed paleosol A horizon. A 'swirl' texture of digested soil material traces the path of the burrow. Admiralty Quarry, Portland. Resin-impregnated thin-section. Plane polarized light. Width of field of view 0.25 mm.

This suggests that the original soil may have been subject to reworking and the plant material winnowed away. The typical paleosol with larger pebbles and carbonaceous marl matrix is present below this top lithified layer but only as an infilling of hollows and fissures in the top of the underlying Hard Cap (Figure 5(a)). In places (e.g. Breston Quarry, British National Grid Reference, SY 6898 6962), the Great Dirt Bed is absent altogether, leaving the Soft Cap overlying the nodular Hard Cap, suggesting topographic control of soil formation and erosion.

The Lower and Basal Dirt Beds

These two Dirt Beds are lithologically very similar with very simple profiles. In places where they are well developed (e.g. the Basal Dirt Bed at God Nore, British National Grid Reference, SY 690 698 and the Lower Dirt Bed in Sheat Quarry, British National Grid Reference, SY 689 698), both consist of an upper black carbonaceous layer (2 cm) underlain by up to 8 cm of brown marl which grades down into a cream-coloured marl at the base. At some localities, e.g. Holworth House (British National Grid Reference, SY 762 816) the middle brown layer grades down into the white marl via a zone of mottling (Figure 5(c)). The boundary between the white marl and the underlying limestone is a gradational one in contrast to the nodular, rubbly zone at the base of the Great Dirt Bed. Often only the middle brown layer is present.

The top of each Dirt Bed has a fairly sharp contact with the overlying micrite and organic material from the soil does not appear to have been reworked into the overlying bed. The Basal Dirt Bed is generally horizontal and persistent, though becoming thin and shaly in places, whereas the Lower Dirt Bed varies in thickness greatly over the irregular relief of the algal stromatolitic mounds of the Skull Cap.

The brown marl shows a faint lamination consisting of short, horizontal streaks of black and orange plant material, brown clay and white carbonate (Figure 6(b),(c)). Small lenses of white, microcrystalline carbonate, up to 2–3 mm long, 300 μm wide, are common in both Dirt Beds. The difference in colour of the marl layers is determined by the amount of carbonaceous material present. The darker layers are composed almost entirely of small fragments of plant material which tend to have a parallel orientation. The matrix consists of 1.5–8 μm grains of carbonate and clay minerals. Sub-angular grains of quartz silt are rare (< 1% matrix) and a few clasts of chert (90–800 μm) enclosing pyrite cubes are present. Bone fragments have also been found. Infilled animal burrows have not been observed but lignitic roots are quite common. Pollen grains (of cycadophytes in particular) were found in abundance in the white marl of the Lower Dirt Bed at Sheat Quarry, along with very fine rootlets which have nodular swellings, possibly similar to modern mycorrhizae (Francis, 1983b, pl. 41, Figure 6).

Although the Lower Dirt Bed profile is now very thin it once supported large conifer trees which are now preserved *in situ*, confirming its identification as a paleosol. The similar morphology of the Basal Dirt Bed implies this is also a paleosol although plant material within it is poorly preserved, apart from an accumulation of fusain at Bacon Hole.

INTERPRETATION OF THE PALEOSOLS

The Great Dirt Bed

Since its formation as a forest soil the Great Dirt Bed has obviously undergone many diagenetic changes, its mineralogy and geochemistry being most susceptible. Compaction will have reduced the thickness of the profile by an unknown amount and destroyed voids and burrows within the marl layer. The upper layers of loose, partially decayed plant debris were probably washed away by the incoming lagoon waters.

Some of the features used for modern soil classification (Buol *et al.* 1973) would be impossible to assess, e.g. base status, ion exchange capacity, moisture level. However, the morphology of the Great Dirt Bed and its horizon development appears relatively unchanged with no subsequent pedogenic alteration.

Therefore, using the scheme of modern soil description (Buol *et al.*, 1973) the Great Dirt Bed can be described as shown below. This is based on typical profiles from several sites.

Horizon		*Description*
?O	1–2 cm$^+$ (top eroded)	Black, organic-rich layer (sometimes absent)
A (or A/C)	10–25 cm	Dark brown/black carbonaceous marl with granular texture, incorporating silicified tree stumps and lignitic roots. Black and white limestone pebbles scattered throughout, often cemented by secondary carbonate. Pebbles at the base grade into C
C (or C ca in places	10 cm$^+$	Rubbly nodular peloidal and intraclastic parent limestone, partly micritized at the top. Algal limestone below

Table 1. Analysis of some constitutents of selected basal Purbeck Dirt Beds (% sand, silt and clay measured with carbonate present).

	Great Dirt Bed (Perryfield Quarry Portland)	Lower Dirt Bed (Perryficld Quarry Portland)	Basal Dirt Bed (Wakeham Quarry Portland)
% sand	40.3	48.9	41.4
% silt	36.0	38.0	40.6
% clay	23.7	13.0	18.1
% $CaCO_3$	50.5	58.9	88.4
% Organic carbon	1.71	0.91	0.34

The profile is very simple with one main undifferentiated A horizon of variable thickness grading into the bedrock below. Where the lower pebbles are cemented together by secondary carbonate the A/C boundary is much sharper (e.g. at the Fossil Forest). There are neither pale, leached B horizons present nor reddened layers of iron accumulation. This description above is that of the fossil soil since in the original soil the thicknesses were probably greater, particularly the upper organic-rich layer (mollic epipedon, Brewer, 1964a) which would have been capped by forest floor litter.

Textural analysis (by sieving) indicates that the Great Dirt Bed is now composed of similar amounts of sand, silt and clay-sized particles and is texturally classified as a loam (Buol *et al.*, 1973) (Table 1), although this texture may have been influenced by diagenetic alteration.

The diagnostic features of the Great Dirt Bed paleosol are (i) its simple A/C profile, (ii) dark, organic-rich solum and (iii) high carbonate content and calcareous parent rock. Using modern soil classification it is identified as a rendzina (New Soil Classification of England and Wales; Curtis *et al.* 1976) as suggested by West (1979). Under the US Comprehensive Soil Classification System (Soil Survey Staff, 1975) it is identified in the Order Mollisol (dark, organic-rich soil), in the sub-order Rendoll, defined as having calcareous parent rock, mollic epipedon, and with up to 40% $CaCO_3$ and pebbles up to 7.5 cm diameter in the A horizon. There are no Great Groups in this sub-order. The Great Dirt Bed rendzina can be further described as lithomorphic (Ragg and Clayden, 1973) as it formed on lithified parent rock, and also as a xero-rendzina due to the presence of secondary carbonate, implying a seasonally arid climate (Kubiena, 1970).

The characters of rendzinas are determined by the nature of the parent rock rather than by climate, that is, they are azonal soils. They are found, for example, on the Chalk in England (Townsend, 1973), on coral limestones in the tropical Seychelles (Lionnet, 1952), on dune-sands in semi-arid South Australia (Stace, 1956) and on Mediterranean limestones (Townsend, 1973). Today they are typically grassland soils and are generally too dry and shallow for forest trees (Duchaufour, 1982). Trees only become established on rendzinas where the profile is deeper and the more acidic forest litter then dissolves the carbonate more rapidly, allowing free iron to accumulate in a B-horizon, forming a brunified rendzina (Duchaufour, 1982). This is not present in the Great Dirt Bed.

The modern relationship between soil type and vegetation clearly cannot be applied to these Jurassic soils, since at that time angiosperms (including grasses) had yet to appear and the vegetation was dominated by the gymnosperms (Wesley, 1973). The rather uniformly warm and equable Mesozoic climate, with its broad climate zones (Frakes, 1979) would also have altered the distribution of plants and soils.

The Lower and Basal Dirt Beds

The identification of these paleosols is not as straight forward as for the Great Dirt Bed because the profiles are thin and not well developed. Where best developed both Dirt Beds have a similar profile, described below (compiled from typical sections from several localities).

Horizon		*Description*
?O	0–2 cm (top eroded)	Dark brown/black carbonaceous layer
A	1–8 cm	Brown marl with streaks of lignite (roots), plant material and clasts of white carbonate. Faint lamination apparent. No limestone pebbles
C	0–4 cm	Pale brown/cream coloured marl with very little organic material. Grades into limestone below. May have a mottled zone at A/C boundary

The presence of *in situ* tree stumps in the Lower Dirt Bed confirms its identification as a paleosol. Even though *in situ* plants have not been found in the Basal Dirt Bed its similar composition and structure to the Lower Dirt Bed (Figure 6(b),(c)) implies that it was also originally a soil. The thinness of the profiles suggest that they have been highly compacted or greatly eroded from once thicker profiles able to support large trees.

Both Dirt Beds are texturally similar to the Great Dirt Bed and classified as loam (Table 1). The carbonate content decreases and the organic matter content increases from the Basal Dirt Bed up to the Great Dirt Bed, reflecting the maturity of the better developed soil and perhaps sparser vegetation in the earlier soils.

Their high carbonate content and the calcareous parent rock suggests that the Lower and Basal Dirt Beds can also be identified as rendzinas, although much more immature forms. The absence of large pebbles and the presence of streaks of carbonate suggests that the underlying bedrock was only partially lithified during soil formation. The slight horizontal orientation of the soil matrix might be related to an alluvial influence and the soil matrix may have been originally transported from very local areas. There are no sedimentary features that suggest the Dirt Beds are true pedoliths (in the sense of Gerasimov, 1971) so as to suggest that their soil matrix has been derived from previously formed soils.

THE PEBBLES IN THE GREAT DIRT BED

The black and white limestone pebbles in this Dirt Bed range in size from those considered part of the S-matrix (< 2 mm) to some of boulder size (> 25 cm, Wentworth Scale). The larger pebbles tend to occur over mounds in the Hard Cap below and some are in fact still attached to the underlying limestone. The white pebbles often exhibit interlocking shapes. Although West (1975) reported an increase in pebble size from the west to the Lulworth area, pebble measurements show that there is no apparent size sorting or grading throughout the Dirt Bed profile or over its outcrop area as a whole. Details of pebble measurements and analyses can be found in Francis (1983(a)).

The blackened pebbles have a smaller size range than the white, only up to 14 cm diameter, and tend to be up to 10% more spherical than the flatter, platy white pebbles. They are generally 'compact' in shape, the white ones being 'bladed' (categories of Sneed and Folk, 1958) (Figure 8). The black pebbles are randomly distributed throughout the profile and on average constituted 30–40% of all pebbles.

The pebbles are lithologically similar to the limestones of the Hard Cap, also noted by West (1975). There is no evidence to suggest that

Figure 7. Calcrete textures in the Great Dirt Bed. (a) Irregular and truncated laminae of calcrete enclosing clast within Great Dirt Bed, Fossil Forest, Lulworth. Width of field of view 1.3 cm. (b) Black and white pebbles enclosed within calcrete matrix, base of Great Dirt Bed, Fossil Forest. The blackened clasts are more rounded than the angular white ones. Width of field of view 1.2 cm. (c) Rounded clast of intraclastic sediment coated with irregular laminae of calcrete. Great Dirt Bed, Fossil Forest, Lulworth. Width of field of view 6 mm. (d) Mottled micrite between blackened clasts, observed within pebble in the Great Dirt Bed, Fossil Forest. Similar mottled textures also observed

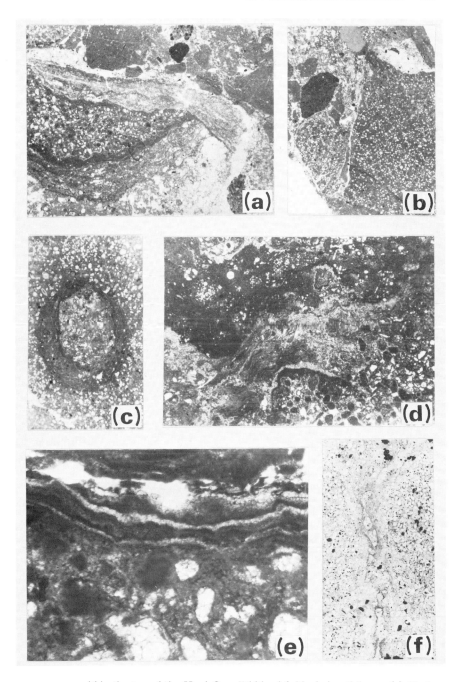

within the top of the Hard Cap. Width of field of view 5.5 mm. (e) Blackened pebble coated with undulating laminae of fibrous calcite and organic matter. Width of field of view 0.9 mm. (f) Trace of a root or burrow in a pebble of oosparite. The walls of the cavity are lined with micrite and infilled with calcite spar. Width of field of view 0.5 cm.

the pebbles were derived from beyond the Lulworth/Portland area or from beds below the Hard Cap, contrasting with the report of Webster (1826) that some were derived from the underlying Portland Stone invoking local uplift and erosion. In the field the 'black' (dark grey) and 'white' (buff-coloured) pebbles are easily distinguishable but in thin-section they are lithologically similar apart from blackening of allochems and/or matrix. Pebble composition includes algal micrite, pelletoid micrites and sparites, micrite with calcitized lenticular gypsum pseudomorphs and micrite containing several generations of intraclasts or oolitically-coated clasts. Many variations of these types are present. Calcitized gypsum pseudomorphs are more common in the black pebbles (as nodular clusters) than in the white pebbles where they have oolitic coatings, suggesting derivation from an older sediment.

Within the pebbles many variations of black and/or white clasts, such as intraclasts and peloids, occur within black or white matrices; for example black clasts are present within a white matrix but the overall pebble colour remains white. Black peloids can be seen within a black matrix. The only combination not observed was the presence of white clasts within a black matrix, suggesting that the blackened sediment was formed before the white.

The blackening of the limestone appears to be due to the inclusion of finely disseminated organic matter throughout the matrix, rather than superficial staining by iron sulphides which occurs in some anaerobic environments (Illing, 1954; Sugden, 1966; Maiklem, 1967). Organic matter constitutes approximately 10−20% of the non-carbonate fraction (20−40%) of a pebble. Crystals of pyrite or other iron minerals were not observed.

Many pebbles, both black and white, have coatings of irregular laminae of calcium carbonate with textures typical of calcrete (Strasser and Davaud, 1982). The pebbles and the clasts within pebbles have layers of finely laminated micrite and fibrous radial calcite 1−5 μm thick, forming rinds up to 2 cm thick (Figure 7(e)). Individual laminae are often stained dark brown due to the inclusion of organic material and drape from one grain to another enclosing several clasts (composite ooliths). The laminae are often truncated by successive layers (Figure 7(a)). Some pebbles have a cupped upper surface and draped laminae on the lower side (Figure 7(c)). Many of the flat platy pebbles are composed entirely of this laminated, fenestral micrite and this same sediment often encloses many original pebbles to form large nodules. Pebbles with calcrete textures are located throughout the Dirt Bed profile but in places (e.g. the Fossil Forest) the base of this paleosol has pebbles cemented by microcrystalline calcite which is not laminated but has a vaguely pelletoid or rather mottled appearance (Figure 7(b)).

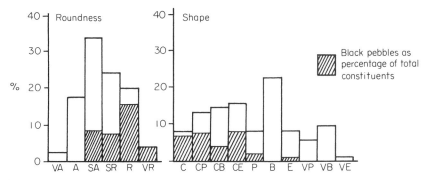

Figure 8. Distribution of shape and roundness of 165 pebbles from the Great Dirt Bed. The roundness classes of Powers (1953) are used and the shape classes of Sneed and Folk (1958). The roundness categories are: VA, very angular; A, angular; SA, sub-angular; SR, subround, R, round; VR, very round. The shape categories are: C, compact; CP, compact platy; CB compact bladed; CE, compact elongate; P, platy; B, bladed; E, elongate; VP, very platy; VB, very bladed; VE, very elongate. The results indicate that the blackened pebbles are more rounded in terms of abrasion and more spherical than the white.

This micritic cement extends down into the top of the Hard Cap to give it a patchy, mottled texture enclosing partially micritized skeletal grains (Figure 7(d)). There is no evidence that this micritization is due to the effects of boring algae (the type described by Bathurst, 1966) or the mottling due to bioturbation. Small isolated fragments of plant material (cuticle, pollen) appear better preserved within this fine-grained carbonate than within the marl matrix. There are also small tubular moulds, 50–100 μm in diameter, often filled with pelletoid micrite or sometimes lignite, which may represent small burrows or root moulds (Figure 7(f)).

The Origin of the Calcrete and Black Pebbles

The pebbles in the Great Dirt Bed have many textures and fabrics similar to those found in ancient and modern calcretes, including laminar crusts (Bernoulli and Wagner, 1971; Robbin and Stipp, 1979), coated grains (James, 1972; Read, 1974; Riding and Wright, 1981), cupped pebbles with accretionary rims (Bretz and Holberg, 1949) and micritized grains and bedrock (Arakel, 1982; Strasser and Davaud, 1982). Calcrete forms by the accumulation of secondary carbonate in soil profiles as nodular or laminar horizons, further developing into thick, indurated layers (Aristarian, 1970). A semi-arid or sub-humid climate with seasonal rainfall and periods of high evaporation is necessary (Boulaine, 1961; Harrison, 1977). Rainwater dissolves calcium carbonate from the soil as it percolates downward but when the water is subsequently evaporated the carbonate remains in the soil, firstly coating pebbles then developing into an impermeable laminar layer on top of which further carbonate is deposited (the K-horizon of Gile *et al.* 1966).

In the Great Dirt Bed calcrete is not only found coating pebbles but many pebbles are formed entirely of laminar calcrete, suggesting that they have formed by the brecciation of laminar layers. This can occur in mature profiles by weathering (Stage 4 of Harrison, 1977) and also by the mechanical destruction of the calcrete by the growth of tree roots—the 'rhizobrecciation' of calcrete by the roots of pine trees in Spain, described by Klappa (1980). Lower layers of calcrete remain intact and act as an aquiclude, retaining a lens of fresh-water above them. A similar store of water in the Great Dirt Bed would have been useful for the Purbeck trees as their water supply appears to have been limited (Francis 1984) and, like modern rendzinas (Fitzpatrick, 1980), the soil probably had little capacity for water storage.

The Lower and Basal Dirt Beds do not contain calcrete nodules, although the white carbonate clasts may represent initial stages of its development.

Blackened clasts are commonly associated with calcrete deposits (Multer and Hoffmeister, 1968; Ward *et al.*, 1970; Wilson, 1975; Riding and Wright, 1981), including the Purbeckian deposits of the Jura (Cotillon, 1960; Bläsi, 1980; Strasser and Davaud, 1983), and are also considered to be indicative of sub-aerially exposed surfaces. Although some blackening may be due to external staining by iron sulphide (Illing, 1954; Sugden, 1966; Maiklem, 1967), the most common cause is due to the inclusion of finely disseminated organic matter throughout the sediment matrix. On the shores of hypersaline lagoons of Isla Mujeres, Yucatan, blackened clasts of aeolianite bedrock and calcrete are mixed with similar white clasts formed by desiccation fracturing, described by Ward *et al.* (1970). The resemblance to the Purbeck pebbles was noted by West (1975). The black sediment originated from black, gelatinous mud and algal mat at the lake margin (Folk, 1967). The black pebbles are smaller and rounder than the larger platy white ones, the difference being due to the greater susceptibility of the black rock to fracturing. This relationship was also observed amongst the Great Dirt Bed pebbles and those from the Jura Purbeck Beds. Strasser and Davaud (1983) found that the blackening resulted from the inclusion of both algal matter and charcoalified higher terrestrial plant material

By comparison, the Great Dirt Bed black pebbles appear to have originated from sediment in a reducing alkaline environment in the shallow hypersaline lagoon, which was blackened by algae, some pyrite and probably terrestrial plant material. Subsequent reworking led to the incorporation of clasts of the black sediment into an unstained matrix. As it became exposed, desiccated and fractured, black pebbles mixed with white clasts on the supratidal flats, later to become incorporated into the developing soil.

DISCUSSION

During minor regressional phases in the early Purbeck the algal sediments, previously formed in the intertidal zone of the Purbeck lagoon (Brown, 1963; West, 1975), became sub-aerially exposed for sufficient periods to allow desiccation of the shoreline and the formation of soils. During the formation of the Great Dirt Bed paleosol the period of exposure was of sufficient length to allow the development of recognizable soil horizons in the soil. However, the absence of distinct horizons in the Lower and Basal Dirt Beds suggests that these soils were exposed for relatively shorter periods.

The blackened clasts at the base of the Great Dirt Bed, by comparison with those on modern hypersaline lake margins (Ward et al., 1970) and other geological occurrences (Strasser and Davaud, 1983), appear to have originated from blackened, organic-rich sediment formed in a reducing environment in the lagoon. As this became exposed it became lithified and desiccated and fractured into clasts which were reworked and mixed with clasts of unblackened sediment on the shoreline, similar to the origin proposed for the blackened pebbles in the Purbeck Beds of the Jura by Strasser and Davaud (1983).

With continued exposure, surface weathering and colonization by plants led to the formation of a soil profile suitable for the establishment of a conifer forest. The rendzina-like character of the Great Dirt Bed paleosol was most strongly influenced by the nature of the calcareous bedrock, resulting in a high carbonate content and containing large limestone pebbles derived from the underlying bedrock. If the bedrock had been somewhat more acidic and the climate slightly more humid, a more typical podsol soil with an illuvial B horizon would probably have developed under the forest, as characteristically associated with forest vegetation today (Ragg and Clayden, 1973). Although modern rendzinas are typically grassland soils (Townsend, 1973), it is apparent that the different vegetational patterns have to be considered when interpreting the origin of fossil rendzinas.

The A-horizon of the Great Dirt Bed appears to have been dominantly formed of plant debris derived from local vegetation on a clay base. The main constituents were conifer twigs and foliage which contained large quantities of resin (Francis, 1983b) and was thus the probable source of the orange resinous blobs in the S-matrix. Some fern and lycopod foliage from the herbaceous undergrowth would also have been included. In such a dry environment the moisture content of this humus would have been lowered to a flammable level and fueled forest fires, burning only dry humus and surface litter, as observed in modern fires by Hill (1982). Lightning probably started such fires as it does in modern forests today (Komarek, 1972). The absence of large deposits of charcoal or charred tree stumps indicates that the fire was not able to damage the large trees completely.

There is little evidence within the paleosol of animal or insect activity. A few burrows (Figure 6(e)) and small bones are present but there are no quantities of faecal pellets which would have resulted from the digestion of the soil by soil organisms, such as the types described by Szabo *et al.* (1964). Such pellets dominate the organic horizons of modern rendzinas, forming characteristic moder fabrics (Kubiena, 1970). Several types of moder fabric have been observed in a Carboniferous rendzina (Wright, 1983), representing the activity of soil micro-arthropods within the O-horizon. The semi-arid and periodically dry Purbeck climate (Francis, 1984) may have deterred soil organisms in the Great Dirt Bed but such features would also have had poor potential for preservation.

The organic material within the A-horizon of the Great Dirt Bed is mostly in a fairly well oxidized and amorphous state. No identifiable cuticles were recovered from the marly S-matrix. The plant debris would have been readily oxidized in such an aerobic environment, though in some modern soils a high carbonate content inhibits breakdown (Fitzpatrick, 1980). The average organic carbon content of the Great Dirt Bed is 1.71% compared to values from modern rendzinas of 5.1% in the Seychelles (Lionnet, 1952) and 4.2% in Australia (Stace, 1956), though the significance of such comparisons is unknown as the low content of the Great Dirt Bed could be related to sparser vegetation or diagenesis. In the Great Dirt Bed at God Nore, (where the S-matrix appears reworked), conifer cuticle is better preserved (Francis, 1983b), suggesting that conditions here were slightly less aerobic.

The seasonal aspect of the climate (Francis, 1984) also had an important influence on the Great Dirt Bed paleosol and the formation of calcrete. This process probably commenced with exposure of the bedrock, as in Florida today (Robbin and Stipp, 1979) and continued as the soil cover accumulated (Multer and Hoffmeister, 1968), deriving carbonate from pebbles within the soil. The pebbles originated by fracturing of the bedrock by soil-forming processes, probably aided by the rhizobrecciation of the bedrock by the roots of the conifers, as occurs today in Spain (Klappa, 1980). The layers of laminar calcrete formed within the soil were also brecciated to form pebbles.

The thinner and less well-defined paleosol profiles of the Basal and Lower Dirt Beds indicate that soil processes were not so developed and the period of exposure probably shorter. The underlying sediment was not well lithified or desiccated prior to their formation preventing inclusion of pebbles in the soil matrices.

The conifers which grew in these paleosols had shallow horizontal root systems, receiving their water supply from fresh water retained in the soil (a seasonally erratic supply) rather than deeper ground water (Francis, 1983b). There is no evidence from fossil tree stumps to suggest that the conifers were adapted for growth with their bases in water, such as buttressed or mangrove-like roots. The absence of any features of waterlogging in the paleosols and their obvious aerobic nature, particularly that of the Great Dirt Bed, is important evidence that the Purbeck cheirolepidiacean conifers grew in well-drained soils even though they are preserved in a sequence of marginal continental deposits. The habitats of the Cheirolepidiaceae are considered to be rather diverse (Alvin, 1982) and although some may have been coastal, mangrove-like plants (Upchurch and Doyle, 1981) others were large forest trees (Alvin, 1983; Francis, 1983b) and, in the case of the Purbeck trees, grew in well-drained soils.

Estimates of the length of time for formation of the paleosols can be obtained from several sources. Rendzinas are essentially immature soils which have not had time to undergo further horizon development. Typical modern rendzinas may develop over a span of 10 000 years (Fitzpatrick, 1980). This order of time may be applicable to the Great Dirt Bed. Similarly laminar calcrete layers in Florida have been estimated to have formed at a rate of 1 cm per 2000–4000 years (Robbin and Stipp, 1979). Calculations based on the thickness of Purbeck calcrete within the Great Dirt Bed are difficult as complete laminar layers are not present and the semi-arid Purbeck climate may have affected the rate of carbonate deposition. The maximum thickness of any pebbles composed of laminar calcrete was found to be approximately 5 cm which, using the formation rate above, is estimated to have taken 10 000–20 000 years to form.

A value for the minimum length of exposure of the soils can be estimated from the life span of the trees, which can be calculated from their growth rings. The average width of the Purbeck rings is 1.13 mm (Francis 1984). Assuming the growth increments are annual, the tree stump of the largest diameter (approximately 1.4 m of the remaining inner heartwood only) had a minimum life span of about 700 years, or perhaps 900–1000 years if the outer sapwood and bark layers are taken into account. Using these criteria the large tree stump rooted in the Lower Dirt Bed at Chalbury Camp lived for at least 500 years, giving an estimate of the minimum length of time for the existence of this thin Lower Dirt Bed paleosol.

The development of each Dirt Bed paleosol was terminated by an influx of highly saline lagoon water and the re-establishment of intertidal conditions with algal sediments (the 'Caps'). The flooding of the forests must have been fairly rapid though gentle enough to leave the soil horizons undisturbed, as very little of the soil material was mixed into the overlying sediment, as noted by Damon (1884). The loose upper layer of dried and partially decayed plant litter (the O-horizon) was washed away and mats of algal-bound sediment coated tree bases, fallen logs and the top of the soil, preventing further erosion. This algal 'capping' was an important factor in the preservation of these paleosols. A similar sub-aerial stromatolitic algal mat (the O-horizon) overlies the pelletoid A-horizon of Wright's (1983) Carboniferous xero-rendzina. Algal horizons in other sequences of marginal continental sediments may prove to be potential sites for the location of further paleosols.

SUMMARY

Rendzina paleosols are preserved within a sequence of marginal continental deposits of algal stromatolitic and pelletoid limestones, constituting the basal part of the Lower Purbeck Formation in Dorset.

The Great Dirt Bed is the most well-developed paleosol exhibiting a characteristically simple A/C rendzina profile, consisting of a dark organic-rich horizon overlying limestone bedrock. The main component of the matrix of the A-horizon is decomposed plant debris. There is little evidence of the activity of soil organisms and a notable absence of a pelletoid moder fabric, the faecal pellet textures typical of modern rendzinas. The upper O-horizon of undecomposed plant litter is also lost. This soil supported a conifer forest of slow-growing shallow-rooted trees, now preserved *in situ* as silicified tree trunks and carbonized roots.

Pebbles derived from the underlying limestone are incorporated in the soil matrix. Some consist of blackened sediment derived from a desiccated and fractured organic-rich deposit previously formed on the margins of the adjacent lagoon. The semi-arid, seasonal Purbeck climate promoted the formation of laminated and mottled deposits of secondary carbonate or calcrete, present as micritized bedrock, laminar rinds and cement around pebbles. This was itself brecciated by the soil processes and probably also by the mechanical action of the tree roots, then to become incorporated into the solum as pebbles.

The Lower and Basal Dirt Beds are immature forms of rendzinas, with similar simple profiles of organic-rich layers with high carbonate content, overlying marl and devoid of large pebbles. The Lower Dirt Bed also supported a forest of conifers and cycadophytes.

Each paleosol is capped by algal limestone, which originated as algal-bound sediment formed when rising saline lagoon water successively inundated the forests. This covered the tree stumps and the top of each soil with a protective layer of sediment, ensuring the rather exceptional preservation of these paleosols with *in situ* tree stumps.

ACKNOWLEDGEMENTS

This paper presents results of a research project undertaken in the Geology/Biology departments at Southampton University and funded by an NERC grant which is gratefully acknowledged. I am grateful for facilities provided there and thank Drs I.M. West and P.J. Edwards for supervision.

I thank Dr G.T. Creber for reviewing this manuscript and Professor W.G. Chaloner and Professor A.J. Smith at Bedford College for facilities provided there during the tenure of an NERC Post-Doctoral Research Fellowship. I thank Lorna Jackson for patient typing.

REFERENCES

Alvin, K.L. 1982. Cheirolepidiaceae: biology, structure and paleoecology. *Rev. Palaeobot. Palynol.* **37**, 71–98.

Alvin, K.L. 1983. Reconstruction of a Lower Cretaceous conifer. *Bot. J. Linn. Soc.* **86**, 169–76.

Anderson, F.W. 1971. The Ostracods, in *The Purbeck Beds of the Weald (England)*, F.W. Anderson and R.A. Bazley (eds). *Bull. geol. Surv. Gt. Brit.* **34**, 27–174.

Arakel, A.V. 1982. Genesis of calcrete in Quaternary soil profiles, Hutt and Leeman lagoons, Western Australia. *J. sedim. Petrol.* **52**, 109–25.

Aristarian, L.F. 1970. Chemical analysis of caliche profiles from the High Plains, New Mexico. *J. Geol.* **78**, 201–12.

Arkell, W.J. 1947. The geology of the country around Weymouth, Swanage, Corfe and Lulworth. *Mem. geol. Surv. UK*, 386 pp.

Bathurst, R.G.C. 1966. Boring algae, micrite envelopes and the lithification of molluscan biosparites. *Geol. J.* **5**, 15–32.

Bernoulli, D. and Wagner, C.W. 1971. Subaerial diagenesis and fossil caliche deposits in the Calcaire Masiccio Formation (Lower Jurassic, Central Apennines, Italy). *N. Jb. Geol. Paläont. Abh.* **138**, 135–49.

Bläsi, H. 1980. *Die Ablagerungsverhaeltrusse im 'Portlandien' des Schweizerischen und Franzoesischen Juras*. Ph.D. Thesis, University of Bern. 151 pp.

Boulaine, J. 1961. Sur la rôle de la végétation dans la formation des carapaces calcaires méditerranéenes. *C. R. Acad. Sci. Paris* **253**, 2568–70.

Bretz, J.H. and Holberg, L. 1949. Caliche in southeastern New Mexico. *J. Geol.* **57**, 491–511.

Brewer, R. 1964a. *Fabric and Mineral Analysis of Soils*. Wiley, New York. 470 pp.

Brewer, R. 1964b. Classification of plasmic fabrics of soil materials, in *Soil Micromorphology*, A. Jongerius (ed.). Elsevier, Amsterdam. 95–108.

Brown, P.R. 1963. Algal limestones and associated sediments in the basal Purbeck of Dorset. *Geol. Mag.* **100**, 565–73.
Buckland, Rev. M. and De La Beche, H.T. 1836. On the geology of the neighbourhood of Weymouth and the adjacent parts of the coast of Dorset. *Trans. geol. Soc. London* **4**, 1–46.
Buol, S.W., Hole, F.D. and McCracken, R.J. 1973. *Soil Genesis and Classification.* Iowa State University Press, Arnes. 360 pp.
Buurman, P. 1980. Palaeosols in the Reading Beds (Paleocene) of Alum Bay, Isle of Wight, UK. *Sedimentology* **27**, 593–606.
Cope, M.J. 1981. Products of natural burning as a component of the dispersed organic matter of sedimentary rocks, in *Organic Maturation Studies and Fossil Fuel Exploration*, J. Brooks (ed.). Academic Press, London. 89–102.
Cope, M.J. and Chaloner, W.G. 1980. Fossil charcoal as evidence of past atmospheric composition. *Nature* **283**, 647–9.
Cotillon, P. 1960. Caractères pétrographiques et genèse des galets noir observés dans une coupe des 'Calcaires blancs' de Provence (Jurassique supérieur-Crétace inferieur). *C. R. Soc. geol. France* **7**, 170–1.
Curtis, L.F., Courtney, F.M. and Trudgill, S. 1976. *Soils in the British Isles.* Longman, New York and London. 364 pp.
Damon, R. 1884. *Geology of Weymouth, Portland and the Coast of Dorsetshire.* 2nd ed., Weymouth. 250 pp.
Duchaufour, P. 1982. *Pedology: Pedogenesis and Classification.* (English translation by T.R. Paton). George Allen and Unwin, London. 448 pp.
Fitton, W.H. 1835. Notice on the junction of Portland and Purbeck strata on the coast of Dorsetshire. *Proc. Geol. Soc.* **2**, 185–7.
Fitton, W.H. 1836. Observations on some of the strata between the Chalk and the Oxford Oolite, in the south-east of England. *Trans. geol. Soc. Lond.* **4** 103–378.
Fitzpatrick, E.A. 1980. *Soils.* Longman, London. 353 pp.
Folk, R.L. 1967. Carbonate sediments of Isla Mujeres, Quintana Roo, Mexico and vicinity, in *New Orleans Geol. Soc. Guidebook*; *Field Trip to Yucatan.* 100–23.
Frakes, L.A. 1979. *Climates through Geologic Time.* Elsevier, Amsterdam. 310 pp.
Francis, J.E. 1983a. *The Fossil Forests of the Basal Purbeck Formation (Upper Jurassic) of Dorset, Southern England.* Ph.D. Thesis (unpublished), University of Southampton.
Francis, J.E. 1983b. The dominant conifer of the Jurassic Purbeck Formation, England. *Palaeontology* **26**, 277–94.
Francis, J.E. 1984. The seasonal environment of the Purbeck (Upper Jurassic) fossil forests. *Palaeogeogr., Palaeoclimatol., Palaeoecol.* **48**, 258–307.
Gerasimov, I.P. 1971. Nature and originality of palaeosols, in *Paleopedology*, D.H. Yaalon (ed.). Int. Soc. Soil. Sci. and Israel Univ. Press, Jerusalem. 15–28.
Gile, L.H., Peterson, F.F. and Grossman, R.B. 1966. Morphological and genetic sequences of carbonate accumulation in desert soils. *Soil Sci.* **101**, 347–59.
Goldbery, R. 1982. Palaeosols of the Lower Jurassic Mishhor and Ardon Formations ('Laterite Derivative Facies'), Makhtesh Ramon, Israel. *Sedimentology* **29**, 669–90.
Gray, W. 1861. On the geology of the Isle of Portland. *Proc. Geol. Ass.* **1**, 128–47.
Hill, R.S. 1982. Rainforest fire in Western Tasmania. *Aust. J. Bot.* **30**, 583–9.
Harris, T.M. 1981. Burnt ferns from the English Wealden. *Proc. Geol. Ass.* **92**, 47–58.
Harrison, R.S. 1977. Caliche profiles: Indicators of near-surface subaerial diagenesis, Barbados, West Indies. *Bull. Can. Petrol. Geol.* **25**, 123–73.
Howitt, F. 1964. Stratigraphy and structure of the Purbeck inliers of Sussex (England). *Q. Jl. geol. Soc. Lond.* **120**, 77–114.
Illing, L.V. 1954. Bahaman calcareous sand. *Bull. Am. Ass. Petrol. Geol.* **38**, 1–95.
James, N.P. 1972. Holocene and Pleistocene calcareous crust (caliche) profiles: Criteria for sub-aerial exposure. *J. sedim. Petrol.* **42**, 817–36.

Klappa, C.F. 1980. Brecciation textures and teepee structures in Quaternary calcrete (caliche) profiles from east Spain; the plant factor in their formation. *Geol. J.* **15**, 81–9.

Komarek, E.V. 1972. Ancient fires. *Proc. Ann. Tall. Timbers Fire Ecol. Conf.* **12**, 219–40.

Kubiena, W.L. 1970. *Micromorphological Features of Soil Geography*. Rutgers University Press, New Brunswick, New Jersey. 254 pp.

Lionnet, J.F.G. 1952. Rendzina soils of coastal flats of the Seychelles *J. Soil Sci.* **3**, 172–81.

McPherson, J.G. 1979. Calcrete (caliche) palaeosols in fluvial redbeds of the Aztec Siltstone (Upper Devonian), southern Victoria Land, Antarctica. *Sedim. Geol.* **22**, 267–85.

Maiklem, W.R. 1967. Black and brown speckled foraminiferal sand from the southern part of the Great Barrier reef. *J. sedim. Petrol.* **37**, 1023–30.

Meyer, R. 1976. Continental sedimentation, soil genesis and marine transgression in the basal beds of the Cretaceous in the east of the Paris Basin. *Sedimentology* **23**, 235–53

Multer, H.G. and Hoffmeister, J.E. 1968. Subaerial laminated crusts of the Florida Keys. *Bull. geol. Soc. Am.* **79**, 183–92.

Norris, G. 1969. Miospores from the Purbeck Beds and marine Upper Jurassic of southern England. *Palaeontology* **12**, 574–620.

Powers, M.C. 1953. A new roundness scale for sedimentary particles. *J. sedim. Petrol.* **23**, 117–9.

Pugh, M.E. 1968. Algae from the Lower Purbeck limestones of Dorset. *Proc. Geol. Ass.* **79**, 513–23.

Ragg, J.M. and Clayden, B. 1973. The classification of some British soils according to the comprehensive system of the United States. *Soil Survey Technical Monograph*, no. 3, Harpenden. 326 pp.

Read, J.F. 1974. Caliche deposits and Quaternary sediments, Edel Province, Shark Bay, Western Australia, in *Evolution and diagenesis of Quaternary carbonate sequences, Shark Bay, Western Australia*, B.W. Logan, J.F. Read, G.M. Hagan, P. Hoffman, R.G. Brown, P.J. Woods, and C.D. Gebelein (eds). *Am. Assoc. Petrol. Geol. Memoir* **22**, 250–80.

Retallack, G.J. 1976. Triassic Palaeosols in the Upper Narrabeen Group of New South Wales. Part 1: Features of the palaeosols *J. geol. Soc. Austr.* **23**, 383–99.

Retallack, G.J. 1981. Fossil soils: indicators of ancient terrestrial environments, in *Palaeobotany, Palaeoecology and Evolution*, K.J. Niklas (ed.). Praeger Publishers, New York. vol. 1, 55–102.

Riding, R. and Wright, V.P. 1981. Palaeosols and tidal flat/lagoon sequences on a Carboniferous carbonate shelf: sedimentary associations of triple disconformities *J. sedim. Petrol.* **51**, 1323–39.

Robbin, D.M. and Stipp, J.J. 1979. Depositional rates of laminated soilstone crusts, Florida Keys. *J. sedim. Petrol.* **49**, 0175–0180.

Sneed, E.D. and Folk, R.L. 1958. Pebbles in the lower Colorado river, Texas: A study in particle morphogenesis. *J. sedim. Petrol.* **66**, 114–49.

Soil Survey Staff 1975. *Soil Taxonomy*. Agricultural Handbook 436. USDA, Washington.

Stace, H.C.T. 1956. Chemical characteristics of Terra Rossas and Rendzinas of South Australia. *J. Soil Sci.* **7**, 280–93.

Strahan, A. 1898. The geology of the Isle of Purbeck and Weymouth. *Mem. geol. Surv. UK.* 1–278.

Strasser, A. and Davaud, E. 1982. Les croûtes calcaires (calcretes) du Purbeckien du Mont-Salève (Haut-Savoie, France). *Eclogae geol. Helv.* **75**, 287–301.

Strasser, A. and Davaud, E. 1983. Black pebbles of the Purbeckian (Swiss and French Jura): lithology, geochemistry and origin. *Eclogae geol. Helv.* **76**, 551–80.

Sugden, W. 1966. Pyrite staining of pellety debris in carbonate sediments from the Middle East and elsewhere. *Geol. Mag.* **103**, 250–6.

Szabo, I., Martou, M. and Parti, G. 1964. Micro-milieu studies in the A-horizon of a mull-like rendsina, in *Soil Micromorphology*, A. Jongerius (ed.). Elsevier, Amsterdam. 33–45.

Townsend, W.N. 1973. *An Introduction to the Scientific Study of the Soil*. Edward Arnold Ltd, London. 299 pp.

Upchurch, G.R. and Doyle, J.A. 1981. Paleoecology of the conifers *Frenelopsis* and *Pseudofrenelopsis* (Cheirolepidiaceae) from the Cretaceous Potomac Group of Maryland and Virginia, in *Geobotany II*, R.C. Romans (ed.). Plenum Press, New York. 167–202.

Vahrameev, V.A. 1970. Range and palaeoecology of Mesozoic conifers, the Cheirolepidiaceae. *Palaeont.* **4**, 12–25.

Ward, W.C., Folk, R.L. and Wilson, J.L. 1970. Blackening of eolianite and caliche adjacent to saline lakes, Isla Mujeres, Quintana Roo, Mexico. *J. sedim. Petrol.* **40**, 548–55.

Webster, T. 1826. Observations on the Purbeck and Portland Beds. *Trans. geol. Soc. Lond. ser. 2*, **2**, 37–44.

Wesley, A. 1973. Jurassic plants, in *Atlas of Palaeobiogeography*, A. Hallam (ed.). Elsevier, Amsterdam. 329–38.

West, I.M. 1975. Evaporites and associated sediments of the basal Purbeck Formation (Upper Jurassic) of Dorset. *Proc. Geol. Ass.* **86**, 205–28.

West, I.M. 1979. Review of evaporite diagenesis in the Purbeck Formation of Southern England. Symposium 'Sédimentation jurassique W. européan' *ASF Publication spéciale* no. 1, 407–15.

Wilson, J.L. 1975. *Carbonate Facies in Geologic History*. Springer-Verlag, New York.

Wright, V.P. 1983. A rendzina from the Lower Carboniferous of South Wales *Sedimentology* **30**, 159–79.

Chapter 5

TECTONIC CONTROL ON ALLUVIAL SEDIMENTATION AS REVEALED BY AN ANCIENT CATENA IN THE CAPELLA FORMATION (EOCENE) OF NORTHERN SPAIN

CHRISTOPHER D. ATKINSON*
Department of Geology, University College of Swansea, Great Britain
*now at, ARCO Exploration and Techology Company, 2300 West Plano Parkway, Plano, Texas 75075, USA.

INTRODUCTION

The detailed description and interpretation of fossil soils ('paleosols') is a comparatively recent, but rapidly expanding innovation in the study of terrestrial sedimentary sequences. In most previous discussions on pre-Quaternary paleosols, emphasis has tended to be placed either purely on their recognition (e.g. Burgess, 1961; Freytet, 1971, 1973; Buurman, 1975), on their usefulness as indicators of paleoclimates (e.g. Allen, 1960, 1973, 1974a; Steel, 1974; Hubert, 1977) or to derive inferences concerning sedimentation rates (Leeder, 1975). Only a few studies have demonstrated that fossil soils, like their recent counterparts, formed an important part of an actively evolving landscape (e.g. Allen, 1974b; Leeder, 1976; Buurman, 1980; Retallack, 1976, 1977, 1983a,b; Bown and Kraus, 1981; Besly and Turner, 1983). In the landscape, soils and soil types vary laterally in response to regional variations in depositional environment and geomorphology. In modern soils, this type of lateral variation is termed a 'catena', the widely accepted definition of which is:

'A sequence of soils of about the same age, derived from similar parent material and occurring under similar climatic conditions, but having different characteristics due to variations in relief and in drainage.' (Steila, 1976).

The purpose of this paper is to illustrate, using an integrated study of paleosol development and alluvial stratigraphy, the presence of an ancient catena sequence from the middle Eocene Capella Formation of the Tremp-Graus basin in northern Spain. Evidence is presented to demonstrate that this catena developed due to tectonic control on alluvial sedimentation produced by thrust-induced differential subsidence in the basin.

The Capella Formation (Garrido-Megias, 1968) comprises an interbedded suite of alluvial channel and vari-coloured overbank deposits (Nijman and Nio, 1975). Previous descriptions have concentrated either on its stratigraphy (Garrido-Megias, 1968) or on its general sedimentological appearance (Nijman and Nio, 1975). The first suggestion that the Formation contained paleosols was made by Buurman (cited in Nijman and Nio, 1975) who recorded the presence of many features indicative of paleopedogenesis (e.g. vivid colouration, intensive bioturbation, rootlet mottling, mineral accumulation horizons), but did not document their origin nor interpret any of the possible soil types. Thus despite the fact that paleosols were known to exist, no major attempt at their description was presented at that time. In this paper it is argued that paleosols are not only common in the Formation, but also that they exhibit pronounced regional variations across the basin.

GEOLOGICAL SETTING

The Tremp-Graus Basin is a relatively narrow and elongate structure rarely exceeding 25 km in width and stretching for nearly 100 km from west to east across the provinces of Huesca and Lleida in northern Spain (Figure 1). The basin developed during Paleocene and Eocene times as a synclinal depression on top of the southwards migrating Cotiella nappe (Seguret, 1970) or Montsec thrust-sheet (Williams, 1985

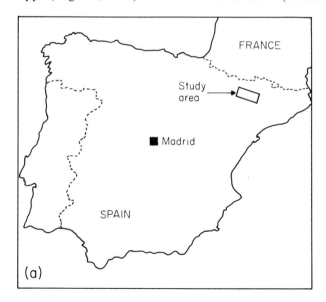

Figure 1. (a) Location of the Tremp-Graus Basin (after Nijman and Nio, 1975).

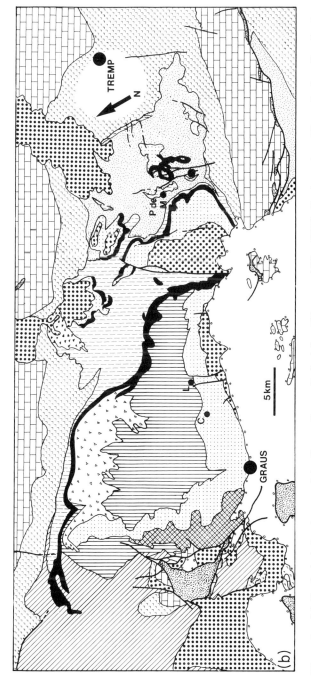

Figure 1. (b) Geological sketch map of the Tremp-Grauss Basin (after Nijman and Nio, 1975). For legend see Figure 2. C = Capella. L = Laguarres. P de M = Puente de Montañana.

press; Williams and Atkinson, in prep.). It thus forms a 'thrust-sheet-top' or 'piggyback' basin in the sense of Ori and Friend (1984) being defined at its base and margins by the flats and ramps of the underlying Montsec thrust sheet. It acted as a major site of sediment accumulation from Paleocene to Oligocene times (Nijman and Nio, 1975). During this period, up to 2000 m of non-marine and marine sediments, arranged into two clastic-wedge sequences (the Montañana and Campodarbe Groups) filled the basin (Figure 2).

The major portion of the non-marine sediments comprising both Groups have been interpreted as the product of several large humid alluvial fans (Atkinson, 1983). Several fan complexes, of both northern and southern derivation, have been recognized within the five main alluvial formations identified in the basin, e.g. the San Esteban, Monllobat, Campanué, Capella and Escanilla Formations (Figure 2). Throughout the basin-filling period the northern fans were of major importance and their relative size increased through time as Pyrenean uplift proceeded. In doing so they became successively offset westwards along the northern basin margin (Atkinson, 1983). The orientation of the fans was such that most sediment dispersal was to the south and south-west. Supplementing these major northern fans were occasional minor fans which developed along the southern basin margin. These fans delivered sediment northwards and resulted from clastic 'backshedding' associated with ramp uplift at the front of the Montsec thrust-sheet (Williams and Atkinson, in prep.). Both the major and minor fan sequences are characterized by very similar sedimentary facies (Figures 3 and 4). Thick, multistorey conglomerates, interpreted to be palaeovalley fills, typify the proximal regions and these grade, via mixed sand and gravel, mid-fan braided/meandering channels, to finer grained, predominantly sandy, meandering/low sinuosity channels in distal fan regions (Atkinson, 1983).

The Capella Formation (upper Ypresian—upper Lutetian) represents the final phases of fan sedimentation within the Montañana Group (Figure 2). The Formation is relatively undeformed and varies in thickness from 150 m in the east at Torsal Gros to in excess of 300 m near the village of Capella in the west. This westwards thickening does not take place gradually, but instead occurs in increments coinciding with the presence of the NNE–SSW trending Ribagorzana and Luzas fault-zones (Figures 1 and 2).

Within the Capella Formation both major northern and minor southern fans have been recognized. Deposits from the northern fans predominate in the western and upper part of the eastern Capella successions whereas southern fan deposits are restricted to the lower portions of the eastern succession (Atkinson, 1983). Throughout

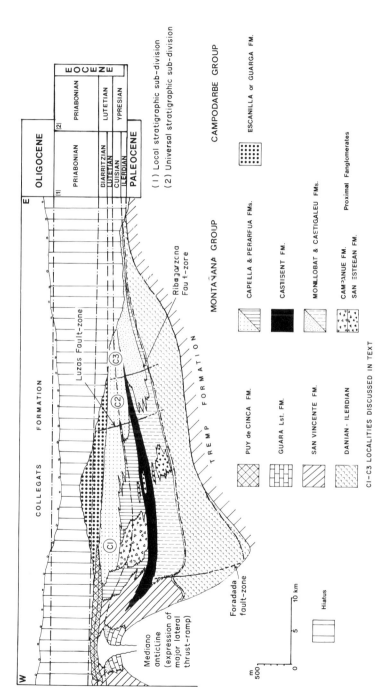

Figure 2. West—east geological cross-section through the Tremp-Graus Basin (modified after Nijman and Nio, 1975).

144   CHAPTER 5

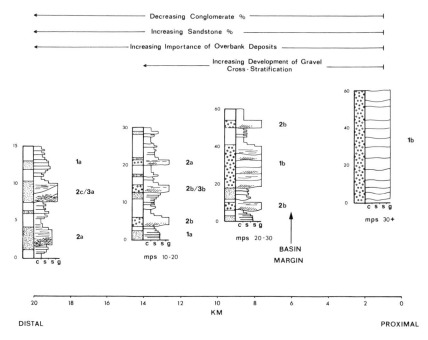

Figure 3. Idealized sedimentary facies of the Montañana and Campodarbe humid alluvial fan sequences. 1a = crevasse/levee sheet sands, 1b = extensive multistorey sheet conglomerates, 2a = simple ribbon bodies, 2b = multi-storey ribbon bodies, 2c = multilateral ribbon bodies, 3a = simple tabular 'point bar' bodies, 3b = multistorey tabular 'point bar' bodies (all terminology modified from Friend et al., 1979).

Capella times, fan activity in the basin was commensurate with the continued emplacement and southwards advance of the Montsec thrust unit (Seguret, 1970; Nijman and Nio, 1975; Atkinson, 1983).

STRATIGRAPHIC SETTING

Previous studies by Garrido-Megias (1968) and Nijman and Nio (1975) suggested that the pronounced eastwards thinning of the Capella Formation was a result of downcutting in the east by a progressive unconformity at the base of the overlying Campodarbe Group (Figure 2). They inferred the presence of this unconformity purely on the basis of the Capella strata reduction, and no biostratigraphic evidence was presented to support their suggestions. However, a large-scale unconformity is not considered feasible since the top of the Capella Formation has remained uneroded as testified by the lateral persistence of a regionally correlatable lacustrine carbonate interval

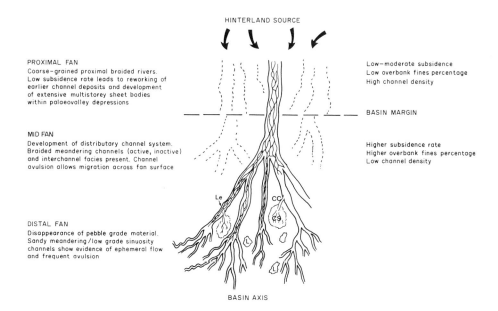

Figure 4. Idealized depositional model for the Montañana and Campodarbe humid alluvial fans. Le = levee deposits, CC = crevasse channels, CS = crevasse splays, L = lacustrine deposits.

(see Figure 5). It is suggested instead that the eastwards thinning of the sediments is a primary depositional feature (see below).

In addition to the Formation thinning eastwards it also becomes stratigraphically older (Figure 2). This is related to the earlier initiation of fan sedimentation in the eastern portion of the basin during late Ypresian times. These early deposits can be traced laterally westwards across the Luzas fault-zone where they grade into brackish and true marine sediments of the Perrarua Formation. By mid-Lutetian times, fan sedimentation advanced further westward and prograded over these earlier Perrarua deposits. The continental/marine transition now lay far to the west near to the line of the Foradada fault-zone (Figure 2).

Therefore the Formation is stratigraphically diachronous with the oldest, but thinnest fan deposits (upper Ypresian—upper Lutetian) occurring in the east of the basin and the youngest, but thickest fan sediments (mid-upper Lutetian) developing in the west (Figure 2). According to the time-scale devised by Mutti *et al.* (1972, their Table II), the Capella succession represents a time interval of nearly 3 million years in the east but only approximately 1 million years in the west.

146   CHAPTER 5

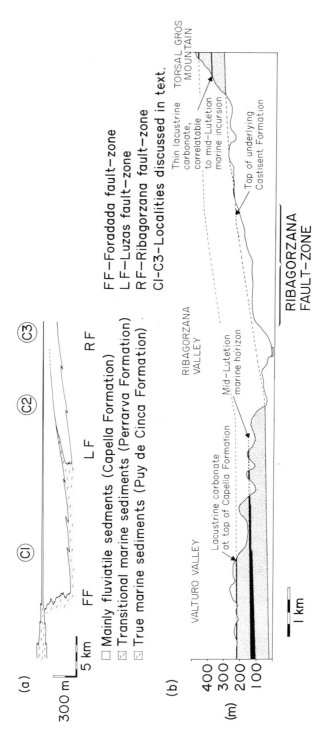

Figure 5. (a) General stratigraphic appearance of the Capella Formation illustrating decrease in thickness to the east and presence of mid-Lutetian marine intercalation and lacustrine carbonate at the base of the overlying Campodarbe Group. (b) Enlarged cross-section from the Valturo valley (locality C2) to the Torsal Gros mountain (locality C3) illustrating the pinching out of the mid-Lutetian marine horizon and the laterally equivalent lacustrine carbonates.

LOCATION OF THE STUDIED SECTIONS

In order to illustrate the catenary form of the Capella paleosols, the Formation was examined at three geographically distinct, but approximately time equivalent locations (labelled C1-C3 in Figure 2). This was made possible by the existence within the Formation of an extensive marine intercalation easily traceable from west to east across the basin (Figure 5). This marine interval, originally recognized by Wasser (1978), and Pronk (1978), was produced by a sea-level rise which occurred in the west of the basin during mid-Lutetian times (Speksnijder and van der Veen, 1978). Using the intercalation as a basal marker horizon and the Formation top as an upper limit, a small portion of the succession dated as mid to late Lutetian was examined at each of the three localities.

FIELD AND LABORATORY PROCEDURES

The three studied sections were examined by means of vertical grain size profiles and larger scale alluvial stratigraphy sections constructed with the aid of photographic panoramas. In order to simplify descriptions of individual channel units, a modified version of the terminology devised by Friend et al. (1979) is used.

Macroscopic description of the paleosols involved the construction of 'key' vertical profiles which demonstrated the typical appearance and characteristics of the paleosols at each locality. Each profile contains a record of paleosol thickness, grain size variation, colour according to the Munsell Color Chart System (Goddard et al., 1948), bioturbation intensity, calcium carbonate content and the presence or absence of nodular mineral accumulation horizons. Samples were collected throughout most profiles and thin section studies and X-ray analyses of the clay fraction were conducted where necessary. Soils terminology used throughout the text is taken from Brewer (1964) and Soil Survey Staff (1975).

CAPELLA FORMATION: SEDIMENTATION AND PALEOSOL DEVELOPMENT

Locality C1—Laguarres—Capella Sequence

Description
This sequence lies to the west of the Luzas fault-zone between the villages of Laguarres and Capella in the western portion of the basin (Figure 1). At this locality all 300 m of exposed Capella sediments lie

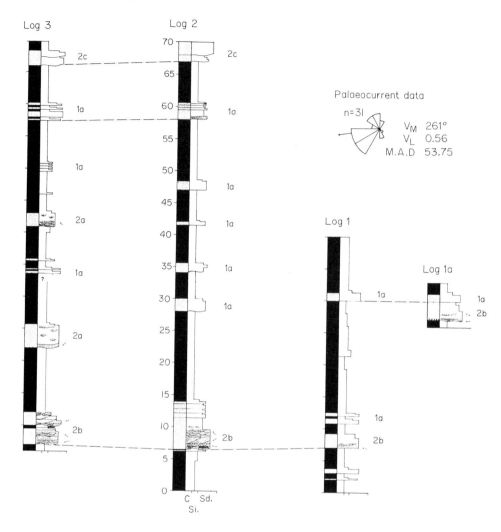

Figure 6. Typical vertical sedimentary profiles and palaeocurrent trends through the Capella Formation at locality C1. Symbols adjacent to channel fills represent channel body type as indicated in Figure 3.

above the mid-Lutetian marker horizon (Figures 2 and 5). The general stratigraphic appearance of the succession is one of low channel density with channel deposits forming only 10–40% and averaging only 20–30% of the exposed sequence. This low density is readily apparent from vertical profiles all of which are clearly dominated by overbank deposits (Figure 6).

Channel bodies are characterized by sand-grade fills and gravel debris is extremely rare. Most display simple or multistorey ribbon

PALEOSOLS AND BASIN TECTONICS   149

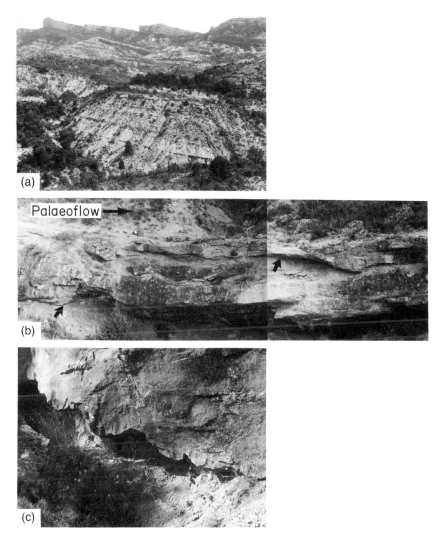

Figure 7. (a) Overall view of the Capella succession at locality C1. Note the laterally persistent thin sheet sands (levee/crevasse deposits) and banded pattern in the overbank deposits indicative of paleosol presence. (b) Oblique section through a multistorey ribbon body (2b) at locality C1. Arrows mark pronounced erosion surfaces at the base of each storey fill. (c) Close up view of irregular scoured base of a simple ribbon (2a) body at locality C1. Scale = 18 cm.

geometries, usually with a thickness of less than 5 m (Figure 7(a) and (b)). Basal erosion surfaces are highly irregular and channel margins are often abrupt and steep (Figure 7(c)). Internally, draping laminations, interpreted as cut and fill structures, reactivation surfaces and repeated grain size variations are common. Upwards through the succession the ribbon bodies become replaced by thicker (5–15 m) and wider units with a sheet-like appearance (Figure 8). These predominantly fining-upwards units again exhibit a multistorey geometry and are typified by the presence of inclined sets of clay/silt draped epsilon or lateral accretion bedding. Occasionally the lateral accretion sets are truncated obliquely by low angle dipping erosion surfaces. Palaeocurrents recorded from both channel types indicate that the main sediment dispersal was towards the west with local deviations to the north and south of this mean.

The volumetrically dominant overbank sediments have a relatively uniform appearance throughout the succession. They consist mainly of pedogenically altered mud-siltsones (see below), the colour of which imparts a yellow-orange appearance to the overbank intervals. The mud-siltstones are punctuated repeatedly by discrete packets of heterogeneous sheet sands which can often be traced laterally into 'wing' features (in the sense of Friend et al., 1979) associated with major ribbon or sheet bodies. These sheet sand packets are interpreted on the basis of their location and distribution to be of either crevasse and/or levee origin.

Figure 8. General view of the upper part of the Capella sequence at locality C1 illustrating the development of thicker, sheet-like sand bodies (arrowed).

Paleosol Characteristics
Pedogenic modification indicated by varicolouration, rootlet mottling and intensive bioturbation occurs repeatedly in the overbank sediments and at the tops and margins of most smaller channel units (Figure 9). The paleosols are remarkably uniform in appearance and consist of cyclic arrangements of grey-yellow-orange horizons traceable over large lateral distances across the outcrop (Figure 7(a)). In detail (Figure 9), few distinct profiles can be recognized and the repeated superimposition indicates that composite pedogenesis, in the sense of Morrison (1978) and Bown and Kraus (1981), has taken place. The colour of the paleosol matrix varies from grey (5Y 5/2) through to moderate yellowish brown (10YR 5/6). In an ideal sequence, matrix percentage is reduced as the intensity of red-yellow mottling increases upwards. Most individual mottles possess diameters in the region of 1–5 cm although occasionally larger mottles with diameters of between 10 20 cm are seen. Generally, coalescence of the mottles results in the production of distinct reddened horizons at certain levels (Figure 9). Red coloration in most mottles tends to concentrate towards the outside of the feature and internally they are bleached grey. Where the soil matrix is grey the mottles remain red throughout.

In thin section, the effects of later calcium carbonate cementation have tended to destroy the finer details of the micromorphology. Where cementation is poorly developed one occasionally sees weakly oriented clay particles within voids ('argillans'—Figure 10(a)) and preferential iron-oxide concentration at the void margins ('neoferrans'—Figure 10(b)). In most cases the voids are filled by later sparry calcite cement (Figure 10(c)). X-ray diffraction analysis reveals that the clay fraction (less than 2 microns) has a high smectite content with subsidiary amounts of kaolinite, illite and chlorite (Figure 9).

Finally, although the host overbank sediments are rich in calcium carbonate (up to 70% detrital limestone debris) there is no macro- or micro-scopic evidence for the existence, either now or in the past, of concretionary calcium carbonate glaebules.

Interpretation
The predominance of ribbon bodies exhibiting many features indicative of rapid channel cutting and filling suggests that the Capella floodplain in this region was characterized for long periods by immature relatively short-lived rivers. Grain size variations and reactivation surfaces imply that the rivers possessed a fluctuating hydrograph and were possibly of ephemeral character (cf. Williams, 1970; Picard and High, 1973). This erratic discharge regime presumably led to frequent overbank flooding as testified by the numerous wing extensions or crevasse and levee

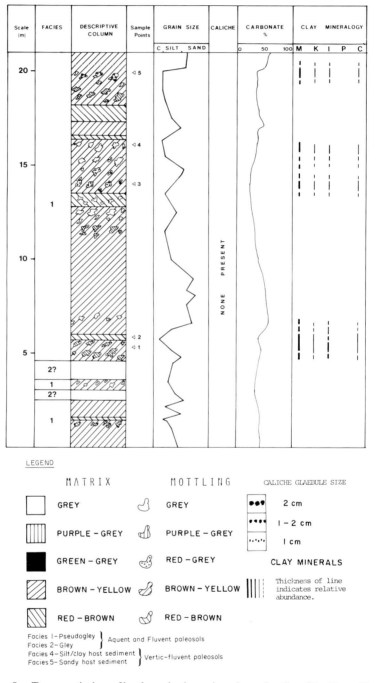

Figure 9. Type vertical profile through the paleosols at locality C1. M = Montmorillonite. K = Kaolinite. I = Illite. P = Palygorskite. C = Chlorite.

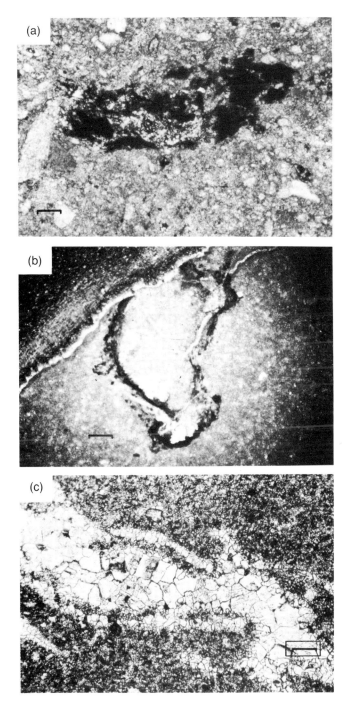

Figure 10. Micromorphological features observed in the paleosols at locality C1. (a) Oriented clay particles within void ('argillan'). (b) Concentration of iron (i.e. haematite) at void margin ('neoferran'). (c) Calcite spar (cement) infilling former root(?) void. Scale bar = 300 μm.

sheet sands. This style of river deposition was eventually replaced by more permanent channel courses which deposited the sheet bodies at the top of the succession. These sheets with their characteristic lateral accretion bedding represent the deposits of well established meandering rivers which developed in this region during late Lutetian times. The heterogeneous nature of the accretion bedding together with the presence of cross-cutting erosion surfaces is very similar to that described from other meandering stream deposits by Elliott (1976) and Puigdefabregas and Van Vliet (1978). These authors suggested that such features would be favoured in meandering streams characterized by large-scale fluctuations in channel discharge. Adopting a similar argument for the Capella rivers it can be assumed, as was the case in the lower part of the sequence, that at least some degree of ephemerality typified their flow. In terms of fan setting both the immature and high sinuosity rivers have been interpreted on the basis of their relatively fine grain size, unstable nature and palaeocurrent distribution to lie within a mid-distal location receiving detritus mainly from northern sources (Atkinson, 1983).

Despite the upward change in river activity and pattern, paleosols in the sequence remain virtually constant in appearance. Their most diagnostic feature is the presence at many levels of well-developed colour mottling. In modern soils mottling of this type occurs along cracks, burrows and root channels and is produced by the redistribution of iron compounds induced by fluctuating water levels (Buurman, 1975, 1980; Fitzpatrick, 1980). The most important processes responsible for this redistribution are alternating periods of oxidation and reduction which are usually accompanied by the destruction of sedimentary features by vegetation growth and animal burrowing (Buurman, 1975; Fitzpatrick, 1980). Soils of this type, otherwise known as hydromorpic soils, are common in floodplain settings and are divided into two main groups: gley and pseudogley soils. Gley soils typify very poorly drained areas and form under the influence of fluctuating groundwater levels. Pseudogley soils develop where drainage improves and arise through the stagnation of downward percolating pluvial water which concentrates after flooding or during a distinct wet season. Recently, Buurman (1980) documented in detail the various attributes of these two soil types. He demonstrated that gley soils are mainly grey in colour since the major part of the soil is reduced and lies below the lowest groundwater table. Oxidation only occurs along cracks or organically produced channels with the result that the soil mottles are characterized by red haematite accumulations at their centres. In contrast, pseudogley soils only become reduced where the downward percolating pluvial water accumulates in the profile. In this reduced horizon iron distribution is similar to that of gley soils whereas below it the soil is unsaturated and potentially oxidized. In this zone, water seeps downwards from the stagnating layer above leading to reduction along the soil channels and concentration of oxidized iron compounds at the channel rim.

The overall character of the Capella paleosols and the distribution of oxidized iron compounds in the soil mottles indicates both gley and pseudogley activity. This is to be expected due to their floodplain setting where accumulation of both rain and floodwater would provide the optimum conditions for the development of hydromorphic soils. Soils of this type, which are relatively immature and lack diagnostic horizons, have only undergone initial soil ripening (Pons and Zonneveld, 1965). These criteria imply that the paleosols represent ancient forms of 'Entisol' (Soil Survey Staff, 1975), i.e. soils which possess poorly developed to absent middle horizons. Since they developed in an obvious floodplain setting where the water-table was either high or constantly fluctuating they can best be classified as representing former Entisols belonging to the Aquent (strong gleying) and Fluvent (pseudogley activity) sub-orders.

The dominance of these Entisol soils is almost certainly a result of the erratic discharge regime of the distal fan rivers. Frequent overbank flooding in this environment already suggested on the basis of numerous crevasse and levee intervals is substantiated by the widespread occurrence of these immature hydromorphic soil types.

LOCALITY C2—VALTURO SEQUENCE

Description
This sequence lies approximately 20 km due east of locality C1, southwest of the village of Puente de Montañana in the Valturo valley (Figures 1 and 2). The exposed succession comprises some 200 m of fluvially dominated sediments which become increasingly more marine influenced westwards as one approaches the Luzas fault-zone (Figure 5). Within this succession the upper 100 m or so above the marker marine horizon are of middle to upper Lutetian age.

Overbank sediments dominate the sequence (Figure 11) with channel units forming only 20–30% of the exposed succession. The main channel bodies possess conglomeratic fills (Figure 12(a)) and have thicknesses of between 3 and 10 m. They display either multistorey ribbon or sheet geometries and virtually all occur as isolated bodies surrounded by overbank deposits (Figure 12(b)). The channels exhibit irregular, erosive bases which are overlain by imbricated to

156 CHAPTER 5

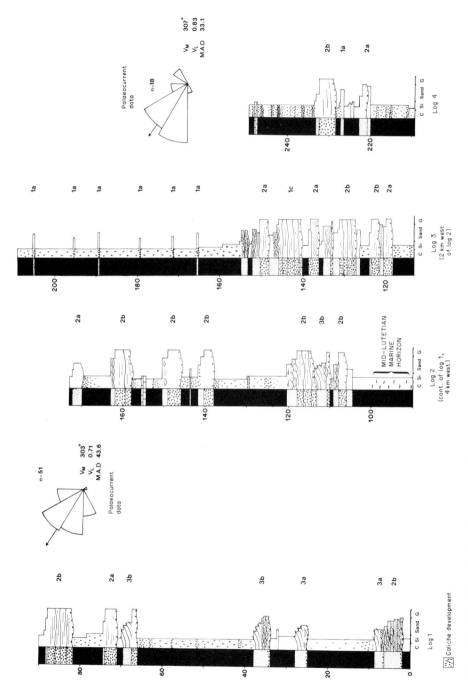

Figure 11. Typical vertical sedimentary profile and palaeocurrent trends through the Capella Formation at locality C2.

PALEOSOLS AND BASIN TECTONICS   157

Figure 12. (a) Detail of pebbly fill within a multistorey/sheet conglomeratic body at locality C2. Note low angle stratification within the gravels and the highly scoured erosive base of the body. Vertical exposure in the photograph = 3−4 m. (b) Typical outcrop architecture of the conglomeratic channel fills seen at locality C2. (c) Small, lenticular channel fill (arrowed) interpreted to be of crevasse channel origin seen within a thick sequence of overbank mud-siltstones and levee/crevasse sheet sands at locality C2.

massive orthoconglomerates diagnostic of waterlain deposition. With the exception of occasional sets of medium to large-scale (foreset height = 0.8–1.2 m) planar cross-stratified gravel (Figure 12(a)), sedimentary structures are rare within the channels. Towards the tops of the bodies abrupt fining occurs to sand grade material, much of which is heavily bioturbated and mottled. Palaeocurrents recorded from the channels reveal that river flow varied throughout the sequence but was predominantly towards the north-west (Figure 11).

Overbank intervals are similar to those described at locality C1, comprising pedogenically altered mud-siltstones punctuated by occasional laterally extensive crevasse and levee sheet-sands. In some of these sheet sand intervals, particularly close to the flanks of major coarse member units, small lenticular channel bodies may occur (Figure 12(c)). These small channels exhibit palaeocurrents at a high angle to those recorded from the major channels nearby and are interpreted as periodically active flood or crevasse channels.

Paleosol Characteristics

All the overbank sediments at this locality are characterized by a strong orange-red coloration produced by intensive paleopedogenic modification. The paleosols occur as superimposed composite profiles (Figure 13). They dominate the overbank intervals and also occur at the tops of major channel units where rootlet mottling can be traced downwards for up to 1.5 m. Where the host overbank sediment is predominantly fine-grained and dominated by silt-clay, paleosol development is more extensive. In these intervals the colour of the paleosol matrix varies from moderate yellowish brown (10YR 5/6) to dark brown (5YR 3/4) and they contain mottling of a similar type to that seen at locality C1. Most of these mottles possess obvious red haematite accumulation at their margins and frequently they increase in number vertically to culminate in obvious reddened horizons.

Although mottling is ubiquitous throughout the profiles the most diagnostic feature of the paleosols is the presence of calcium carbonate nodules (Figures 13 and 14(a)). The nodules are disorthic (Brewer, 1964; Wieder and Yaalon, 1974) in that they possess sharp boundaries and can easily be removed from the enclosing matrix. Most individual nodules have a diameter of 0.1 to 1.5 cm. They usually increase in number and size upwards through the paleosol forming thin coalesced zones where amalgamation of two or three nodules produces irregular bulbous features up to 3 cm in diameter. The nodules, although common, never form more than 50% by volume of the total paleosol matrix.

In thin section the microfabric of the nodules consists of a 'crystic

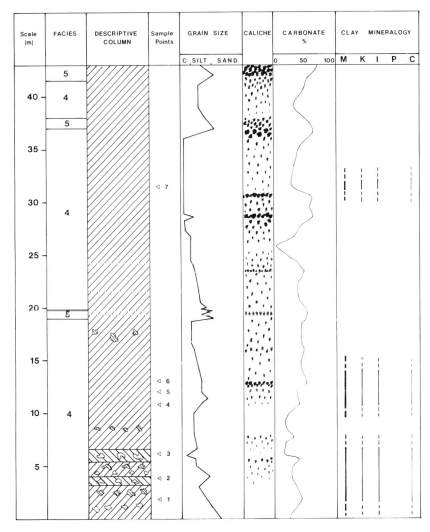

Figure 13. Type vertical profile through the paleosols at locality C2. Legend as in Figure 9.

plasmic fabric' (Brewer, 1964) in which 'crystallaria' and 'septaria' voids are seen (Figure 14(b)). The matrix in which the nodules lie exhibits an anisotropic 'argillasepic' fabric with occasional argillan and neoferran structures. Fractured quartz grains and evidence of displacive calcite cement growth is common (Figure 14(c)). X-ray determination of clay mineralogy reveals that both nodules and matrix contain a very high smectite content with subsidiary illite, kaolinite and chlorite. Total calcium carbonate content varies from 16–55% where

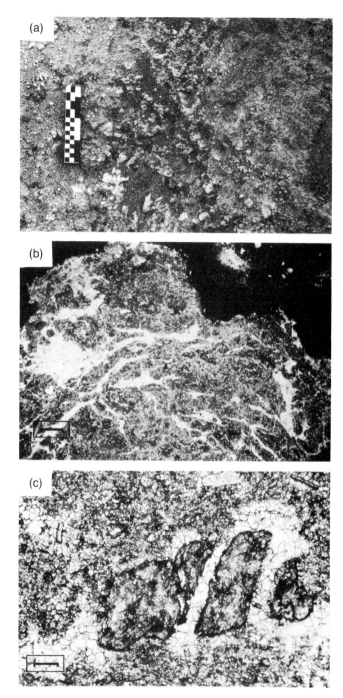

Figure 14. Macro- and micromorphological appearance of the paleosols at locality C2. (a) Caliche nodules with typical disorthic form (scale bar divisions = 1 cm, 2 cm, 5 cm). (b) Caliche nodule in thin section illustrating the presence of concentric crystallaria and radial 'septaria' voids. (c) Fractured quartz grain (dark) separated by displacive calcite cement spar. Scale bar in (b) = 300 μm and in (c) = 125 μm.

nodules are scarce to in excess of 80% where horizons of coalesced nodules are present (Figure 13 and Atkinson, 1983).

Interpretation
The characteristics of the channel units suggests that the mid-late Lutetian floodplains of this region were dominated by stable, coarse-grained, bedload dominated rivers. Within the channels bedload sediments were deposited and shaped into both longitudinal and transverse bar forms now represented by the large-scale sets of planar cross-stratification seen in the channel fills. The channels are assumed to have possessed a minimal sinuosity and at low discharges flow diversion is likely to have occurred around that bars generating a braided channel pattern. The nature of these braided channels implies deposition on relatively steep gradients characteristic of mid to proximal fan settings (Atkinson, 1983). Palaeocurrents and clast provenance demonstrate that the rivers were derived from both southern and northern fan sources, the latter of which became dominant through time.

The proximal setting and apparent stability of the rivers were important factors favouring the development of fairly mature soils in the overbank sequences. Unlike locality C1 the presence of nodular accumulations of calcium carbonate implies the existence of soil horizonation with the establishment of C or even B horizons (Fitzpatrick, 1980). Soils characterized by the progressive crystallization of calcium carbonate are termed caliche or calcrete and are widespread in many recent soils (see Goudie, 1973, for review) and ancient sedimentary sequences (e.g. Steel, 1974; Leeder, 1975; Watts, 1980; Bown and Kraus, 1981; Retallack, 1983a,b). For more information concerning their mode of formation the reader is referred to Goudie (1973) and the paper by Allen (this volume).

In the Capella paleosols the isolated and disorthic form of the soil glaebules is strongly reminiscent of the types 1 and 2 caliches described by Steel (1974). He regarded these types to represent relatively immature caliches analogous to the stage II forms of Gile *et al.* (1966) and the young-mature sequences of Reeves (1970). The presence of large amounts of disorthic glaebules suggests that at the time of formation the soils were subject to the process of 'pedoturbation', i.e. displacement and churning in the solum (see Buol *et al.*, 1973; Wieder and Yaalon, 1974; Wright, 1982). Pedoturbation commonly occurs by a combination of burrowing and plant growth (biopedoturbation), the shrinkage and swelling of clay minerals (argillipedoturbation) and/or soil creep. Within the Capella paleosols the ubiquitous presence of bioturbation and rootlet mottling implies that biopedoturbation almost certainly occurred. In addition their enrichment by smectite clays, whose properties of water retention make them particularly susceptable to seasonal shrink-swell, suggests that argillipedoturbation was also favoured. Proof that shrinkage and swelling did take place is provided by (i) the blocky nature of the paleosol mottles, and (ii) the presence of crystallaria and septaria fractures within the caliche glaebules. In modern soils blocky ped fabrics and internal glaebule fracturing are both interpreted to result from periodic wetting and drying of the soil leading to alternating phases of soil expansion and contraction (see Buol et al., 1973; Fitzpatrick, 1980).

Soils characterized by the above features of calcium carbonate precipitation, pedoturbation, shrink-swell and more importantly high levels of swelling clay are similar to present day tropical Vertisols (Soil Survey Staff, 1975). These soils develop preferentially under seasonal climates which experience strong periodic contrasts in moisture supply (Steila, 1976). They possess deep, wide cracks which remain open for most of the year, are characterized by a swelling clay content of greater than 30% and frequently display precipitation of calcium carbonate in the lower parts of their profile (Soil Survey Staff, 1975). The resulting caliche is relatively immature consisting of small isolated glaebules (Blokhuis et al., 1968; Allen, 1973, this volume) and resembles the types seen in the paleosols at this locality. Unfortunately, in contrast to modern Vertisols, it is not known if actual original percentages of smectitic clay were greater than 30% in the Capella examples. In addition they appear to lack other diagnostic features of true Vertisols such as slickensided ped surfaces and pseudoanticlinal joint planes (Al-Rawi et al., 1968; Buol et al., 1973; Allen, 1974a; Goldbery; 1982). The absence of these features, coupled with the knowledge that the paleosols developed in sediments of fluvial origin, suggest that they cannot be classified *a priori* in the Vertisol Order. Rather, they most probably represent a type of Fluvent soil which developed in host sediments with a composition particularly suited to the formation of vertic characters. On this basis the paleosols are probably better classified in the Entisol soil order as an ancient type of Vertic-Fluvent soil (Soil Survey Staff, 1975).

LOCALITY C3—TORSAL GROSS SEQUENCE

Description
This locality lies approximately 8 km east of C2 and is close to the eastern-most limit of the Capella Formation in the Tremp-Graus basin (Figures 1 and 2). The total thickness of the formation is difficult to

assess, but is nowhere greater than 140 m (Figure 15). Unfortunately, in contrast to the previous localities, the mid-Lutetian marine horizon is not present. Instead its approximate position was suggested by Pronk (1978) to be correlatable with a zone of extensive lacustrine carbonates approximately midway through the succession (Figure 15(a)). Although this suggestion lacks palaeontological support detailed correlation studies and field mapping by the author have tended to confirm this correlation. Nevertheless, since doubts still remain as to the exact position of the marker level its use at this locality cannot be advocated. Instead, because the Capella sequence is relatively uniform throughout, the characteristics of the whole formation are incorporated in the following discussion. It must be remembered, however, that the true mid-upper Lutetian interval only probably represents a small portion at the top of the overall sequence.

Vertical profiles and the alluvial stratigraphy of the sequence (Figure 15(b)) reveal an overbank fine member dominated succession in which the percentage of channel deposits varies from 10–25% averaging approximately 15%. The main channel-fills occur as discrete isolated units (Figure 16(a)) characterized by a conglomeratic-fill in which mudstone cobbles and reworked caliche nodule intraclasts are common. They exhibit both multistorey ribbon and sheet geometries and are very similar in internal composition to the conglomerate bodies described at the previous locality. They are thus similarly interpreted as the deposits of bedload-dominated, braided rivers. Palaeocurrent measurements indicate that river flow was predominantly towards the south and south-west.

The overbank intervals again possess many similarities with those described from the Valturo sequence. Crevasse channel and crevasse splay/levee sands are easily recognizable (Figure 16(b)) although not as numerous as at the two previous locations. The main difference is in the appearance of the mud-siltstone intervals which are vividly coloured and characterized by a diagnostic and often intense reddening.

Paleosol Characteristics
The effects of paleopedogenic modification are seen extensively at this locality. Most, if not all, channel units possess intensely mottled upper boundaries and in some cases several smaller bodies (< 2 m thickness) are completely penetrated by rootlet traces giving them a highly blocky and irregular appearance (Figure 16(c)).

In the mud-siltstone intervals the paleosols are similar in appearance to those described from the Valturo valley. The major differences are the presence of thicker and more numerous reddened horizons and the

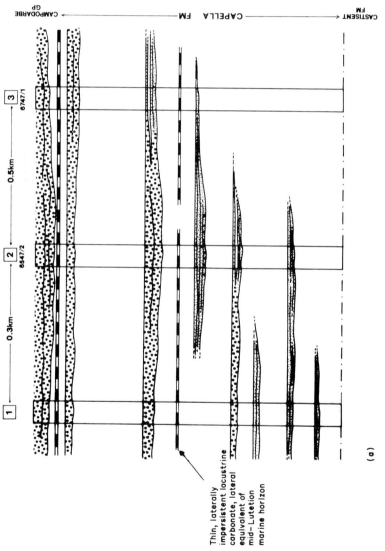

Figure 15. (a) Simplified alluvial stratigraphy of the Capella sequence at locality C3.

abundance throughout the succession of caliche nodules. All the caliche horizons, although still nodular, are in a much more advanced stage of formation than any of the profiles in the Valturo sequence (Figure 15). Coalesced horizons with nodular accumulations up to 3 cm in diameter occur repeatedly and most intervals resemble stage III caliches of Gile *et al.* (1966) and the advanced type 2 profiles of Steel (1974). They represent the optimum form of caliche development seen anywhere in the Capella outcrop.

Interpretation
The gross similarity in appearance between the deposits of this locality and the previous one suggests that the nature of the Capella floodplain in both areas was also very similar. On the basis of this evidence it is concluded that during Capella times the Torsal Gros region was dominated by a stable floodplain on which coarse-grained braided channels acted as the main drainage paths. The dominance of channels of this type again implies deposition on steep gradients in a mid to

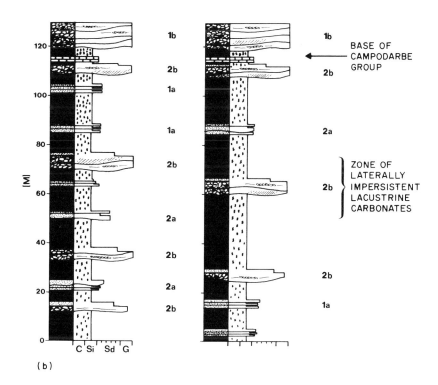

Figure 15. (b) Vertical sedimentary profiles through the Capella succession at locality C3.

Figure 16. (a) Overall view of the Capella sequence at locality C3. Arrows indicate the level of the lacustrine carbonates interpreted to be equivalent to the mid-Lutetian marine intercalation. Note channel body in centre of section. (b) Typical appearance of the overbank sediments at locality C3 in which intensely mottled sheet sands are overlain by a sequence of caliche paleosols (scale bar = 18 cm). (c) Close up view of a levee/crevasse sheet sand displaying intense paleopedogenic modification (field of view is 4 m wide).

proximal fan setting. Unlike the Valturo channels, palaeocurrents and clast lithology indicate that these rivers were sourced solely from the north.

The decreased density of channel deposits is interpreted as a response to less intense river activity than in the Valturo region. This decreased activity is also suggested by an increase in the maturity of the paleosols. Although the paleosol sequences are strongly reminiscent of the previously described Vertic-Fluvent types, the caliche they contain is clearly in a much more advanced state of formation. They resemble present day 'Torrert' soils which are a type of Vertisol occurring in arid climatic regions (Soil Survey Staff, 1975; Steila, 1976). However, there is no direct evidence in the Torsal Gros sequence supporting the existence of a demonstrably arid climate, e.g. desiccation cracks, aeolian sediments, etc. and thus the paleosols are considered more likely to be a hypermature variant of the Vertic-Fluvent soils described above.

ORIGIN AND CAUSES OF PALEOSOL VARIATION

The above descriptions demonstrate that a strong west to east variation in soil type existed on the Capella floodplain during late Lutetian times. In the west, relatively poorly developed, strongly gleyed Aquent and Fluvent soils predominated on a floodplain characterized by unstable, and probably ephemeral, distal fan channels. In contrast, further eastwards, better drained soils of Vertisol affinity became increasingly more important as more permanent, coarse-grained braided channels of mid to proximal fan setting were encountered. This gradation in soil type must have been promoted by changes in the controls governing soil formation across the Capella floodplain.

Studies on modern soils have shown that there are five main controls governing soil formation: climate, parent material, biological activity, topographic relief and the length of time of soil formation (Bridges, 1978; Fitzpatrick, 1980; Retallack, 1983a,b). Variations in one or more of these factors can cause soil character to change and may lead ultimately to the production of a different soil type (Steila, 1976; Retallack, 1983a,b). In the case of the Capella paleosols climate, parent material and biological activity were all too uniform to be considered as the likely cause influencing soil variation. Fluctuations in regional climate would be on far too large a scale to explain the relatively rapid lateral transition seen in the basin. Likewise, since all the floodplain sediments were derived from virtually identical sources, rich in Mesozoic carbonates, changes in the gross composition of parent material can also be discounted. Similarly, biological activity represented by bioturbation and rootlet mottling also varies little throughout the region. Although the apparent intensity of rootlet activity increases in the eastern soils this is thought to be a consequence rather than a cause of soil change. It is merely felt to reflect the increased maturity of these soils in which the mottling colours are more vivid and thus the rootlet traces more obvious.

In contrast to the above conclusions there is good evidence suggesting that the final two factors, topographic relief and length of time of soil formation, did vary across the basin. Evidence for the existence of a depositional floodplain topography is based firstly on the prevailing humid fan environment, and secondly, on the local palaeogeography. Studies in modern environments have shown that a preprequisite for humid fan sedimentation is the presence of an uplifted source terrain at the margin of the depobasin (Denny, 1967; Heward, 1978). The nature of this setting implies the existence of a depositional gradient from proximal, elevated regions at the fan apex to distal, low lying zones at the fan toe. In the Capella deposits, sequences of proximal fan affinity related to the larger northern fans tend to be restricted to the east of the basin, whereas distal fan sequences are concentrated in the west. On the basis of this information it is logical to assume that at the time of deposition the western floodplain lay at a lower topographic level than that in the east. This suggestion is substantiated by the local palaeogeography which reveals that during late Lutetian times the marine environment, and hence sea-level, lay at the western margin of the basin close to the Foradada fault-line (Figures 1 and 2). Presumably, as in the present day, floodplain elevation increased inland from this coastline as the river gradients steepened in an upstream direction towards the fan hinterlands.

The presence of a depositional gradient contributed to soil formation by influencing the local drainage conditions across the Capella floodplain. In the west the low elevation of the fan surface, coupled with the nearness of the marine base-level, contributed to poor floodplain drainage and promoted the development of gleyed Aquent and Fluvent soils. In contrast, the higher proximal portions of the fan surface possessed a deeper water-table enhancing floodplain drainage. In these areas increased soil drying occurred prompting seasonal shrink-swell and amplifying caliche formation. These features contributed singificantly to the establishment of more mature soils with Vertisol-like characteristics.

The final factor, time, represented by relative floodplain aggradation rates is felt to have been the major control influencing soil change across the basin. Studies on the length of time of soil development have shown that soil maturity increases the longer the time a soil remains at or near the sediment surface (Allen, 1974a; Leeder, 1975; Bown and Kraus, 1981; Kraus and Bown, this volume). Soil maturity is thus inversely related to the local sedimentation rate, and the greater this rate, the lower the potential maturity of the developing soils.

In the Capella Formation estimates of floodplain aggradation rate have been calculated by dividing the thickness of preserved sediment at each locality by the length of time the sediments took to accumulate (Table 1). This relatively simple form of calculation, whilst ignoring the effects of possible post-depositional compaction, can be considered as giving a standardized approximation of the average rates of sedimentation prevailing at the time of deposition. The calculations suggest that local floodplain aggradation rates could have been up to six times greater in the western part of the basin (Table 1).

Supplementing this information the average length of time required to develop successive, composite soil horizons has also been estimated (Table 1). The calculations reveal that at locality C1 values were probably in the order of 14 000 years whereas further to the east at locality C2 values as high as 40 000 years may have prevailed. It must be remembered that these values are estimates and include not only the time taken for soil formation, but also for overbank sediment accumulation and the establishment, at certain times, of river channels (see Kraus and Bown, this volume). According to evidence from other studies (e.g. Bown and Kraus, 1981; Retallack, 1983a,b) formation of individual hydromorphic soil profiles can occur very quickly, possibly in the order of 100 years or less, where optimum conditions favourable to their development prevail. In contrast, the caliche-bearing Vertic-Fluvent soils almost certainly took a longer time to evolve. Whilst there are many problems with using pedogenic carbonate to assess the length of time of pedogenesis (see Allen, this volume), the superabundance of $CaCO_3$ in the host floodplain sediments of the Capella Formation implies they formed relatively quickly in comparison to the estimates of 10^3-10^4 years for caliche in other ancient sequences (e.g. Allen, 1974a,b; Leeder, 1975). As inferred by Whiteman (1971), providing optimum conditions are present, there is no reason why pedogenic carbonate may not form in as little as a few 10's to 100's of years. Assuming the above values are in the right order of magnitude for the Capella paleosols then on a geological scale soil formation in the basin may have occurred over a relatively short time period. However, this is not considered to be the case since the above values are derived from modern analogues and represent solely the time needed to establish a single soil profile. Because the Capella paleosols are composite profiles they represent many incremental periods of soil formation. As such their formation must have taken a much longer

Table 1. Rates of floodplain aggradation and soil reoccurence.

	Length of time for accumulation of sequence[1] (millions of years)	Thickness of sequence (metres)	Average floodplain aggradation rate (metres/year)	Average number of composite paleosol intervals	Average time between soil forming episodes (years).
Locality C1 Laguarres-Capella	1	300	0.3	72	14 000
Locality C2 Valturo valley	1	c.120	0.12	26	40 000
Locality C3 Torsal Gros[2]	3	c.150	0.046	no data	no data

[1] Based on time correlation of Mutti et al. (1972)
[2] Torsal Gros estimates are based on the whole Capella Formation.

period of time than indicated above and presumably represents the majority of the re-occurrence time calculated in Table 1. This conclusion agrees well with the ideas of Kraus and Bown in this volume who suggest that sedimentary stasis indicated by paleosols occupies much of the time interval recorded by overbank deposition.

Unfortunately, all the above measurements are based on the belief that the floodplain aggrades in a uniform manner. However, it is more probable that floodplain aggradation is highly irregular (see Kraus and Bown, this volume). Bridge and Leeder (1979) documented that sediments delivered to the floodplains of modern rivers do not accumulate in a steady-state fashion (see their Table 1). They indicated that most overbank floods were of variable magnitude and that periods of local erosion and reworking frequently followed depositional phases. The interplay between deposition and erosion has a pronounced effect on soil development. Where deposition predominates and relatively thick floodplain sequences containing numerous crevasse and levee units result, soil formation is expected to be retarded. In contrast, where flood derived sediments are thin and/or are reduced by floodplain erosion, net deposition decreases, sedimentary stasis is favoured and soil formation is expected to proceed to a more advanced state.

In the Capella Formation, well developed crevasse and levee deposits are found in the west of the basin in association with short-lived distal fan channels. This information has been used to suggest that discharge fluctuations, overbank flooding and channel avulsion were a common feature of these unstable rivers. This presumably contributed significantly to the relatively high rates of overall floodplain sedimentation recorded in this region (Table 1). Under these conditions immature Aquent and Fluvent soils would be strongly favoured. Meanwhile, in the more proximal eastern fan sequences, the importance of crevasse and levee sands diminishes, suggesting that overbank flooding from the braided channels was more restricted. These conditions generated lower floodplain aggradation rates favouring more mature soils of Vertisol character.

If different aggradation rates did typify the western and eastern fan floodplains the simulation studies on alluvial stratigraphy by Allen (1978) and Bridge and Leeder (1979) predict that they should be characterized by different channel densities and overbank sediment percentages. According to their models, if aggradation rates are relatively high successive channel fills are rapidly buried and become encased in thick overbank sequences. This results in an alluvial stratigraphy typified by a relatively low channel density (see Figure 8 of Allen, 1978). On the other hand if rates are low, channel burial proceeds more slowly with the result that successive channel fills stack closer together and the preservation potential of overbank sediments is severely reduced. Under these conditions the resulting alluvial stratigraphy is characterized by a relatively high density of channel bodies (see Figure 4 of Bridge and Leeder, 1979).

In the Capella Formation estimates of channel density are remarkably uniform (approximately 20%) throughout the basin although a slight decrease occurs in the east. This behaviour is the exact opposite of that predicted by the simulation studies of Allen (1978) and Bridge and Leeder (1979), since slightly lower channel densities occur in a region with a lower calculated floodplain aggradation rate. This effect may be explained by the fact that the simulation studies only modelled the evolution of a single river channel, whereas in the fan environment of the Capella Formation numerous channels were presumably active at the same time (Figure 4). Downstream splitting of the major proximal trunk rivers creates a large, distributary network of smaller channels in the distal fan reaches. Therefore, although floodplain aggradation rate is higher in the west, the increased number of distal fan rivers means that overall channel density would remain relatively high and similar, if not greater, than that seen further to the east.

In summary, it is suggested that the soil change recorded across the Capella floodplain was produced by differences not in climate, biological activity or parent material, but by changes in floodplain topography and local floodplain aggradation rates related to varying river activity and regime. These two factors strongly influenced local drainage characteristics which subsequently caused the variation in paleosol type across the basin. Adopting terminology from modern environments (Steila, 1976; Bridges, 1978) this type of lateral soil variation produced under conditions of uniform climate and parent material can be considered as representing a good example of ancient 'catena' formation.

ORIGIN OF DIFFERENTIAL SEDIMENTATION RATES ACROSS THE BASIN

Although it has been argued that the western floodplains were characterized by higher floodplain aggradation rates no cause or reason for this behaviour has as yet been proposed. In general there are two main mechanisms which could lead to increased fluvial aggradation: (1) eustatic sea-level rise which would lead to decreased river gradients in near coastal areas promoting increased sedimentation rates and thus floodplain aggradation as rivers attempted to re-attain grade, and (2) increased basin subsidence promoted by tectonic activity which would allow an increased thickness of fluvial sediments to accumulate thus generating a high floodplain aggradation rate. Obviously, if floodplain sedimentation did not keep pace with basin subsidence and the marine environment lay close-by, a relative sea-level rise could take place with similar consequences to (1) above.

Facies distribution and palaeogeographic reconstructions of the basin indicate that during late Lutetian times the coastline remained stable along the N–S line of the Foradada fault-zone (Figure 17). This implies that relative sea-level must have been rising at a rate comparable to sediment delivery since neither major transgression nor regression took place. The question arises as to whether this sea-level rise was eustatic or related to a local increase in basin subsidence which also promoted increased sediment accumulation. According to the world sea-level curves of Vail *et al.* (1977), eustatic sea-level fluctuations were common during Eocene times. These are amply illustrated in the Pyrenean region by such well defined events as the lower Eocene 'Ilerdian transgression' and the upper Eocene 'Biarritzian transgression' (Nijman and Nio, 1975; Puigdefabregas, 1975). Both these transgressions are recorded throughout northern Spain and have been interpreted to be of eustatic origin (Puigdefabregas, 1975; Nio, 1982, pers. comm.).

In contrast, the scale of the Capella sea-level rise is much smaller and restricted solely to the Tremp-Graus region. This suggests it is of more localized origin and probably more likely related to accelerated basin subsidence rather than extensive eustacy. Differential basin subsidence could result from the unstable tectonic setting of the basin on top of the southwards advancing Montsec thrust-sheet. Direct evidence of tectonic control on sedimentation is suggested by the important role the Foradada, Luzas and Ribagorzana fault-zones played in controlling not only the incremental thickening of the Capella sediments but also the position of environmental transitions, e.g. continental to brackish water/marine to true marine (Figure 17).

In an independent study (Williams and Atkinson, in prep.), these fault-zones have been interpreted as the surficial expressions of lateral thrust surfaces at depth ('lateral ramps' in the sense of Butler, 1982). In the east of the basin (Luzas-Ribagorzana region) the thrusts propagated in response to the formation of a piggy-back style culmination structure. Stratal duplication in the culmination (Figure 17) caused upward elevation of the eastern part of the basin leading to relatively low subsidence rates at the time of deposition of the Capella sediments. Further westwards (west of the Luzas fault-zone) the effects of the culmination feature diminished with the result that higher rates of subsidence accentuated by local thrust slice loading, characterized the basin.

The effects of this differential tectonic activity may have further enhanced the topographic relief across the Capella floodplain. It could be expected that preferential eastern upthrow on the fault-zones would have tended to increase the elevation of the floodplain in an easterly direction. Thus, in addition in the topographic gradient inherent in the fan model, regional tectonics may have also contributed to the establishment of a basin wide floodplain relief.

CONCLUSIONS

An examination of paleosols within a stratigraphically equivalent portion of the Capella Formation has demonstrated a pronounced variation in soil type across the Tremp-Graus basin. The variation involves a transition from immature, hydromophic Aquent and Fluvent soils developing in association with unstable, distal fan channels to more mature soils of Vertisol-like character which typify more proximal fan environments dominated by relatively stable channels. There is

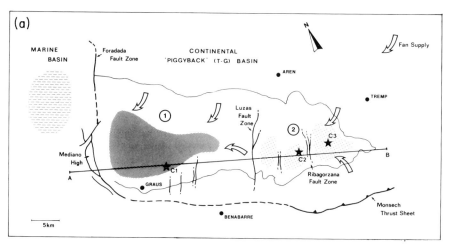

Figure 17. (a) Structural elements and relationship to Capella sedimentation in the Tremp-Graus Basin. Position of cross-section seen in Figure 17(b) is indicated by the line A–B. Zone ①. Region of relatively high basin subsidence rate and high floodplain aggradation rate. Floodplain characterized by unstable (ephemeral), rapidly avulsing distal fan rivers. Immature Aquent and Fluvent soils favoured. Zone ②. Region of lower basin subsidence rate caused by upward elevation of the floodplain produced during sub-basinal culmination uplift. Floodplain tends to be higher and better drained. Fairly stable (probably incised) mid-proximal fan, braided rivers and mature Vertisol soil development.

PALEOSOLS AND BASIN TECTONICS   175

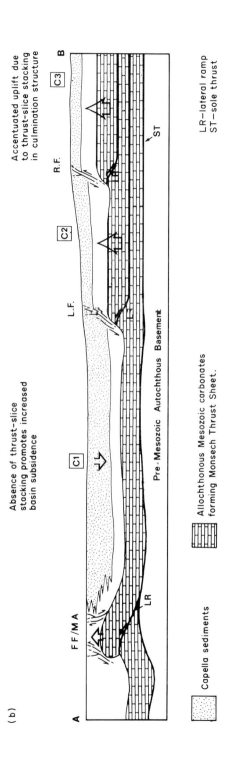

Figure 17. (b) Schematic cross-section transverse to propagation direction of the Montsec thrust sheet demonstrating how culmination uplift below the east of the basin generates lateral thrust ramps and ultimately synsedimentary faulting at the depositional surface. Note the cross-section is not to scale.

strong evidence to suggest that the main influences on soil change were an increase in relief and decrease in floodplain sedimentation rate from west to east across the basin. Changes in soil type produced by such a mechanism, where climate and host sediment remain constant, are referred to in modern landscapes as a catena.

In the Capella Formation the existence of an ancient catena aids significantly in the recognition of tectonic control on alluvial sedimentation. In the west, the presence of relatively immature soils in a succession characterized by a high floodplain aggradation rate, rapidly avulsing unstable channels and a low relief are taken to reflect the presence of a relatively high basin subsidence rate. Further east, increasing soil maturity in a succession dominated by relatively lower floodplain aggradation rates, larger, more stable channels and increased relative elevation implies that much lower subsidence rates affected the basin.

Similar conclusions were adopted by Steel (1974) to explain the increased maturity of caliche profiles in condensed sequences from the Permo-Triassic in Scotland. Whilst Steel could only infer that differential tectonic subsidence was the cause of the soil change, this study had independent evidence which indicated that this was most probably the case. Changes in basin subsidence rate are interpreted to result directly from synsedimentary fault activity. These faults (Foradada, Luzas and Ribagorzana) record the presence at depth, of lateral thrust surfaces which delineate a culmination structure below the eastern portion of the Tremp-Graus basin. In the east, where culmination uplift was most pronounced, relatively low subsidence rates prevailed and mature paleosols dominated a reduced sedimentary sequence. In contrast further west, the influence of the culmination decreased and comparatively higher rates of subsidence affected the basin. This resulted in a preponderance of immature paleosols in a much thicker sedimentary succession.

Finally, according to some paleosol workers (e.g. Valentine and Dalrymple, 1976) recognition of catenas is the only proper method one can employ to prove the existence of ancient soils. In this study a paleosol defined catena not only helps to elucidate the variable nature of former continental landscapes but also aids significantly in understanding the controls operating on the landscape.

ACKNOWLEDGEMENTS

The material in this paper forms part of the author's doctoral research which was supervised by Professor T. Elliott at the University College of Swansea, UK. It was originally presented as a poster display at the

XIth International Congress on Sedimentology held in Hamilton, Ontario during August 1982. Funding for the research was provided by a Post-graduate Studentship Award from the Shell International Petroleum Company whom I gratefully acknowledge. Thanks are extended to Trevor Elliott, Paul Wright and Peter Friend for constructive reviews of the manuscript, to Peter Buurman for stimulating my interest in paleosols and to Djin Nio and Graham Williams for useful discussions on Pyrenean geology.

REFERENCES

Allen, J.R.L. 1960. Cornstone. *Geol. Mag.* **97**, 43–8.
Allen, J.R.L. 1973. Compressional structures (patterned ground) in Devonian pedogenic limestones. *Nature, Phys. Sci.* **243**, 84–6.
Allen, J.R.L. 1974a. Studies in fluviatile sedimentation: implications of pedogenic carbonate units, Lower Old Red Sandstone. Anglo-Welsh outcrop. *Geol. J.* **9**, 181–208.
Allen, J.R.L. 1974b. Geomorphology of Siluro Devonian alluvial plains. *Nature, Phys. Sci.* **245**, 644–5.
Allen, J.R.L. 1978. Studies in fluviatile sedimentation: an exploratory qualitative model for the architecture of avulsion controlled alluvial suites. *Sedim. Geol.* **21**, 129–47.
Al-Rawi, G.J., Sys, C. and Laruelle, J. 1968. Pedogenic evolution of the soils of the Mesopotamian floodplain. *Pedologie* **18**, 63–109.
Atkinson, C.D. 1983. *Comparative Sequences of Ancient Fluviatile Deposition in the Tertiary South Pyrenian Basin, Northern Spain.* Unpublished Ph.D Thesis, University of Wales, UK. 350 pp.
Besly, B. and Turner, P. 1983. Origin of red beds in a moist tropical climate (Etruria Formation, Upper Carboniferous, U.K.), in *Residual Deposits: Surface Related Weathering Processes and Materials*, R.C.L. Wilson (ed). Published for the Geological Society of London by Blackwell Scientific Publications, Oxford. 131–47.
Blokhius, W.A., Pape, T. and Slager, S. 1968. Morphology and distribution of pedogenic carbonate in some Vertisols of the Sudan. *Geoderma* **2**, 173–200.
Bown, T.M. and Kraus, M.J. 1981. Lower Eocene alluvial paleosols (Willwood Formation, northwest Wyoming, U.S.A.) and their significance for paleoecology, paleoclimatology and basin analysis. *Palaeogeogr., Palaeoclimatol., Palaeoecol.* **34**, 1–30.
Brewer, R. 1964. *Fabric and Mineral Analysis of Soils.* Wiley, New York. 470 pp.
Bridge, J.S. and Leeder, M.R. 1979. A simulation model for alluvial stratigraphy. *Sedimentology* **26**, 617–44.
Bridges, E.M. 1978. *World Soils.* Cambridge University Press, Cambridge. 128 pp.
Buol, S.W., Hole, F.D. and McCraken, R.J. 1973. *Soil Genesis and Classification.* Iowa State University Press, Ames, Iowa. 360 pp.
Burgess, I.S. 1961. Fossils soils of the Upper Old Red Sandstone of South Ayrshire. *Trans. geol. Soc, Glasgow* **24**, 138–53.
Butler, R.W.H. 1982. The terminology of structures in thrust belts. *J. Struct. Geol.* **4**, 239–45.
Buurman, P. 1975. Possibilities of palaeopedology, *Sedimentology* **22**, 289–98.
Buurman, P. 1980. Palaeosols in the Reading Beds (Paleocene) of Alum Bay, Isle of Wight, U.K. *Sedimentology* **27**, 593–606.
Denny, C.S. 1967. Fans and pediments. *Am. J. Sci.* **265**, 81–105.

Elliott, T. 1976. The morphology, magnitude and regime of a Carboniferous fluvial distributary channel. *J. sedim. Petrol.* **46**, 70–6.

Fitzpatrick, E.A. 1980. *Soils. Their Formation, Classification and Distribution.* Longman Group Ltd., London. 353 pp.

Freytet, P. 1971. Palaéosols résiduels et paléosols alluviaux hydromorphes associes aux depots fluviatiles dans le Crétacé supérieur et l'Éocène basal du Languedoc. *Rev. Geogr. Phys. Geol. Dyn.* **2(13)**, 245–68.

Freytet, P. 1973. Petrography and palaeoenvironment of continental carbonate deposits with particular reference to the Upper Cretaceous and Lower Eocene of Languedoc (southern France). *Sedim. Geol.* **10**, 25–60.

Friend, P.F., Slater, M.J. and Williams, R.C. 1979. Vertical and lateral building of river sandstone bodies, Ebro basin, Spain. *J. geol. Soc. Lond.* **136**, 39–46.

Garrido-Megias, A. 1968. Sobre la estratigrafia de los conglomerades de Campanue (Santa Liestra) y formaciones superiores del Eocene (extremo occidental de la cuenca Tremp-Graus, Pirineo Central). *Acta. Geol. Hisp.* **3(2)**, 39–43.

Gile, L.H., Peterson, F.F. and Grossmann, R.B. 1966. Morphological and genetic sequences of carbonate accumulation in desert soils. *Soil Sci.* **100**, 347–60.

Goddard, E.N., Trask, P.D., De Ford, R.K., Rove, O.N., Singewald, J.T. and Overbeck, R.M. 1948. *Rock-Color Chart.* Geol. Soc. Am., Boulder, Colorado.

Goldbery, R. 1982. Structural analysis of soil microrelief in paleosols of the Lower Jurassic 'Laterite-Derivative Facies' (Mishhor and Ardon Formations) Makhtesh Ramon, Israel. *Sedim. Geol.* **31**, 119–40.

Goudie, A. 1973. *Duricrusts in the Tropical and Sub-Tropical Environment.* Clarendon Press, Oxford. 174 pp.

Heward, A.P. 1978. Alluvial fan sequence and megasequence models: with examples from the Westphalian D—Stephanian B coalfields, N. Spain, in *Fluvial Sedimentology*, A.D. Miall (ed.). *Can. Soc. Pet. Geol. Mem.* **5**, 668–702.

Hubert, J.F. 1977. Paleosol caliche in the New Haven Arkose, Connecticut: record of semiaridity in Late Triassic-early Jurassic time. *Geology* **5**, 302–4.

Leeder, M.R. 1975. Pedogenic carbonates and floodplain sediment accretion rates: a quantitative model for alluvial arid-zone lithofacies. *Geol. Mag.* **112**, 257–70.

Leeder, M.R. 1976. Palaeogeographic significance of pedogenic carbonates in the topmost Old Red Sandstone of the Scottish Border Basin. *Geol. J.* **11**, 21–8.

Morrison, R.B. 1978. Quaternary soil stratigraphy, in *Quaternary Soils*, W.C. Mahaney (ed.). Geo Abstracts, Norwich. 77–108.

Mutti, E., Luterbacher, H.P., Ferrer, J. and Rossel, J. 1972. Schema stratigrafico e linemento di facies del Paleogene marine della zona Centrale Sudpirnaica tra Tremp e Pamplona. *Mem. Soc. Italia XI* 391–416.

Nijman, W.J. and Nio, S.D. 1975. The Eocene Montañana Delta, in *Guidebook to Excursion 19, part B, IXth Int. Sed. Congr. Nice, France*, J. Rosell and C. Puidefabregas (eds). 56 pp.

Ori, G.G. and Friend, P.F. 1984. Sedimentary basins formed and carried piggy back on active thrust sheets. *Geology* **12**, 475–8.

Picard, M.D. and High, L.R. Jnr. 1973. *Sedimentary Structures of Ephemeral Streams. Developments in Sedimentology 17.* Elsevier, Amsterdam. 223 pp.

Pons, L.J. and Zonneveld, I.S. 1965. *Soil Ripening and Soil Classification.* Inst. Land. Recl. Improv. (ILRI) Publ. 13. 128 pp.

Pronk, J.W. 1978. *Een Sedimentologische Interpretatie van de Capella Formatie ten oosten can de Rio Isabena, Huesca, Spanje.* Unpublished Doctoral Thesis (MSc), University of Utrecht, The Netherlands.

Puigdefabregas, C.T. 1975. *La Sedimentacion Molasica en la Cuenca de Jaca.* Mono. del Inst. de Estudios Pirenaicos, 104. 188 pp.

Puigdefabregas, C.T. and Vliet, A. van 1978. Meandering stream deposits from the

Tertiary of the Southern Pyrenees, in *Fluvial Sedimentology*, A.D. Miall (ed.). *Can. Soc. Petrol. Geol. Mem.* **5**, 469–85.
Reeves, C.C., Jnr. 1970. Origin, classification and geologic history of caliche on the Southern High Plains, Texas and Eastern New Mexico. *J. Geol.* **78**, 352–62.
Retallack, G.J. 1976. Triassic palaeosols in the upper Harrabeen Group of New South Wales, 1. Features of the paleosols. *J. Geol. Soc. Aust.* **23**, 383–99.
Retallack, G.J. 1977. Triassic palaeosols in the upper Harrabeen Group of New South Wales, 2. Classification and reconstruction. *J. Geol. Soc. Aust.* **24**, 19–36.
Retallack, G.J. 1983a. A palaeopedological approach to the interpretation of terrestrial sedimentary rocks. The mid-Tertiary fossil soils of Badlands National Park, South Dakota. *Bull. geol. Soc. Am.* **94**, 823–40.
Retallack, G.J. 1983b. *Late Eocene and Oligocene Paleosols from Badlands National Park, South Dakota*. Geol. Soc. Am. Spec. Paper, 193. 82 pp.
Seguret, M. 1970. *Étude Tectonique de Nappes et Series Decolles de la Partie Centrale du Versant Sud de Pyrenees*. Univ. Sci. Technology du Languedoc, Pub. 2. 155 pp.
Soil Survey Staff 1975. *Soil Taxonomy*. Agric. Handbook 436, USDA, Washington.
Speksnijder, A. and Veen, T.B. van der 1978. *Interaktie Tussen Alluvial Fan—Fluviatile en Lagunaire Sedimentatie in Het Eocene Nabij Graus (Huesca), Zuidelijke Pyreneen, Spanje*. Unpublished Doctoral Thesis (M.Sc), University of Utrecht, The Netherlands.
Steel, R.J. 1974. Cornstone (fossil caliche)—its origin, stratigraphic and sedimentological importance in the New Red Sandstone, western Scotland. *J. Geol.* **82**, 351–69.
Steila, D. 1976. *The Geography of Soils*. Prentice Hall, New Jersey ??? pp.
Vail, P.R., Mitchum, R.M. Jnr., Todd, R.G., Widmier, J.M., Thompson, S. III, Sangres, J.B., Bubb, J.N. and Hatteild, W.G. 1977. Seismic stratigraphy and global changes of sea-level, part 4, Global cycles of relative changes of sea-level, in *Seismic Stratigraphy—Applications to Hydrocarbon Exploration*, C.E. Payton (ed.). *Am. Ass. Pet. Geol. Mem.* **26**, 83–97.
Valentine K.W.G. and Dalrymple, J.B. 1976. Quaternary buried paleosols, a critical review. *Quat. Res.* **6**, 209–20.
Wasser, G.G.M. 1978. *De Capella Formatie tussen Graus en Laguarres*, Unpublished Doctoral Thesis (M.Sc.), University of Utrecht, The Netherlands.
Watts, N.L. 1980. Quaternary pedogenic calcretes from the Kalahari (southern Africa): mineralogy, genesis and diagenesis. *Sedimentology* **27**, 661–86.
Whiteman, A.J. 1971. *The Geology of the Sudan Republic*. Oxford University Press, London. 290 pp.
Wieder, M. and Yaalon, D.H. 1974. Effect of matrix composition on carbonate nodule crystallisation. *Geoderma* **11**, 95–121.
Williams, G.D., 1985. Thrust Tectonics in the South Central Pyrenees. *J. Struct. Geol.* **7**, 11–17.
Williams, G.D. and Atkinson, C.D., in prep. Thrust tectonics and sedimentation patterns of the Montsec Thrust Sheet, South Central Pyrenees.
Williams, G.E. 1970. Flood deposits of the sand bed ephemeral streams of Central Australia. *Sedimentology* **17**, 1–40.
Wright, V.P. 1982. Calcrete paleosols from the Lower Carboniferous, Llanelly Formation, South Wales. *Sedim. Geol.* **33**, 1–33.

Chapter 6

PALEOSOLS AND TIME RESOLUTION IN ALLUVIAL STRATIGRAPHY

MARY J. KRAUS
Department of Geological Sciences, University of Colorado, USA

THOMAS M. BOWN
US Geological Survey, Colorado, USA

INTRODUCTION

Traditionally, the utility of paleosols in stratigraphic and palaeoenvironmental studies has been championed by students of the Quaternary. Because individual soils are commonly morphologically distinctive and areally widespread, they have been good stratigraphic markers for subdividing Quaternary deposits and are useful for local correlations (e.g. Richmond, 1962; Morrison, 1964, 1967; Mahaney, 1978) as well as regional correlations (Morrison and Frye, 1965; Birkeland et al., 1971; Mahaney, 1978). The methods and principles of Quaternary soil stratigraphy are well established and were adequately summarized by Richmond (1962) and Morrison (1967, 1978). However, in spite of several important works in the last decade, examination of the rich record of Tertiary and older paleosols has tended to lag far behind in terms of recognition, methodology, and applications.

Paleosols and soils were accorded formal stratigraphic status by the American Commission on Stratigraphic Nomenclature (1961) following the suggestion of Richmond and Frye (1957). Though the fundamental soil-stratigraphic unit in this scheme is the 'soil', Morrison (1967) suggested that 'geosol' be substituted for 'soil' to avoid confusion concerning the precise meaning of the latter term. He defined the geosol as '...a weathering profile that formed at (immediately beneath) and essentially parallel with the land surface, has physical characteristics and stratigraphic relations that permit its consistent recognition and mapping, and whose stratigraphic interval in the rockstratigraphic sequence (relationship to immediately older and younger deposits) is known quite definitely' (Morrison, 1967: 9–10). Because paleosols are subject to lateral variability in their physical attributes, this definition was intended to emphasize the consistent stratigraphic relations of the soil in a stratigraphic sequence. Morrison recognized both composite geosols, produced by two or more episodes of pedogenesis, and multistorey geosols, consisting of two or more distinct profiles that are vertically superposed but which exhibit little or no overlap between successive profiles. In the recent (1983) edition of the North American Stratigraphic Code, the geosol was adopted as the basic 'pedostratigraphic' unit. INQUA-ISSS, however, chose the 'pedoderm' as their basic soil stratigraphic unit (Parsons, 1981), a term first proposed by Brewer et al. (1970). Among the ways in which the two concepts differ are the following: a geosol is defined as all or part of a buried soil that can occur at any position in a stratigraphic sequence; whereas a pedoderm consists of a complete surficial soil which may be exhumed, relict, or buried.

In addition to their utility in physical correlation, paleosols have been used as boundary units for Quaternary glacial and interglacial stages and they have thus, properly or improperly, been accorded time-stratigraphic status (Morrison and Frye, 1965; Morrison, 1967; Schultz and Stout, 1980). Some Quaternary paleosols are also potentially useful for the absolute dating of deposits because they occasionally contain material capable of C-14 dating (Mahaney and Fahey, 1980; Evans, 1982).

Though the North American paleosol literature is dominated by examples from the Quaternary record, geologists have not ignored the potential of buried paleosols in unraveling pre-Quaternary stratigraphy (see, e.g. references in Richmond and Frye, 1957 and Evans, 1982). However, with Tertiary and older deposits, the tendency has been for geologists to recognize solitary, highly distinctive, and well developed paleosols or paleosol horizons that are believed to mark major unconformities (e.g. Ritzma, 1955, 1957; Schultz, et al., 1955; Pettyjohn, 1966; Abbott et al., 1976; Mabesoone and Lobo, 1980; Schultz and Stout, 1980) or at least relatively lengthy pauses in a more or less continuous record of sediment accumulation (e.g. Allen, 1974; Behrensmeyer and Tauxe, 1982). The association of paleosols with unconformities in some areas has tended to overemphasize pedogenic modification as a process that requires geologically lengthy periods of landscape stability (non-deposition and non-erosion). For paleosols formed on bedrock parent materials this is probably true, although these records for pre-Quaternary paleosols are relatively uncommon. For example, the Eocene (?) Interior Paleosol of South Dakota might record as much as several million years of landscape stability (Retallack, 1983a,b), and silica/iron-aluminum duricrusts of middle and low latitudes (Dury, 1971; Goudie, 1973) also formed over long time spans on stable landscapes.

Quaternary paleosol studies also have tended to support the ingrained concept that pedogenesis is coincident with periods of landscape stability that alternate with longer, more normal periods of more or less continuous deposition or erosion (e.g. Frye, 1949; Jungerius, 1976). For example, in their description of the late Quaternary history of deposits in South Australia, Williams and Polach (1971) distinguished discrete periods of deposition ranging from 2000 to 8000 years alternating with discrete periods of non-deposition and pedogenesis that ranged from several thousand to 6000 years. Pawluk (1978) suggested that in situations where there is a surface of continual accumulation of detritus, surface horizons will be thick and weakly developed if accretion is uniform, or the stratigraphic sequence will consist of weakly developed composite soils under variable rates of sediment accumulation. If erosion occurs relatively rapidly, horizons tend to be thin and poorly developed: 'Only on relatively stable land surfaces is the residence time of the soil body sufficiently long so that reorganization of geological material fully reflects the expression of the pedogenic setting' (Pawluk, 1978; p. 70).

The authors believe that development of soils in all alluvial regimes is a normal and expected phenomenon and that most ancient alluvial deposits contain numerous superposed ancient soil profiles. Recently, geologists have identified several thick alluvial sequences that consist largely of stacked paleosols, which may be either multistorey or composite in character (e.g. Chalyshev, 1969; Freytet, 1973; Retallack, 1977a,b; Bown, 1979; Bown and Kraus, 1981a,b, in press; Kraus and Bown, 1982; Freytet and Plaziat, 1982; Retallack, 1983a,b). The genesis of these stratigraphically arranged successions of paleosols, often consisting of several tens to more than a thousand superposed soils, cannot be satisfactorily explained by correlating all of these paleosols with discrete and relatively long-lived periods of landscape stability in sedimentary regimes in which either active erosion or sedimentation occupies geologically significant fractions of the geologic time represented by the rock column.

In this paper, the authors will focus on thick sequences of superposed alluvial paleosols and their relevance to concepts of time stratigraphy. It is emphasized that episodes of actual sediment accumulation and actual erosion are highly sporadic in the alluvial regime and in sum occupy very little of the geologic time represented by any column of rocks of fluvial origin. In fact, the correct interpretation of an alluvial record is conceptually one of long intervals of stasis punctuated by very brief intervals of deposition or erosion. Alluvial paleosols are potentially useful in understanding the episodic nature of fluvial sedimentation and erosion and, consequently, the manner in which alluvial successions develop. In other sections of this paper, examples from the lower Eocene Willwood Formation of Wyoming and the Upper Triassic Chinle Formation of Arizona are drawn upon to illustrate the utility and importance of thick, superposed sequences of paleosols to:

(1) recognizing episodic and geologically significant base-level fluctuations in the fluvial record;
(2) biostratigraphy and evolutionary studies; and
(3) determining rates of sediment accumulation.

STRATIGRAPHIC CONCEPTS

Wheeler (1958, 1964a,b) emphasized the relative roles of erosion, non-deposition, and sedimentation in generating stratigraphic sections. He suggested (1964a, p. 600) that 'Many inadequacies of concept and practice stem from the popular notion that stratigraphy is the science of past sedimentation, to the exclusion of degradation...'. More recently, the highly episodic nature of sedimentation and consequent incompleteness of the sedimentary record have been examined by Sadler (1981) and Sadler and Dingus (1982).

Building on the work of Blackwelder (1909) and the base-level concept of Barrell (1917), Wheeler constructed schematic diagrams of restored stratigraphic sequences to illustrate the 'space time "volume"' that is missing at unconformities; that is, he restored sequences to completion by converting the thickness of a vertical section to time and by schematically depicting the strata lost during periods of degradation and non-deposition. These 'lost' strata he termed the 'degradational vacuity' and the 'hiatus' respectively.

Wheeler (1964a,b) envisioned the Earth's surface divided into innumerable areas where either sedimentation or erosion occurs. Base-level is the imaginary surface separating these areas; areas undergoing sediment accumulation lie below base-level, whereas areas undergoing degradation lie above it. Change from sedimentation to erosion in any area, or *vice-versa* involves, respectively, a rise or fall of base-level in that area. Although Wheeler (1964a, p. 603) observed that base-level '...intersects the lithic surface at all points of equilibrium...' (equilibrium = no sedimentation or erosion), he overemphasized the roles of sedimentation and erosion by making no provision for that part of the time restored section in which nothing, neither erosion nor sedimentation, happens.

Wheeler's concepts of base-level and the restoration of stratigraphic sections can be integrated with current notions of alluvial stratigraphy and paleosol development. Sandy fluvial systems deposit two basic lithofacies—a coarse facies of sands and gravels that accumulate within channels, and a fine facies of fine sands, silts, and clays that are deposited in interchannel areas. Though pedogenic modification of channel sands has been recognized (Bown and Kraus, 1981a, in press; Bown *et al.*, 1982), alluvial paleosols have been most commonly described from overbank deposits of mud, silt, and clay (e.g. Allen, 1974; McPherson, 1979, 1980; Bown, 1979; Bown and Kraus, 1981a; Goldbery, 1982; Retallack, 1983a,b), and these volumetrically dominant floodplain sediments comprise the bulk of alluvial soil parent materials. In their study of streams in the United States, Wolman and Leopold (1957) found that the frequency of overbank flooding for most streams was generally once every one or two years. Flood events are usually short-lived phenomena, though prolonged flooding up to two months duration has been recorded (Kesel *et al.*, 1974; Goulding, 1980). Fine sediment settles from water that is either standing or flowing outside the channel. Scouring can occur in some areas, but Jahns (1947), in his study of the Connecticut River floods, observed that significant scouring is restricted to interchannel areas through which unusually large volumes of flood waters pass. Bridge and Leeder (1979) compiled previously published data demonstrating that major seasonal floods typically deposit only millimetres or centimetres of alluvium, and both the thickness and grain size of sediments decrease markedly away from the channel belt into the interchannel areas (Alexander and Prior, 1971; Kesel *et al.*, 1974).

The episodic nature of floodplain sedimentation and soil formation can be illustrated by means of a schematic time restored alluvial stratigraphic column, patterned after the sections of Wheeler (1958). Figure 1 depicts the relative proportions of time occupied by sediment accumulation and soil formation based on an example from a net aggradational system, the lower Eocene Willwood Formation. Hypothetically, during the course of a single year, the average interchannel area was inundated once and underwent either erosion or deposition reflecting its position with respect to base-level. Small increments of sediment were added over time to the sedimentary sequence (Figure 1, column A). The thickness of these additions depended partly on proximity of the interchannel area to the channel belt as well as on flood volumes, sediment availability, and type of floodplain vegetation (e.g. Wolman and Leopold, 1957; Wolman and Eiler, 1958). During most of the year (probably exceeding 95% in most systems), flow was confined to the channel and *neither deposition nor erosion of the floodplain occurred*. During these lengthy periods of stasis, pedogenic processes modified the newly deposited alluvium as

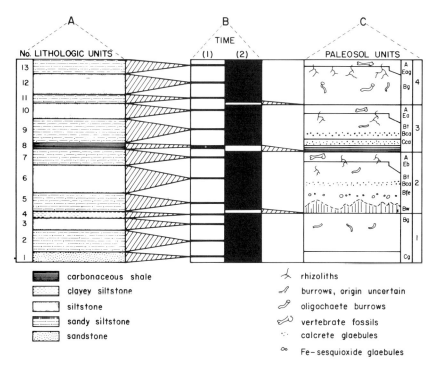

Figure 1. Diagram showing hypothetical fluvial lithologic units 1–13 (column A) and hypothetical paleosol units 1–4 developed on them (column C), and relative proportions of time occupied by deposition (solid portions of column B_1) and soil formation (solid portions of column B_2).

well as all previously deposited sediment that remained near the surface. Column B (Figure 1) shows in black the relative proportions of time elapsed during active sediment accumulation (1) and soil formation (2). This illustration is schematic and does not include unknown amounts of time elapsed during accumulation of sediments or formation of paleosols that may have been eroded at some time during formation of this column. Though the relative durations of time required to form each of four paleosols are depicted as proportional to thickness, they are also unknown.

Accumulation of alluvial sequences also reflects the effects of longer term aggradational or degradational trends superimposed on the yearly and local patterns of net sedimentation and net erosion. Intrabasinal factors, including lateral channel migration and avulsion, will gradually or dramatically alter the proximity of floodplain areas to the channel belt and thus increase or decrease the thicknesses of alluvium deposited locally each year. Over time, these processes will produce variability in local floodplain accretion rates. Over even longer, geologically important periods of time, external controls such as climate and tectonic activity will determine whether an alluvial system undergoes net degradation or net aggradation (e.g. Allen, 1978; Bridge and Leeder, 1979; Miall, 1980). During degradation, part or all of the stratigraphic sequence will be eroded and an unconformity of different magnitude in different places is produced. Figure 2 illustrates a paleosol sequence cut by an unconformity with a complementary time restored section. Soils 1 through 6 were deposited and developed during an aggradational regime. This was followed by a degradational regime during which soils 5 and 6 and part of soil 4 were eroded. Nonetheless, during most of the time represented by the unconformity, the floodplain coincided with base-level, there was neither active deposition nor active erosion and sediment at or near the surface was altered pedogenically. Following a return to net aggradational conditions, soils 7 and 8 were deposited and pedogenically developed.

Sedimentation and erosion are highly episodic and together occupy very little of the time represented by any alluvial sequence. Therefore, it matters not whether the fluvial system was in a net aggradational or a net degradational regime because stasis, and its corollary pedogenesis, are the dominant processes in developing an alluvial sequence and occupy the vastly greater proportion of time (Figure 3). During most of the time the sequences illustrated accumulated (Figures 2 and 3), the floodplain surface and base-level were coincident and soils formed on the alluvial parent materials. Short-lived depositional events occasionally added small increments of alluvium to the floodplain, although scour events at times reversed the process of building. Thin veneers of new alluvium were incorporated into the developing soil profile by soil accretion during ensuing periods of non-deposition and non-erosion.

Prolonged flooding of the interchannel areas and buildup of relatively thick deposits upon developing soils can arrest their formation because the soils become situated beneath the level to which the near surficial pedogenic processes can act on them. In these instances, soil formation begins anew on the fresh sediment and the newly-forming soil is separated from the now buried paleosols by a thickness of unaltered alluvial parent materials. It is in this way that a column of many superposed paleosols is produced in alluvial successions. It follows that most rocks recording accumulation of ancient alluvial sediments should contain numerous superposed soil profiles, and the authors' studies of pre-Quaternary alluvial rocks from several areas of the world demonstrate satisfactorily that such appears to be the case. Indeed, we have often found paleosol development in these rocks to be so pervasive that it is commonly difficult to locate unaltered parent materials upon

PALEOSOLS AND ALLUVIAL STRATIGRAPHY 187

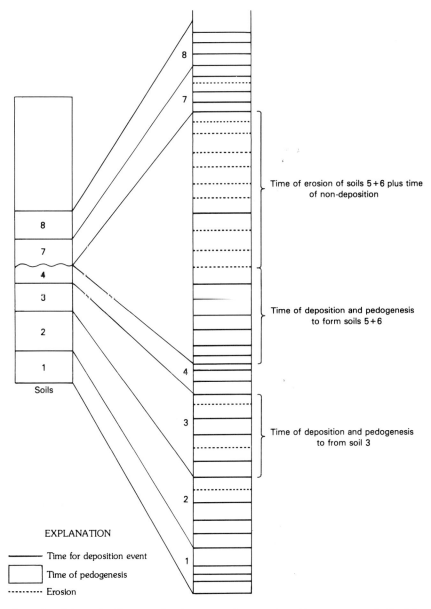

Figure 2. Schematic diagram showing part of a stratigraphic sequence of alluvial paleosols in the lower Eocene Willwood Formation (left column) and a time restored column for the same sequence (right column). Additional discussion in text.

which the ancient soils developed. In the cases of the lower Eocene Willwood Formation of Wyoming, the Oligocene Jebel Qatrani Formation of Egypt, and the Upper Triassic Chinle Formation of Arizona, sediment accumulation in parts of the column appears to have been gradual and incremental enough to allow development of very thick soil profiles. Even relatively thick additions of sediment were incorporated into existing profiles, or were differentially modified and their own distinctive characteristics superimposed on the horizons of buried profiles (Bown and Kraus, 1981a, Figure 10).

PALAEOTOPOGRAPHY AND THE ALLUVIAL RECORD

Large-scale sediment filled scours are widespread in many ancient alluvial sequences and result essentially from two processes. The first of these is channel migration which can be either small scale if caused by bank recession or by cutoffs of meandering channels, or dramatic and large scale if effected by avulsion. Avulsion occurs during floods and is generally associated with the normal evolution of an aggrading stream system. The course of a stream is abandoned in favour of a new course which is established at a lower topographic level near or adjacent to its previous course. The new channel is established by shallow or deep scour into the underlying unconsolidated sediment; this new channel is gradually filled by the accumulation of new influxes of detritus. Sediment filled scours that form by avulsion or smaller scale relocation of part of the stream course are typically floored with relatively coarse material (gravel and sand) that reflects the re-establishment of a channel in a continuously aggrading regime.

A second process that generates large-scale sediment filled scours, but one that has received less attention in the sedimentologic literature, is erosion due to the shift from a dominantly aggrading stream regime to one characterized by net degradation. This shift reflects the lowering of local or regional base-levels and removal of sediment from channel and overbank deposits alike. Re-establishment of aggrading conditions causes these degradational scours to be filled; however, in now rapidly aggrading systems, most of these scours will be floored by fine sediment (sandy mud and mud) as they are transgressed by the laterally expanding and vertically building sequence of new overbank deposits.

At discrete, stratigraphically restricted intervals in the Upper Triassic Chinle Formation of Arizona (Kraus and Middleton, 1984) and in the lower Eocene Willwood Formation of Wyoming (Bown, 1984), channel sandstones and overbank mudstones were extensively

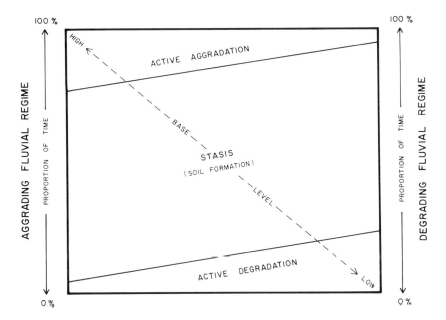

Figure 3. Diagram showing relative proportions of time occupied by active aggradation, active degradation, and stasis (soil formation) and their relationships to Wheeler's concepts of base-level in both aggrading (left) and degrading (right) fluvial regimes. Percents of time shown are for purposes of illustration only; actual percentage of time represented by stasis is probably much larger than depicted.

dissected during episodes of degradation and the resulting palaeotopographies were later filled by younger Chinle and Willwood floodplain sediments, respectively (Figures 4 and 5). These scours range from a few metres to several tens of metres in depth and from a few tens of metres to several kilometres in breadth. The sediment filled scours, which can invariably be traced laterally into relatively thick, multi-storey sand bodies, attest to episodes of areally widespread gullying followed by return to aggrading conditions.

In the Chinle and Willwood examples, both the scoured mudstones and the scour-fill sediments are varicoloured due to translocation and concentration of iron minerals during ancient pedogenesis. Colour differences between paleosol horizons developed on the underlying (dissected) and overlying (scour-fill) sediments commonly enhance recognition of the gullied intervals (Figures 6 and 7). In the Chinle Formation, paleosols developed on scour-fill mudstones and sandy mudstones are relatively immature and their drab coloration and pervasive gley mottles reflect formation on damp or periodically damp

Figure 4. Two superposed sediment filled scours in Petrified Forest Member of Chinle Formation, Petrified Forest National Park, north-east Arizona, showing sequence of thin, immature alluvial paleosols (a) cut by scour (b) that was in turn filled by dipping levee deposits (c) which are characterized also by immature paleosol development. Following deposition of the levee deposits (c), those deposits underwent gullying, producing another scour (d), that was in turn filled with younger overbank materials (e). Development of the mature compound spodic paleosol (f) records a relatively long period of stasis reflecting a return to near equilibrium after fill of the local alluvial topography. Hill in foreground is approximately 4 m high.

sediment. At other localities, formation of a succession of immature soils in the scours under conditions of better soil drainage (or relatively lower accumulation rates) is indicated by a sequence of alternating thin grey or pale red or orange mudstones and sandy mudstones (Figure 4). In contrast, paleosols developed on the eroded floodplain deposits that surround the scours are more mature and possess deep brick red and purple labile accumulation horizons that attest to relatively prolonged development on better drained sediments. The difference in maturity between the paleosols in scour-fill deposits and other floodplain paleosols reflects their relative topographic positions during soil formation, as well as possible permeability differences between the

Figure 5. Sandstone floored, sediment filled scour in lower Eocene Willwood Formation of north-west Wyoming, in which a sequence of relatively mature spodic paleosols (a) are truncated by the scour (b), which is filled with sediments upon which are developed a sequence of relatively less mature spodosols and gleysols (c). The scour is about 15 m deep, of which 10 m is depicted here.

sediments in the scours (generally poorly sorted) and those of older dissected floodplain deposits (generally well sorted).

Similar relations between paleosols within and extraneous to scour-fill deposits are seen in the lower Eocene Willwood Formation of Wyoming, where extensive degradational regime scour systems are developed sporadically throughout the 420–500 m levels of the formation over an area of at least 250 km^2. In the Willwood examples, as in those of the Chinle Formation, paleosol development in the mudstone floored scours is relatively immature compared to that on truncated floodplain soils. Scour-fill deposits are typified by orange and mottled grey and orange soil labile accumulation horizons rich in hydrated iron oxides, whereas floodplain deposits adjacent to and cut by scours possess relatively mature spodic paleosols characterized by brick red and purple B horizons rich in dehydrated iron and aluminium oxides. Once again, it appears likely that soil maturity differences are related to the relative topographic positions and parent material permeabilities of the evolving soil profiles and, perhaps, to a relatively high rate of scour fill.

Figure 6. Sediment filled scour in lower part of lower Eocene Willwood Formation, showing enhancement of the scour (a) by the abutting of relatively immature paleosols in the scour (b) against relatively mature paleosol horizons surrounding the scour (c). Orientation of scour from upper left to lower right. Spodic horizon at (d) is approximately 1.0 m in thickness.

Some mudstones filling scours in the Chinle and Willwood Formations are distinctive in that they dip more steeply than do the scoured mudstones (Figures 4 and 8). The dipping mudstones are commonly made up of alternating pale red and grey bands of immature soils and originate at the lateral margins of extensive sheet sandstones. Dips vary from about 10° to nearly 90°, and the direction of dip always diverges approximately 90° from palaeocurrents determined from laterally adjacent channel systems. The red and grey banded mudstones that dip away from adjacent sheet sandstones constitute a special type of scour fill sediment and may represent immature paleosol development on aggrading levee deposits. The dipping paleosols are coincident with the accretion surfaces of the evolving levees, and the unusually steep dips in some areas may have resulted from compaction of levee deposits against channel deposits during loading.

Dissected stratigraphic horizons, which reflect geologically long term base-level fluctuations, are illustrated by a time restored section for part of the Chinle Formation (Figure 9). The factors responsible for initiating base-level changes during Chinle deposition are uncertain but might include tectonic activity, climatic changes, or both. Initially, aggradational conditions reigned, parent sediments were deposited,

Figure 7. Superposed scours in Petrified Forest Member of Chinle Formation, northeast Arizona. Scour (a) is sandstone floored and reflects downcutting of overbank deposits by stream and fill of scour with proximal channel deposits following stream relocation. Scour (b) is mudstone floored and reflects gullying of the proximal channel deposits, followed by deposition of floodplain mud in the scour. Scour (a) was formed and filled during an aggradational regime; whereas scour (b) was formed during a degradational phase, but was filled following return to aggradational conditions. The more mature paleosol spodic horizons at top and base of hill are darker than the illuviated horizons of less mature paleosols developed on the proximal channel deposits and highlight definition of the two scours. Person shows scale.

and Chinle soils 1 to 8 were formed. Lowering of base-level initiated a degradational interval, during which time floodplain sediments were gullied, causing erosion of soils 6, 7, and 8, and a highly dissected palaeotopography was formed. In Figure 9, a typical scoured interval is represented by the unconformity between paleosols 5 and 9. The time restored section for the dissected surface emphasizes that during part of the time represented by the scour, sediment gradually accreted and soils formed, but those conditions were followed by a degradational regime in which short-lived intervals of erosion alternated with longer periods during which sediment exposed at the surface was pedogenically modified. Following a return to aggradational conditions, the dissected Triassic terrain was gradually blanketed by additional sediment that

194   CHAPTER 6

Figure 8. Relatively steeply dipping levee deposits (a), and gently dipping overbank mudstones (b), filling scours (c), cut into sequence of moderately well-developed spodic paleosols (d) in middle part of Willwood Formation. Levee and overbank deposits (a and b, respectively) are alternating bands of orange and grey mudstones (e), fine sandstones (f), and carbonaceous shale (g). From bottom of gully to top of large hill in right foreground is approximately 12 m.

was then pedogenically modified (paleosols 9 to 14, Figure 9).

Because base-level fluctuations are common in the geologic record, the authors expect that many other alluvial sequences contain dissected horizons similar to those described from the Chinle and Willwood Formation. Pedogenically modified mudstones that fill such scours are valuable indicators of climatically and/or tectonically controlled episodes of degradation and can be useful in reconstructing the palaeogeomorphologic history of continental basins.

SECTION COMPLETENESS, PALEOSOLS AND THE FOSSIL RECORD

Problems of relative section completeness and time resolution are important to palaeontologic studies because the proper resolution of

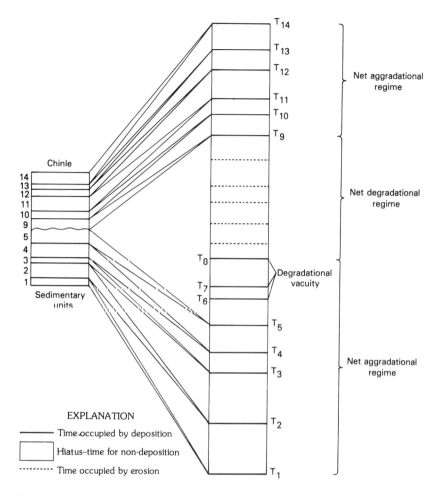

Figure 9. Schematic restored time section for a column of sedimentary units in the Triassic Chinle Formation of north-east Arizona, and its relation to net aggradational and net degradational regimes. Additional discussion in text.

the amount and distribution of time in restored sections of fossiliferous sediments imposes constraints on biostratigraphy and on the nature and tempo of evolutionary changes in fossil organisms. Following the seminal work of Sadler (1981), a spate of articles in the palaeontologic literature has attempted to relate concepts of stratigraphic completeness to the fossil record and to current thought concerning evolutionary mechanisms; principally the tenets of punctuated equilibria and phyletic gradualism and the evidence for either (e.g. Behrensmeyer, 1982a,b; Dingus and Sadler, 1982; Gingerich, 1982; Schindel, 1982). Comprehensive review of these topics is beyond the scope of this chapter; however, a brief introduction showing how the varied constructs of section completeness and modern evolutionary concepts might relate to thick, time averaged fossiliferous sections that contain numerous superposed paleosols follows.

Though Sadler (1981) and Dingus and Sadler (1982) have rightly stressed that lengths of time represented by a section are important criteria in determining relative *lithostratigraphic* completeness, their corollary implication, that lengths of time represented by a section are also instrumental in establishing relative *biostratigraphic* completeness, deserves additional consideration. Dingus and Sadler (1982, p. 400) stated 'Since a biostratigraphic record cannot be resolved more finely than the lithostratigraphic record containing it, these estimates (of average or expected completeness) usually represent the maximum amount of temporal precision attributable to a given biostratigraphic pattern.' It would seem logical that 'incomplete' parts of the rock record that are not represented by sediment also lack the fossils requisite for biostratigraphic and evolutionary analyses. But is this always true?

Gingerich (1974, 1976, 1982) emphasized the necessity of a relatively densely spaced, section-controlled stratigraphic record of fossils in determining evolutionary rates, modes, and patterns in early Eocene mammals from the Willwood Formation of north-west Wyoming. Gingerich concluded that most of the evolutionary patterns that emerged tend to support phyletic gradualism, rather than punctuated equilibria, as the principal means of speciation in these mammals. Using the time span and thickness for the Willwood Formation provided by Gingerich (1976) and Schankler (1980), Sadler and Dingus (1982) and Dingus and Sadler (1982) presented estimates of expected lithostratigraphic completeness for the Willwood Formation as functions of selected time intervals varying by orders of magnitude from 10 to 10^6 years. From these estimates they concluded that the Willwood section is expected to be complete between 10^5 and 10^6 year levels of temporal resolution, and that '...if complete sections are required to distinguish between punctuation and gradualism, (for the Willwood section) we should not expect to resolve punctuations lasting less than hundreds of thousands of years.' (Dingus and Sadler, 1982, p. 408). In addition, the last two workers cited Bown (1979) who recognized the numerous gaps in the Willwood section by identification of '...extensive development of paleosols.' Again, in application of their time averaging concept to the alluvial Willwood Formation and its contained fossil vertebrate faunules, stress was laid on biostratigraphic incompleteness as judged from lithostratigraphic incompleteness.

Incompleteness in the alluvial record can arise from erosion or stasis (non-deposition), and stasis is nearly always associated, in the fluvial regime, with soil formation. In reconstructing alluvial sedimentary history by means of a time restored section, it is necessary to distinguish time that is simply not represented by accumulation of sediments (non-deposition and/or erosion in the marine realm; erosion in the alluvial realm) from time that is not represented by rocks but is represented by episodes of ancient soil formation. Though an interval or intervals of time might not be represented by deposition of a column of sediment, that time *is preserved* in the alluvial record as a period of soil formation. Development of a methodology to accurately or relatively date formation time for pre-Quaternary paleosols by their maturity will allow restoration of much of the time restored alluvial section. In other words, in an alluvial sequence containing paleosols, the only time that is 'missing' (not represented) is that time elapsed during erosion and that time lost from the rocks and paleosols that were removed by erosion. As observed above, the cumulative time during which erosion occurred is, geologically speaking, very small. In most well exposed alluvial sequences, the body of sediment lost to local erosion can be restored by examining sections lateral to the scours. Very widespread, basinwide episodes of erosion (not yet known in the Willwood Formation or the Chinle Formation) naturally remain a problem.

We submit that the Willwood Formation, which contains approximately 500–1200 superposed paleosols (Bown and Kraus, 1981a), is a very complete alluvial sequence, and that probably the only time that is missing is that represented by scours in local sections. As shown by Bown (1984), this time can be accounted for adequately in most instances by correlation with adjacent sections. Dingus and Sadler (1982) are correct in evaluating the Willwood paleosols as intervals of non-sedimentation and thereby as gaps in the sediment accumulation history of the Willwood Formation. They are mistaken in inferring that they also represent gaps in the biostratigraphic record because the most significant accumulations of fossil vertebrates in the Willwood Formation occur in specific horizons in the upper sola of these paleosols.

Bown and Kraus (1981b) provided taphonomic evidence that fossil vertebrate remains from Willwood paleosols are scavenged time-lag accumulations that probably developed over the time of formation of the paleosols. The time of formation, determined by analogy with modern soils of the same general classification and maturity, is of the order of 2×10^3 to 1.4×10^4 years. These figures have no value as absolute age determinations of Willwood paleosols, but if analogy with modern soils of similar types is useful in estimating properties of ancient soils, they must certainly be within the proper order of magnitude.

In summary, it is emphasized that for biostratigraphic and evolutionary studies, it is not the time represented by sediments, rather it is the time represented by the fossils that is important. The Willwood Formation is a fluvial unit that, like most other alluvial successions, is relatively incomplete from the standpoint of sediments and lithostratigraphy. As Gingerich (1982) has pointed out, from the standpoint of lithostratigraphic completeness, the Clark's Fork Basin Willwood sequence, containing a higher proportion of sediment per unit time, is expected to be more complete than the central basin Willwood Formation. Like many other alluvial deposits, the Willwood Formation is relatively complete with respect to time because of its rich representation of paleosols. Unlike most other alluvial deposits, it possesses what we believe to be a relatively complete fossil vertebrate succession. Therefore, in disagreement with Dingus and Sadler's (1982, p. 400) argument, the Willwood Formation is an example of a unit in which the biostratigraphic record can be resolved more finely than can the lithostratigraphic record containing it. To the extent that paleosol concentrations of vertebrate fossils in the Willwood Formation reflect accumulation of these remains over a time period comparable to the periods of formation of the discrete paleosols in which the fossils are contained, the individual vertebrate fossil accumulations in the Willwood Formation can be discerned at the 2000 to 14 000 year levels of discrimination, if the paleosols can be sampled individually on a reliable basis. In the southern Bighorn Basin of Wyoming, Willwood exposures and spacing of fossiliferous portions of Willwood paleosols are such that individual paleosols generally can be sampled for their fossil vertebrate contents with confidence. In areas of steeper topography, samples generally can be determined to originate from a stratigraphic interval of about 5–10 m. The greatest number of discrete paleosols in a 10 m interval of sediments recorded by us in more than 1600 m of 24 separate partial sections of the Willwood Formation is four. Therefore the worst levels of faunal resolution likely to be contained in our samples from the southern Bighorn Basin Willwood badlands are on the order of approximately 56 000 years. This figure is probably well within the time limits of resolution necessary to distinguish punctuated and gradual evolution in the Willwood mammal record.

Thus fossil soils provide us with a method of evaluating stratigraphic completeness for a body of alluvial rocks that represents far less time than a formation, member, or even most series of units as they might be grouped in the field. Using the Willwood Formation as an example, it has been shown that fossiliferous paleosols can even be used to establish more complete records of time than are commonly afforded by the rocks in which they are contained.

PALEOSOLS AND SEDIMENT ACCUMULATION RATES

Fluvial systems, and the nature of their deposits, are governed by interrelated controls which are both extrabasinal and intrabasinal in character (Beerbower, 1964; Miall, 1980). Examples of extrabasinal controls include climate in the source area and depositional site, source area lithology, and tectonic activity; whereas intrabasinal mechanisms are inherent to the stream system and include the type and frequency of channel relocation. Recent studies, including both computer simulations (e.g. Allen, 1978, 1979; Leeder, 1978; Bridge and Leeder, 1979; Read and Dean, 1982) and field projects (e.g. Puigdefabregas and Van Vliet, 1978; Flores, 1981; Galloway, 1981; Read and Dean, 1982), have focused on understanding the development of alluvial sequences in response to different intrabasinal and extrabasinal factors.

Changes in intrabasinal or extrabasinal controls may be mirrored by changes in the rate of sediment accumulation through time. Allen (1978), for example, predicted that decrease in the rate of basin subsidence will be accompanied by decrease in the rate of sediment accumulation because of increased reworking of floodplain deposits as channels migrate. When subsidence and sediment accumulation are rapid, alluvial suites will be dominated by fine-grained overbank deposits with single or weakly multistorey sand bodies. When subsidence and sediment accumulation are slow, thick and laterally extensive sand bodies can develop. Thus independent evidence for rates of sediment accumulation and changes in those rates through a section are potentially useful in elucidating the geologic history of alluvial successions.

For many alluvial sequences, radiometric dates are rare or are widely spaced stratigraphically. Temporal resolution through palaeomagnetic dating is still relatively coarse. Consequently, sediment accumulation rates are long term because they are calculated for sections that span relatively long periods (on the order of 10^6 to 10^7 years). The interpretation of ancient fluvial sequences would be facilitated if sections could be subdivided into units spanning shorter periods (on the order of 10^4 to 10^5 years) and accumulation rates determined for those subdivisions. Because most alluvial sequences are comprised of multistorey paleosols, fossil soils are a potentially powerful method of evaluating short-term sediment accumulation rates.

The interrelationship between rate of sediment accumulation and rate of soil maturation has been examined by Bown and Kraus (1981a) and Retallack (1983a,b, 1984). Bown and Kraus (1981a) divided the total time represented by a thick sequence of paleosols in the Willwood Formation by the number of paleosols present in the section to estimate the average rate of soil formation. Their calculations assumed that no erosional unconformities truncated the section and that most paleosols were composites of two or even three epsiodes of pedogenesis. The rate of soil formation determined by this method represents the average time required to deposit and pedogenically modify small subdivisions of the alluvial sequence.

Cognizant of many of the possibilities of utilizing paleosols for evaluating stratigraphic completeness, Retallack (1983a,b, 1984) calculated sediment accumulation rates directly from paleosols in the White River Group of South Dakota. Retallack (1) estimated the *minimum* amount of time needed to develop different paleosol types in his section, based on modern analogs; (2) computed the total time represented by all paleosols in a section of known thickness; and (3) divided section thickness by the total soil time to produce a sediment accumulation rate for the section. Sedimentation rates determined by this technique were approximately 20 to 60 times more rapid than rates calculated for the same section on the basis of radiometric and paleomagnetic data. For example, the middle portion of the White River section, the Scenic Member of the Brule Formation, was deposited during a span of 1.7 m.y. (Prothero *et al.*, 1982) and its calculated rate of sediment accumulation over this time interval is 0.023 mm/yr (Retallack, 1983b, p. 837–8). The rate of sediment accumulation based on fossil soils in the Scenic Member is roughly 20 times larger or 0.47 mm/yr (Retallack, 1983b, p. 838). This difference largely reflects assumptions underlying the calculations, which include: (1) the paleosols are not composite and have undergone little or no erosion; (2) significant time gaps occur between rather than within those parts of the section that were analysed; and (3) then upon maturation of each paleosol it was buried by additional sediment. Despite problems arising from such assumptions, this method produces short-term sediment accumulation rates which Retallack (1984) suggested can be used confidently to compare and contrast different parts of a fluvial section.

Although the authors applaud his insight, the authors believe that development of a rigorous methodology for dating ages of maturation of different Tertiary and older soils is necessary, though lacking. Furthermore, for analysing the genesis of alluvial sequences and changes in tectonic and climatic setting etc., average or maximum times required to form different soils are probably more useful than minimum times. These points can be illustrated by re-examining the Scenic Member of the White River section, discussed above.

The thickness of the Scenic Member is 38.4 m, as estimated directly from the measured stratigraphic column in Figure 5 (Retallack, 1983b). Section thickness divided by the soil derived sediment accumulation rate (0.47 mm/yr, see above) indicates that a minimum of 81 700 years was required to deposit the sediment *and* to pedogenically alter it. Thus 81 700 years is the geologic time occupied by both active aggradation plus stasis and pedogenesis. Dividing 81 700 years by the entire time spanned by the Scenic Member (1.7 m.y.) reveals that only 5% of the total time over which the Scenic Member accumulated is represented in its preserved depositional and pedogenic history (Figure 10). A schematic time restored stratigraphic section for the Scenic Member, derived from the time estimates of Retallack (1983a,b), is shown in Figure 10. The time required to deposit and develop 10 soil profiles is illustrated by the heavy black lines in the time restored column. The time lines have been drawn to scale to occupy 5% of the total length of the time column.

Retallack (1984) proposed using minimum times of soil development for two basic reasons: following an initial relatively rapid period of maturation, soils attain equilibrium and continue to develop more slowly, and climatic fluctuations can cause periods of rapid development interspersed with periods of negligible pedogenesis. If only 5% of the time represented by the Scenic Member can be accounted for using minimum maturation times, it suggests one or more of the following: (1) erosion removed much sediment and there should be unconformities such as those described above for Chinle and Willwood sediments (Kraus and Middleton, 1984; Bown, 1984); (2) the paleosols represent multiple stages of alteration (Bown and Kraus, 1981a); or (3) the minimum estimates for rates of soil maturation used are too low. Because in studying the development of any alluvial sequence the time occupied by non-deposition (pedogenesis or erosion) is at least as important as that occupied by deposition, these three factors cannot be ignored. Thus for alluvial basin analysis, average or maximum rates of soil maturation are probably more useful because they provide a better picture of the large amount of non-depositional time that is represented by small segments of fluvial sequences, i.e. paleosols.

SUMMARY

Floodplain environments undergo relatively infrequent episodes of sedimentation or erosion; the cumulation of the time of these episodes

202   CHAPTER 6

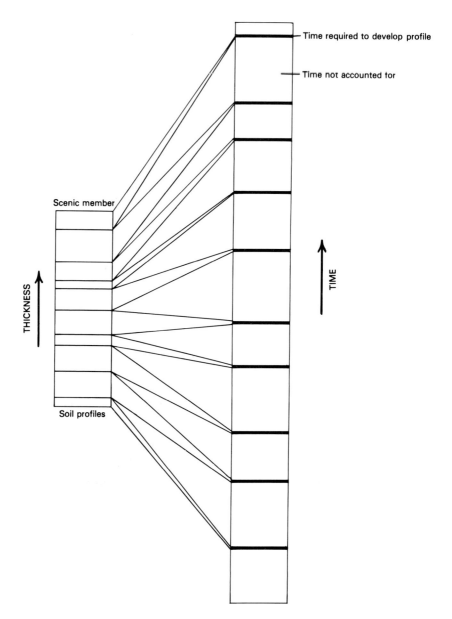

Figure 10. Generalized diagram of a sequence of paleosols in the Scenic Member of the Brule Formation, Badlands National Park, South Dakota (left column), and a time restored column for the same sequence (right column) based on Retallack (1983b). Additional discussion in text.

is geologically limited for any alluvial sequence and the normal state is one of stasis, or non-erosion and non-deposition. During stasis, pedogenic modification of alluvial parent material occurs. Consequently, most ancient successions of overbank deposits consist of sequences of composite or multistorey paleosols; this concept is strengthened by our examination of Tertiary and older alluvial sediments, as well as by the observations of other researchers. Because alluvial sequences containing paleosols preserve not only the record of depositional events but also the combined record of intervals of non-deposition and non-erosion, their study is essential to alluvial time resolution and thereby to elucidating the complete geologic history of alluvial sediments. Our studies demonstrate that fossil soils are valuable indicators of base-level changes in continental basins, and thus they are potentially useful in evaluating climatic changes or tectonic events as well as the palaeogeomorphologic history of a basin. In the future, with more sophisticated and rigorous techniques for evaluating soil maturation times, paleosols will become an even more powerful tool in continental basin analysis because they will facilitate the relatively accurate estimations of short-term sediment accumulation rates. Because most alluvial sequences consist of stacked paleosols, alluvial rocks provide a relatively complete record of short-term sediment accumulation rates, changes in which reflect intrabasinal or extrabasinal controls, or a combination of both. Fossils in fossil-bearing paleosol sequences can represent relatively more time than the sediments that encase them.

ACKNOWLEDGMENTS

We thank P.A. Allen, L.T. Middleton, G. Retallack, and S.W. Wing for technical review of this paper. We are grateful to the Superintendent, Chief Ranger, and other personnel of Petrified Forest National Park, Arizona for their generous cooperation in allowing us to pursue geologic studies of the Chinle Formation in the Park.

Research on the Chinle Formation in Petrified Forest National Park was supported by Research Grant #2904−84 from the National Geographic Society. Kraus' research in the Bighorn Basin was supported by a Faculty Development Award from the University of Colorado and by National Science Foundation Grant EAR-8319421.

REFERENCES

Abbott, P.L., Minch, J.A. and Peterson, G.L. 1976. Pre-Eocene paleosol south of Tijuana, Baja California, Mexico. *J. sedim. Petrol.* **46**, 355−61.

Allen, J.R.L. 1974. Studies in fluviatile sedimentation: implications of pedogenetic carbonate units, Lower Old Red Sandstone, Anglo-Welsh Outcrop. *Geol. J.* **9**, 181–208.

Allen, J.R.L. 1978. Studies in fluviatile sedimentation: an exploratory quantitative model for the architecture of avulsion-controlled alluvial suites. *Sedim. Geol.* **21**, 129–47.

Allen, J.R.L. 1979. Studies in fluviatile sedimentation: an elementary geometrical model for the connectedness of avulsion-related channel sand bodies. *Sedim. Geol.* **24**, 253–67.

Alexander, C.S. and Prior, J.C. 1971. Holocene sedimentation rates in overbank deposits in the Black Bottom of the Lower Ohio River, Southern Illinois. *Am. J. Sci.* **270**. 361–72.

American Commission on Stratigraphic Nomenclature 1961. Code of stratigraphic nomenclature. *Bull. Am. Ass. Petrol. Geol.* **45**, 645–65.

Barrell, J. 1917. Rhythms and the measurement of geologic time. *Bull. Geol. Soc. Am.* **28**, 745–904.

Beerbower, J.R. 1964. Cyclothems and cyclic depositional mechanisms in alluvial plain sedimentation, in *Symposium on Cyclic Sedimentation*, D.F. Merriam (ed.). *Bull. Geol. Surv. Kansas* **169**, 31–42.

Behrensmeyer, A.K. 1982a. Time sampling intervals in the vertebrate fossil record. *Proc. Third N. Am. Paleontol. Conv.* **1**, 41–5.

Behrensmeyer, A.K. 1982b. Time resolution in fluvial vertebrate assemblages. *Paleobiology* **8**, 211–27.

Behrensmeyer, A.K. and Tauxe, L. 1982. Isochronous fluvial systems in Miocene deposits of Northern Pakistan. *Sedimentology* **29**, 331–52.

Birkeland, P.W., Crandell, D.R. and Richmond, G.M. 1971. Status of correlation of Quaternary stratigraphic units in the western conterminous United States, *Quat. Res.* **1**, 208–27.

Blackwelder, E. 1909. The valuation of unconformities *J. Geol.* **17**, 289–99.

Bown, T.M. 1979. *Geology and Mammalian Paleontology of the Sand Creek Facies, Lower Willwood Formation (lower Eocene), Washakie County, Wyoming*. Geol. Surv. Wyo. Mem. 2. 155 pp.

Bown, T.M. 1984. Biostratigraphic significance of baselevel changes during deposition of the Willwood Formation (lower Eocene), Bighorn Basin, Wyoming. *Geol. Soc. Am. Abs. with Prog.* **16**, 216.

Bown, T.M. and Kraus, M.J. 1981a. Lower Eocene alluvial paleosols (Willwood Formation, northwest Wyoming, U.S.A.) and their significance for paleoecology, paleoclimatology and basin analysis. *Palaeogeogr., Palaeoclimatol., Palaeoecol.* **34**, 1–30.

Bown, T.M. and Kraus, M.J. 1981b. Vertebrate fossil-bearing paleosol units (Willwood Formation, northwest Wyoming, U.S.A.); implications for taphonomy, biostratigraphy and assemblage analysis, *Palaeogeogr., Palaeoclimatol., Palaeoecol.* **34**, 31–56.

Bown, T.M. and Kraus, M.J., in press. Geologic and paleoenvironmental conspectus of the Oligocene Jebel Qatrani Formation and adjacent rocks, Fayum Depression, Egypt. *Ann. Geol. Surv. Egypt.*

Bown, T.M., Kraus, M.J., Wing, S.L., Fleagle, J.G., Tiffney, B.H., Simons, E.L. and Vondra, C.F. 1982. The Fayum primate forest revisited. *J. Human Evol.* **11**, 603–32.

Brewer, R., Crook, K.A.W. and Speight, J.G. 1970. Proposal for soil stratigraphic units in the Australian Stratigraphic Code. *J. Geol. Soc. Austr.* **17**, 103–9.

Bridge, J.S. and Leeder, M.R. 1979. A simulation model of alluvial stratigraphy *Sedimentology* **26**, 617–44.

Chalyshev, V.I. 1969. A discovery of fossil soils in the Permian Triassic. *Dokl. Akad. Sci. USSR* **182**, 53–6.

Dingus, L. and Sadler, P.M. 1982. The effects of stratigraphic completeness on estimates of evolutionary rates. *Syst. Zool.* **31**, 400–12.

Dury, G.H. 1971. Relict deep weathering and duricrusting in relation to the palaeoenvironments of middle latitudes. *Geogr. J.* **137**, 511–22.

Evans, L.J. 1982. Dating methods of Pleistocene deposits and their problems: VII. Paleosols. *Geosci. Can.* **9**, 155–60.

Flores, R.M. 1981. Coal deposition in fluvial paleoenvironments of the Paleocene Tongue River Member of the Fort Union Formation, Powder River area, Powder River Basin, Wyoming and Montana, in *Recent and Ancient Nonmarine Depositional Environments: Models for Exploration*, F.G. Ethridge and R.M. Flores, (eds). *Soc. Econ. Paleontol. Mineral. Spec. Publ.* **31**, 169–90.

Freytet, P. 1973. Petrography and paleoenvironment of continental carbonate deposits with particular reference to the Upper Cretaceous and Lower Eocene of Languedoc (southern France). *Sedim. Geol.* **10**, 25–60.

Freytet, P. and Plaziat, J-C. 1982. Continental carbonate sedimentation and pedogenesis— Late Cretaceous and Early Tertiary of southern France. *Contrib. Sediment.* **12**, 212 pp.

Frye, J.C. 1949. Use of fossil soils in Kansas Pleistocene stratigraphy. *Trans. Kansas Acad. Sci.* **52**, 478–82.

Galloway, W.F. 1981. Depositional architecture of Cenozoic Gulf coastal plain fluvial systems, in *Recent and Ancient Nonmarine Depositional Environments. Models for Exploration*, F.G. Ethridge and R.M. Flores, (eds). *Soc. Econ. Paleontol. Mineral. Spec. Publ.* **31**, 127–56.

Gingerich, P.D. 1974. Stratigraphic record of early Eocene *Hyopsodus* and the geometry of mammalian phylogeny. *Nature* **248**, 107–9.

Gingerich, P.D. 1976. Paleontology and phylogeny: patterns of evolution at the species level in Early Tertiary mammals. *Am. J. Sci.* **276**, 1–28.

Gingerich, P.D. 1982. Time resolution in mammalian evolution: smapling, lineages, and faunal turnover. *Proc. Third N. Am. Paleontol. Conv.* **1**, 205–10.

Goldbery, R. 1982. Palaeosols of the Lower Jurassic Mishhor and Ardon Formations ('Laterite Derivative Facies'). *Sedimentology* **29**, 669–704.

Goudie, A. 1973. *Duricrust in Tropical and Subtropical Landscapes*. Clarendon Press, Oxford. 174 pp.

Goulding, M. 1980. *The Fishes and the Forest, Explorations in Amazonian Natural History*. University of California Press, Berkeley. 280 pp.

Jahns, R.H. 1947. *Geological Features of the Connecticut Valley, Massachusetts, as Related to Recent Floods*. US Geol. Surv. Water Supp. Irrig. Pap. 996. 45 pp.

Jungerius, P.D. 1976. Quaternary landscape development of the Rio Magdalena Basin between Neiva and Bogota (Colombia). *Palaeogeogr., Palaeoclimatol., Palaeoecol.* **19**, 89–137.

Kesel, R.H., Dunne, K.C., McDonald, R.C., Allison, K.R. and Spicer, B.E. 1974. Lateral erosion and overbank deposition on the Mississippi River in Louisiana caused by 1973 flood. *Geology* **1**, 461–4.

Kraus, M.J. and Bown, T.M. 1982. Alluvial paleosols: recognition and significance for paleoenvironmental reconstruction and basin analysis. *11th Int. Congr. Sediment. Abs.*, 13.

Kraus, M.J. and Middleton, L.T. 1984. Dissected paleotopography and base level fluctuations in an ancient fluvial sequence. *Geol. Soc. Am. Abs. with prog.* **16**, 227.

Leeder, M.R. 1978. A quantitative stratigraphic model for alluvium, with special reference to channel deposit density and interconnectedness, in *Fluvial Sedimentology*, A.D. Miall, (ed.). *Can. Soc. Petrol. Geol. Mem.* **5**, 587–96.

Mabesoone, J.M. and Lobo, R.C. 1980. Paleosols as stratigraphic indicators for the Cenozoic history of northeastern Brazil. *Catena* **7**, 67–78.

Mahaney, W.C. 1978. Late-Quaternary stratigraphy and soils in the Wind River Mountains, western Wyoming, in *Quaternary Soils*, W.C. Mahaney, (ed.). Geo Abstracts, Norwich. 223–64.

Mahaney, W.C. and Fahey, B.D. 1980. Morphology, composition and age of a buried paleosol, Front Range, Colorado, USA. *Geoderma* **23**, 209–18.

McPherson, J.G. 1979. Calcrete (caliche) palaeosols in fluvial redbeds of the Aztec Siltstone (Upper Devonian), southern Victoria Land, Antarctica. *Sedim. Geol.* **22**, 267–85.

McPherson, J.G. 1980. Genesis of variegated redbeds in the fluvial Aztec Siltstone (Late Devonian), southern Victoria Land, Antarctica. *Sedim. Geol.* **27**, 119–42.

Miall, A.D. 1980. Cyclicity and the facies model concept in fluvial deposits. *Bull. Can. Petrol. Geol.* **28**, 59–80.

Morrison, R.B. 1964. *Lake Lahontan; Geology of southern Carson Desert, Nevada.* US Geol. Surv. Prof. Pap. 401. 156 pp.

Morrison, R.B. 1967. Principles of Quaternary soil stratigraphy, in *Means of Correlation of Quaternary Successions*, R.B. Morrison and H.E. Wright, (eds). Int. Assoc. Quat. Res. (INQUA), VII Cong., 1965, Proc. 9, 1–69.

Morrison, R.B. 1978. Quaternary soil stratigraphy—concepts, methods, and problems, in *Quaternary Soils*, W.C. Mahaney, (ed.). Geo Abstracts, Norwich. 77–108.

Morrison, R.B. and Frye, J.C. 1965. *Correlation of the Middle and Late Quaternary Successions of the Lake Lahontan, Lake Bonneville, Rocky Mountain (Wasatch Range), Southern Great Plains, and Eastern Midwest Areas.* Nev. Bureau of Mines Rep. 9, 45 pp.

North American Commission on Stratigraphic Nomenclature 1983. North American Stratigraphic Code. *Bull. Am. Assoc. Petrol. Geol.* **67**, 841–75.

Parsons, R.B. 1981. Proposed soil-stratigraphic guide, in Internat. Union Quat. Res. and Int. Soc. Soil Sci. INQUA Comm. 6 and ISSS Comm. 5 Working Group. *Pedology Rep.* 6–12.

Pawluk, S. 1978. The pedogenic profile in the stratigraphic section, in *Quaternary Soils*, W.C. Mahaney, (ed.). Geo Abstracts, Norwich. 61–75.

Pettyjohn, W.A. 1966. Eocene paleosol in the northern Great Plains, *US Geol. Surv. Prof. Pap.* **550C**, 61–5.

Prothero, D.R., Denham, C.R. and Farmer, H.G. 1982. Oligocene calibration of the magnetic polarity time scale. *Geology* **10**, 650–3.

Puigdefabregas, C. and Van Vliet, A. 1978. Meandering stream deposits from the Tertiary of the southern Pyrenees, in *Fluvial Sedimentology*, A.D. Miall, (ed.). *Can. Soc. Petrol. Geol. Mem.* **5**, 469–85.

Read, W.A. and Dean, J.M. 1982. Quantitative relationships between numbers of fluvial cycles, bulk lithological composition and net subsidence in a Scottish Namurian basin. *Sedimentology* **29**, 181–200.

Retallack, G.J. 1977a. Triassic paleosols in the upper Narrabeen Group of New South Wales, 1. Features of the paleosols. *J. Geol. Soc. Austr.* **23**, 383–99.

Retallack, G.J. 1977b. Triassic palaeosols in the upper Narrabeen Group of New South Wales, 2. Classification and reconstruction. *J. Geol. Soc. Austr.* **24**, 19–36.

Retallack, G.J. 1983a. *Late Eocene and Oligocene Paleosols from Badlands National Park, South Dakota.* Geol. Soc. Am. Spec. Pap. 193, 82 pp.

Retallack, G.J. 1983b. A paleopedological approach to the interpretation of terrestrial sedimentary rocks: the mid-Tertiary fossil soils of Badlands National Park, South Dakota. *Bull. Geol. Soc. Am.* **94**, 823–40.

Retallack, G.J. 1984. Completeness of the rock and fossil record: some estimates using fossil soils. *Paleobiology* **10**, 59–78.

Richmond, G.M. 1962. *Quaternary Stratigraphy of the La Sal Mountains, Utah*. US Geol. Surv. Prof. Pap. 324. 135 pp.
Richmond, G.M. and Frye, J.C. 1957. Status of soils in stratigraphic nomenclature. *Bull. Am. Assoc. Petrol. Geol.* **41**, 758–63.
Ritzma, H.R. 1955. Late Cretaceous and Early Cenozoic structural pattern, southern Rock Springs Uplift Wyoming. *Wyo. Geol. Assoc., 10th Ann. Field Conf. Guidebook*, 135–7.
Ritzma, H.R. 1957. Fossil soil at base of Paleocene, southwestern Wind River Basin, Wyoming. *Wyo. Geol. Assoc., 12th Ann. Field Conf. Guidebook*, 165–6.
Sadler, P.M. 1981. Sediment accumulation rates and the completeness of stratigraphic sections. *J. Geol.* **89**, 569–84.
Sadler, P.M. and Dingus, L.W. 1982. Expected completeness of sedimentary sections: estimating a time-scale dependent, limiting factor in the resolution of the fossil record. *Proc. Third N. Am. Paleontol. Conv.* **2**, 461–4.
Schankler, D.M. 1980. Faunal zonation of the Willwood Formation in the central Bighorn Basin, Wyoming. *Univ. Michigan Mus. Paleont. Papers on Paleont.* **24**, 99–114.
Schindel, D.E. 1982. Resolution analysis: a new approach to the gaps in the fossil record. *Paleobiology* **8**, 340–53.
Schultz, C.B. and Stout, T.M. 1980. Ancient soils and climatic changes in the central Great Plains. *Trans. Neb. Acad. Sci.* **8**, 187–205.
Schultz, C.B., Tanner, L.G. and Harvey, C. 1955. Paleosols of the Oligocene of Nebraska. *Bull. Univ. Neb. State Mus.* **4**, 1–16.
Wheeler, H.E. 1958. Time-stratigraphy. *Bull. Am. Assoc. Petrol. Geol.* **42**, 1047–63.
Wheeler, H.E. 1964a. Baselevel, lithosphere surface, and time-stratigraphy. *Bull. Geol. Soc. Am.* **75**, 599–610.
Wheeler, H.E. 1964b. Baselevel transit cycle. *Kansas Geol. Surv. Bull.* **169**, 623–9.
Williams, G.E. and Polach, H.A. 1971. Radiocarbon dating of arid-zone calcareous paleosols. *Bull. Geol. Soc. Am.* **82**, 3069–86.
Wolman, M.G. and Eiler, J.P. 1958. Reconnaissance study of erosion and deposition produced by the flood of August 1955 in Connecticut. *Trans. Am. geophys. Un.* **39**, 1–14.
Wolman, M.G. and Leopold, L.B. 1957. River flood plains: some observations on their formation. *US Geol. Surv. Prof. Pap.* **282-C**, 87–107.

Chapter 7

RECOGNITION OF PALEOSOLS IN QUATERNARY PERIGLACIAL AND VOLCANIC ENVIRONMENTS IN NEW ZEALAND

I. B. CAMPBELL

New Zealand Soil Bureau, Lower Hutt, New Zealand

INTRODUCTION

Paleosols have long been recognized as important features of New Zealand soils and Quaternary deposits. They occur within many present day soils and influence their properties, while their presence has been very useful in constructing the New Zealand record of Quaternary events. In the earliest New Zealand soil surveys, for example, Grange (1931), Taylor (1933), Grange et al. (1939) and Pohlen et al. (1947) recognized in the North Island the existence of paleosols buried by accumulations of volcanic ash, and they characterized and traced them on the basis of their properties which included colour, texture, mineral content, and the degree of weathering. Later in the South Island, paleosols were identified in late Quaternary loess by Raeside (1964) and Young (1964), these being thought to represent the remains of soils that were buried during glacial episodes.

A brief review of the nature of paleosols in New Zealand and their classification was given by Gibbs (1971). He emphasized the need to distinguish the features of a paleosol or a buried soil from relict soils which are features within the solum attributable to changes in soil forming conditions. Gibbs also highlighted some of the problems in identifying paleosols and showed how their recognition varied according to the age and nature of the materials in which they are formed.

New Zealand is admirably suited for the occurrence of paleosols because periodic glaciation, volcanism, and continuing tectonic activity (Figure 1) have given rise to repeated periods of accumulation and weathering, during which soils have formed and been subsequently buried. The uplift and erosion accompanying tectonic activity has, however, been of such intensity that most soils formed during earlier weathering cycles have been destroyed. Consequently New Zealand paleosols, along with the landscapes in which they occur, are mostly of late Quaternary age.

Paleosols as stratigraphic entities have rarely been studied in New Zealand as they seldom occur as discrete recognizable bodies over wide areas. Knowledge about them has generally been derived incidently to the study of soils, tephra chronology, glacial stratigraphy or geomorphology.

Over the last decade or so, however, there has been a growing interest in paleosols because of their usefulness in Quaternary and pedological studies, and aspects of their recognition and processes involved in their ageing have been reported.

The sporadic accumulation of volcanic deposits has occurred widely over the central North Island (Figure 1), and the paleosols associated with these intermittently deposited tephras contrast quite markedly with the paleosols of the South Island which are largely a product of late Quaternary glacial or periglacial activity and the cyclic accumulation and weathering of associated deposits such as loess. In general, the paleosol sequences of the periglacial deposits give a reasonably clear indication of the succession of late Quaternary time units, although the events are not accurately dated owing to the paucity of datable materials. The reverse is true, however, of the paleosols in the tephra, which although well dated, do not give such a clear picture of the late Quaternary because of the extreme fragmentation of time by repeated eruptions.

In this paper, the recent work which considers the identification and relationships of the paleosols in glacially and climatically related deposits, and the paleosols formed through the intermittent deposition of tephras, will be reviewed.

PALEOSOLS OF GLACIALLY AND PERIGLACIALLY RELATED DEPOSITS

Glaciation in New Zealand was confined almost completely to the South Island (Figure 1). Within the glaciated areas paleosols are rare because the extensive erosion and reshaping of the landscapes that resulted from direct glacial action destroyed the soils formed in the former weathering cycle. Beyond the zone of direct glacial action, in what is described as the periglacial zone, many areas were strongly influenced, especially in a geomorphic sense, by the associated cold climate. Periglacial activity primarily consisted of surface erosion and lowering by frost action, with the downslope transport of angular frost shattered debris and its subsequent deposition on lower slopes. Successive cycles

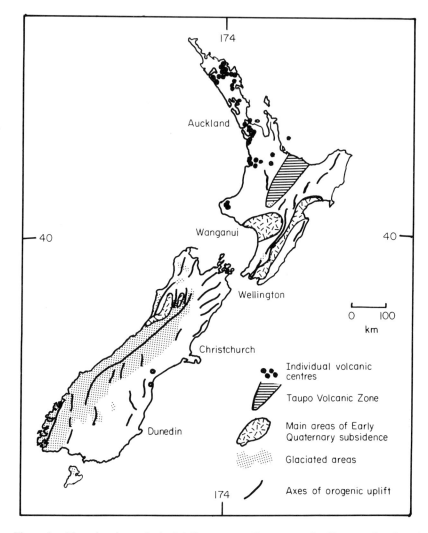

Figure 1. Map showing principal influences on Quaternary landform and paleosol development. Paleosols are most commonly found in the non-glaciated areas where loess is deposited and where there are thick accumulations of tephra (redrawn from Suggate, 1978).

of warming and cooling during the Quaternary have resulted in alternating periods of weathering, downcutting and rejuvenation of much of the hill and mountain land in New Zealand, more especially in the cooler South Island. The intensity of the cycles, however, has varied from place to place. In the higher mountain areas, for example, removal of frost shattered debris was predominant. In lower altitude areas, where the periglacial climate was less severe, removal and accumulation were more balanced and paleosols and weathering products are often interbedded with coarse textured aggradational material derived from erosion during cooler climatic periods. In the areas where periglacial activity has been less intensive, the products of glacial and periglacial erosion are frequently found redistributed as loess. In these latter areas where the destructive effects of periglacial activity have been limited, the loess remains as thick, cool climate-related deposits and it is within them that paleosols are commonly found. Paleosols that have been described from loess and periglacial slope deposits are outlined below.

Paleosols in South Island Loess Deposits

Loess occurs extensively on the eastern side of the South Island, to a limited extent in western South Island districts, and in the south and central areas of the North Island (Figure 2). These deposits have been described in a number of studies. Raeside (1964) outlined the distribution of loess in the South Island and described the features of several deposits. The loess at Timaru (Figure 2) was taken as the type section and it was characterized by six distinct layers, each with a fossil soil more or less clearly expressed. An important attribute of each layer was a very compact horizon which was considered to be a hard pan formed during the period of loess accumulation in cold glacial periods. Biological activity and active soil formation, which Raeside believed occurred during interglacial and interstadial periods, were thought to have caused fragmentation of the compact horizon or pan and is now expressed as a series of paleosols. These were recognized by the presence of root traces, worm casts, humus staining, and colloidal coatings. The paleosols of late Quaternary age (between loesses 1 & 2 and 2 & 3, Otira Glaciation), (Suggate 1965, Table 1) showed a moderate degree of development, while the paleosol between loesses 3 & 4 showed a greater degree of weathering and was considered to represent the weathering of an interglacial stage. Paleosols occurring between loesses 4 & 5 and 5 & 6 were considered to represent the weathering during interstadials of the penultimate glaciation (Waimean Glaciation, Table 1). Raeside considered that the features developed in each paleosol at the time of its formation (including the horizons and coarse subsoil jointing pattern) were essentially fossilized, having been sealed from further weathering by the addition of the succeeding loess layer. He also recognized that where the loess was deposited in a different climatic region from the type section (i.e. higher rainfall), a similar sequence of beds could be recognized but the paleosols were less readily identifiable because of the absence of massive prismatic structured horizons within the loess.

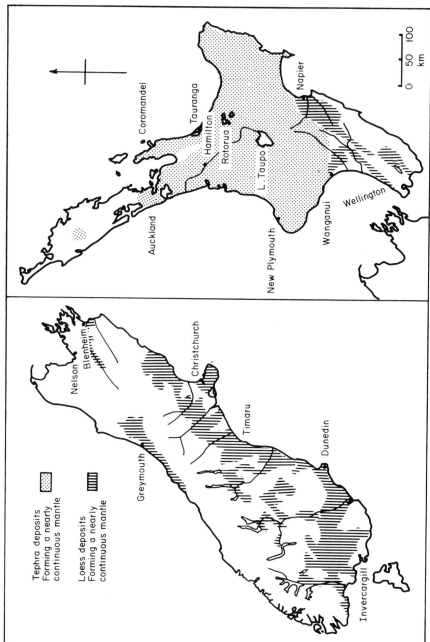

Figure 2. Areas of loess and tephra deposits in New Zealand.

A detailed study of loess deposits over an area of about 150 km² in north-east Otago (100 km north of Dunedin) was carried out by Young (1964). He recognized five loesses (Figure 3) separated by interstadially weathered paleosols, and a more strongly weathered paleosol interpreted as the product of interglacial weathering (Oturian Interglacial, Table 1). The paleosols were again identified by the presence of root traces, worm casts, colour changes, and in some instances, secondary carbonate deposits. Young (1967) also reported on the occurrence of loess deposits from the West Coast of the South Island but he suggested that distinctive expression of paleosols was lacking because of the nature of the weathering that occurs under very high rainfall conditions there.

Paleosols occurring in loess deposits of the southern South Island region were described by Bruce (1973a,b) and Bruce *et al.* (1973) who recognized that the loess, while originating from differing source areas, had nevertheless formed a regional deposit characterized by a number of consistently identifiable layers. The loess cover as a whole was defined as Stewarts Claim Formation and was considered to represent a sequence of glacially related events spanning Otiran to Aranuian

Glacial Stage	Interglacial Stage	Ice Advance	Ice Retreat
	ARANUIAN		Major retreat
OTIRA		Kumara 3_2	
			Minor retreat
		Kumara 3_1	
			Interstadial
		Kumara 2_2	
			Interstadial
		Kumara 2_1	
	OTURI		Major retreat
WAIMEA		Kumara$_1$	
	TERANGIAN		Major retreat
WAIMAUNGAN			

Table 1. New Zealand subdivisions of the Late Quaternary, (Suggate 1965).

214 CHAPTER 7

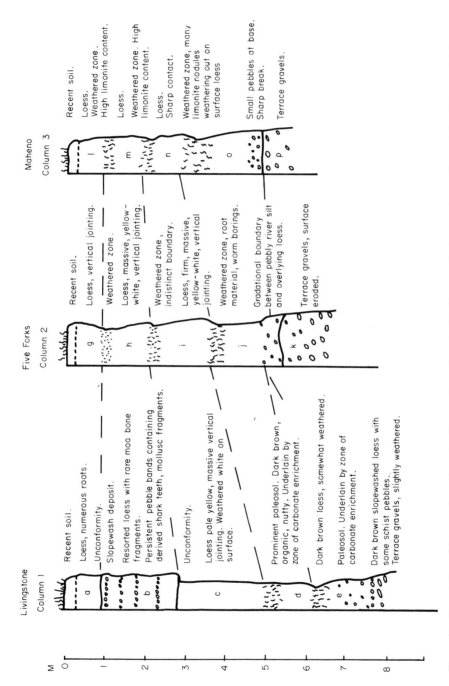

Figure 3. Correlation of loess and weathered units seen in sections in the Oamaru district, north-east Otago (after Young, 1964).

time (Table 1). At its type locality, it is represented by four members, with the uppermost, the Yellow Loess Member (including Yellow A and Yellow B), successively overlying Brown A Loess Member, Brown B Loess Member and the oldest, Brown C Loess Member. The paleosols within the individual members were identified largely on morphological features including structure, colour, compaction, and presence of concretions. Bruce considered, however, that the paleosols were incomplete and represented mainly subsolum horizons, the upper horizons having been removed by pedosphere stripping. In most places the paleosols had features comparable with the soil at the present surface, the blocky structured upper horizon of the paleosol being similar to that of the modified fragipan of the soil at the surface. This, along with concentrations of small stones lying on the surface of the paleosol, was taken as evidence that soil erosion had taken place prior to deposition of the succeeding loess. Lateral variation in the paleosols, similar to that occurring in the present soil because of climatic and soil forming differences across the landscape, was also recognized, with the paleosols of higher rainfall areas (1200 mm rainfall) being less distinctly separated than those of drier areas (650 mm rainfall). The paleosols were considered by Bruce (1973a) to represent interstadial weathering of the loess deposited in the preceeding glacial period. The paleosol capping Brown Loess C Member was considerably more weathered and showed greater pedological development than the other paleosols while also showing some regional variation in properties. Bruce considered that this was probably developed during a full interglacial period rather than an interstadial. Bruce (1973b) also showed that the loesses and their paleosols could be traced across terrace sequences of differing ages, the lower terraces having fewest loesses and paleosols and the higher terraces having a more complete sequence (Figure 4).

Loess deposits and associated paleosols of the mid-latitude eastern region of the South Island were discussed by both Ives (1973) and Griffiths (1973). Ives used paleosols as a means of helping to establish the regional correlation of loess and compared chemical properties (including CEC, BS% & Ca%) of paleosol horizons with those of modern soils. In a number of sections on Banks Peninsula near Christchurch, Griffiths (1973) identified paleosols mainly by colour, mottling, texture, consistence, and structural differences. As with the paleosols described by Bruce (1973a,b), those of the Banks Peninsula region were very similar to the present soil at the surface while the age of the loess sequence and the paleosols probably span a time interval from the last glacial to the penultimate glacial period.

The loess deposits at Timaru, South Canterbury, and the age relationships of the paleosols within them (previously suggested by

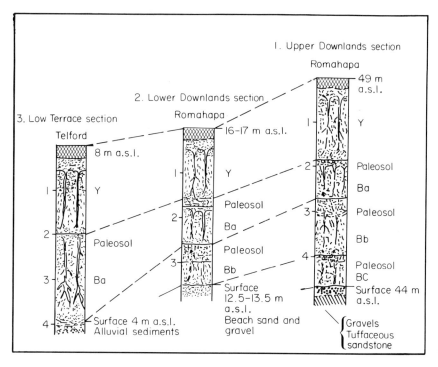

Figure 4. Correlation of paleosols in loess layers from terraces in the southern South Island region (after Bruce, 1973b).

Raeside 1964) were re-interpreted by Tonkin et al. (1974) on the basis of ^{14}C dates from bore-holes through loess on the Timaru downlands. Six loess members grouped in two formations were identified along with five paleosols that were delineated by colour, texture, consistence, structure, mottles, casts, and concretions. The upper loess member was separated from the next loess by a buried peat which gave dates for fulvic acid, humic acid, and residue material from 11 800 to 31 000 years and paleosol 1 was considered to have formed during this period.

The Upper Loess 1 Formation (containing loesses 1 & 4) was considered to equate with the whole of the Dashing Rocks sequence (Figure 5) previously described by Raeside (1964) (which he considered may have covered 2 glaciations). The Lower Loess 2 Formation contained Members 5 & 6. The dates were later revised after the peat was subjected to seven different chemical treatments, including alkali-pyrophosphate, organic solvents, acid hydrolyses and combined treatments (Goh et al. 1978) and a date of 49 000 BP for the buried peat overlying Loess Member 2 was suggested.

Runge et al. (1974) had earlier studied the paleosol sequence

Figure 5. Loess and paleosol sequence described by Raeside (1964) at Dashing Rocks, Timaru. Fragipans forming paleosubsoils are formed in the prominent prismatic structured layers. The more weathered zones are represented by finer structured paleohorizons shown by less steeply sloping layers. (Photograph: Q. Christie.)

collected from the bore-holes at Timaru with a view to identifying the paleosols using analyses of phosphorus, manganese, bulk density and clay contents. They showed that there were systematic differences in the phosphorus fractions, extractable manganese percentage, and bulk densities associated with the paleosols of each loess member. Clay differences were less distinctive, however, and did not accurately identify the paleosols.

The revised ^{14}C dates given by Goh *et al.* (1978) make the correlation of the various paleosols very uncertain. Firstly, the weathering data of Runge *et al.* (1974), give no indication that any of the paleosols represent the weathering of a full interglacial period (Oturian) and the paleosols are said to have features which closely resemble those found in the present soil which is the product of about 12 000 years of weathering. Markedly different features could be expected if an interglacially weathered paleosol were present as the Oturian was thought to span about 20 000 yrs (Figure 6). Secondly, the influx of volcanic glass in Loess 3 of the Dashing Rocks section as reported by Raeside (1964), together with its absence in underlying Loess 4 suggests the possibility that Loess 3 could date from about 20 000 yrs, the glass representing the appearance of Kawakawa Tephra (Vucetich and Kohn, 1973; Vucetich and Howorth, 1976; Campbell, 1979) which is now known to have fallen over a wide part of the South Island (Campbell, in prep.). If the ^{14}C dates are correct then it would seem that some younger loess depositional events may be missing in the Timaru downlands bore-hole section.

PALEOSOLS IN NORTH ISLAND LOESS DEPOSITS

Loess deposits and associated paleosols of North Island regions were outlined by Cowie and Milne (1973) and Milne (1973a,b). During glacial periods the cold climate produced large amounts of periglacial debris in North Island hill and mountain areas due to extensive frost shattering. The removal of this by rivers during the thaw periods resulted in marked river aggradation followed by river downcutting in the warmer interstadial or interglacial periods when the debris to water ratio in rivers changed. During the periods of heavy river loading, the aggrading surfaces provided much fine material as a result of the cold climate physical weathering and this was deposited as loess on the older terraces which were being progressively uplifted. The loesses and paleosol sequences in the North Island are thus closely associated with a terrace system of climatic and tectonic origin. The relationship between loess deposits, paleosols, and terraces is illustrated in Figure 6.

The paleosols that occurred in the loess beds were not defined with clearly identifiable A, B, or C, horizons but were recognized mainly by two morphological features, grey veining and oxide nodules. The grey veining, by analogy with the present day soils, is generally found in the lower B and C horizons. The zone above the reticulate veins is occupied by a gleyed oxide-mottled clayey zone that was thought to have included the fossilized A and B horizons. The second diagnostic feature was a zone of oxide nodules which appeared to be confined to the base of a loess layer. In addition, penetration resistance was found to be low in the upper part of the paleosol and increased towards the base of the loess layer. The relative degrees of weathering occurring within each layer were considered to be subjective because of possible further weathering after burial. Like Raeside (1964), facies were recognized in which weathering had occurred under both higher and lower moisture regimes.

PALEOSOLS OF PERIGLACIAL SLOPE AND COLLUVIAL DEPOSITS

Paleosols of hill and mountain periglacial environments are comparatively widespread, although they tend to be very discontinuous in their occurrence. They have been described in a number of reports.

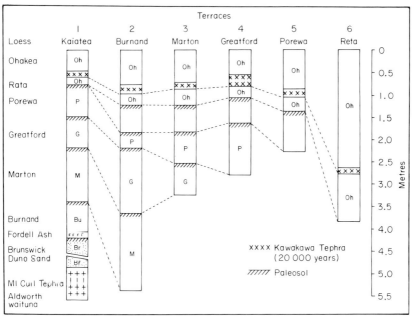

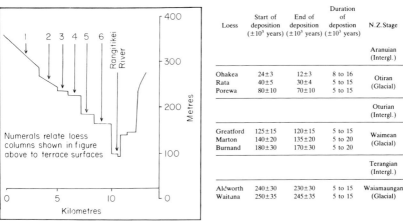

Figure 6. Correlation columns showing paleosols in loesses and tephra in Rangitikei Valley with loess ages and relationships to terraces (after Cowie and Milne, 1973).

Erosional and sedimentary features typical of periglacial conditions in hill and mountain lands were described by Stevens (1957) from the Wellington district. He identified surfaces which had been planed by erosion and in which fossil gullies had formed and subsequently been filled with predominantly angular frost shattered debris. The cyclic nature of the erosion, accumulation and weathering processes were recognized along with some palaeo-pedological features including iron pans of earlier weathering profiles. Raeside (1964) also noted the widespread occurrence in the Dunedin district of stratified slope deposits containing coarse angular fragments along with boulder-filled gullies, and he related them to cooler climatic periods. Near Mt. Cargill five distinct layers were identified, the lowermost part of each containing abundant coarse debris and the upper part a greater proportion of loess. Raeside considered that a soil had formed on each layer but that truncation of the paleosols had occurred.

Another study which recognized paleosols in periglacial slope deposits was by Wilde (1972). He described an age sequence of parent materials, and the soils which were formed on them, in Central Otago (west of Dunedin) and found that five cycles of erosion and deposition had occurred in association with late Pleistocene glacial events. Where the paleosols could be recognized they were distinguished by their colour, texture and compaction. Similarly, Quaternary deposits and surfaces on the Otago Peninsula were described by Leslie (1973) where loess, colluvium and slope deposits were interbedded with, and in places separated by, paleosols. Some of the paleosols were found to have been truncated by erosion or buried by slope drift.

In another investigation in Central Otago layers of angular frost-shattered periglacial slope detritus, which were separated by loess layers each with a paleosol, were described by Leamy and Burke (1973). The deposits, exposed in a fan, consisted of four layers of coarse debris which had been transported downslope after intense physical disintegration under periglacial conditions. Each has a loess layer above which was thought to have been derived during the late part of a cold stadial interval and prior to the commencement of soil formation during the warm interstadial period. The paleosols which were developed on the loess layers and the lowermost gravel were identified by soil morphological criteria, although the morphology tended to be variable through the area in which the paleosols occurred. Leamy and Burke (1973) also used results of soil chemical analyses, principally phosphorus fractionations, to help identify the individual paleosols, which they considered ranged in age from Otiran (glacial) for paleosols 1–4 to Oturian (interglacial) for the lowermost paleosol.

A series of geomorphic surfaces with associated deposits of loess and colluvium, together with related paleosols, were described by Laffan and Cutler (1977) from the Wither Hills in Marlborough, north-eastern South Island (Figure 2). Seven ground surfaces, designated K_1 to K_7 were recognized largely by lithological differences, while the paleosols (Figure 7) were identified by soil structure differences and

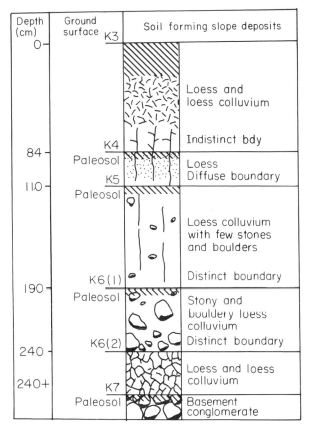

Figure 7. Diagramatic representation of slope deposits, ground surfaces and associated paleosols of the Wither type section, north eastern South Island. K_1 and K_2 ground surfaces only occur on the lower slopes. The K_3 ground surface is mostly the present soil forming surface (after Laffan and Cutler, 1977.)

evidence of faunal activities. The paleosol on the K_6 surface was thought to represent a truncated strongly weathered soil (Oturian paleosol) judged by the presence of moderately weathered stones and thick continuous cutans on ped surfaces.

Paleosols occurring in some sections in the Wellington region were described by Leamy (1964). These were Pleistocene and Holocene deposits of windblown sand, volcanic ash, and loess on greywacke which in places were overlain by periglacial slope deposits. As well as using soil morphological criteria to separate up to four paleosols, he used clay mineralogical differences to help identify and distinguish individual paleosols. There appeared to be no correlation among the three paleosol sequences which were considered to span both glacial and interglacial conditions.

Recent observations by the author in northern South Island districts suggests that late Quaternary periglacial activity has been widespread, but that its influence has been variable across the landscape so that the preservation of paleosols has been quite patchy. Many hill and mountain slopes have been subjected to intensive cold-climate gully dissection followed by infilling with angular fossil scree deposits (Pearce et al. 1983). On the uppermost slopes the regolith and soils are shallow and are of late Otiran to Holocene age. On mid slopes, the regolith may be either deep fossil scree deposits or shallow scree and slope drift (occasionally interbedded with 20 000 yr old Kawakawa tephra (Campbell 1979)) overlying weathered bedrock or interstadially weathered paleosol remnants (Figure 8). Lower slopes may have deep, bedded layers of colluvial deposits with paleosol remnants which possibly date from Terangian or earlier time. The paleosols are seldom clearly defined except on gently sloping surfaces such as fans, and generally comprise weathered regolith material from which the original soil has been substantially removed.

Within the hill and mountain areas of New Zealand therefore, repeated erosional and weathering events have given rise to the occurrence of quite widespread late Last Glaciation slope deposits which are occasionally interbedded with loess, but which seldom have complete paleosols owing to truncation by erosion. The paleosol remnants are difficult to identify and trace as entities because of their sporadic occurrence.

PALEOSOLS OF VOLCANIC DEPOSITS

Deposits of volcanic materials, principally tephra, cover either partially or completely a considerable portion of the North Island (Figure 2), and they have been erupted intermittently throughout the last 0.5 million years, chiefly from eight volcanic centres (Figure 9).

Unlike the South Island where the paleosols are primarily related to fluctuations in late Quaternary climate and provide a key to the establishment of late Quaternary chronology, the paleosols within the tephra deposits are less clearly linked to Quaternary climatic events. They are nevertheless of considerable value in reconstructing the tephra sequences because the bedding within many deposits, especially the older formations, often cannot be distinguished by variations in colour, grain size, mineral content or consistence, and may only be recognized through the weathering and the paleosols that have formed. Ward (1967) observed that weathering may be so advanced that the original ash has been transformed to soil, and in such cases, when the usual basis for recognizing stratigraphic units is not available, the

Figure 8. Periglacial scree, loess and tephra forming truncated paleosols in slope deposits near Nelson. From the bottom up the interpretation is as follows: The coarsely jointed material (5) is a truncated interstadially weathered paleosol from loess. The angular debris above (4) represents cold climate accumulation (Kumara 3_1). The material above (2) is loess with weak paleosol features and it encloses 20 000 yr old Kawakawa Tephra (3). At the top is late glacial (Kumara 3_2) slope detritus with a Holocene soil.

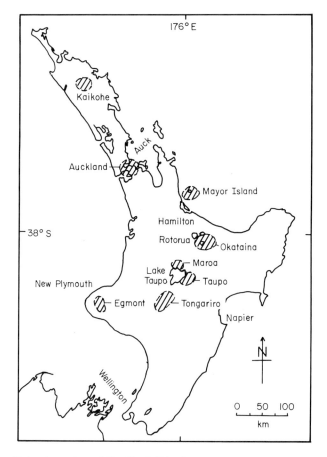

Figure 9. Volcanic centres of the North Island.

paleosols which record periods of non-deposition can be used as an acceptable criteria for mapping tephra units.

The principal tephra formations that are recognized and which to a considerable extent have been mapped and defined through paleosol identification are given in Table 2. This is derived from the large amount of work that has been published over the last two decades and which is briefly summarized below.

OLDER PALEOSOLS FROM TEPHRA

Some of the oldest paleosols recognized are in tephras of the Kauroa Ash Formation (Table 2) (Ward 1967) which is a multiple-bedded, early tephra that occurs widely in the North Island but more particularly in the South Auckland to Hamilton region. Ward recognized 17 mainly rhyolitic ash beds, the upper member of which had a well developed paleosol with clay veins and manganese nodules similar to those found in the modern soil where the tephra is exposed at the surface. Paleotopsoils were also recognized by their structure along with the presence of soil plant and organism impressions. The beds and paleosols were generally characterized by strong weathering, red to yellowish colours and high clay contents, particularly halloysite. The beds above the Kauroa Ash Formation were defined by Ward (1967) as the Hamilton Ash Formation which, like the beds below, were strongly weathered and in places showed distinctive paleosol development in some members. Subsequent work has shown that the upper paleosols are formed in much younger tephras from the Maroa-Taupo Volcanic Centre. Also, beds which Ward considered were a single stratigraphic unit but with differing paleosol expressions were later thought to be separate tephras (Pain 1975). Vucetich et al. (1978) identified the lowermost member of the Hamilton Ash Formation as a separate tephra, called the Ohinewai Ash Member, which although well weathered was contrasted sharply by the underlying strongly weathered paleosol of the Kauroa Ash Formation.

Older paleosols are also known from the early tephras from the Egmont Volcanic Centre (Figure 9), these having been mapped and described by Neall (1972) as the New Plymouth Ashes and Buried Soils. The paleosols are formed in andesitic tephra deposits and, like those of the Kauroa and Hamilton Ash Formations, individual tephras are not distinctive and are only identified by the paleosol developed at the surface of each layer (Figure 10). The paleosols were most frequently recognized by their well developed blocky structure, as well as the darker colours which are best developed beneath lapilli horizons. The clearest paleosols also show organic matter accumulations and increased abundance of root channels but the weakly developed paleosols have only weakly developed blocky structures. Neall contrasted the New Plymouth Ashes with their paleosols, the former being pale brown to yellow when dry, appearing to be little weathered and resistant to erosion, being massive with weak prismatic structure, and having sharp lower boundaries. The buried soils on the other hand have pinkish grey colours, well developed prismatic structure (50–100 cm deep prisms) and break to coarse blocky or fine subangular blocky structure. They generally pass gradationally into underlying unweathered ash. The buried soil at the top of the New Plymouth Ashes is believed to represent a considerable time interval as the paleosols contain substantial quantities fo halloysite, whereas above the paleosol, the ashes and paleosols are mainly allophanic.

226 CHAPTER 7

Table 2. Principal tephra formations and radiocarbon ages from deposits of the North Island volcanic centres (compiled from various sources).

Years BP	OKATAINA CENTRE	MAROA-TAUPO CENTRE	TONGARIRO CENTRE	EGMONT CENTRE	OTHER or UNKNOWN CENTRES	Years BP
250	Tarawera Formation (64 BP) (3 members)			Tahurangi Formation (250 BP)		250
500				Burrell Formation (325 BP (4 members)) Newall Formation (450 BP (4 members))	Rangitoto Ash (750 BP) (Auckland Centre)	500
750	Kaharoa Ash (c. 650 BP)					750
1000						1000
1500		Taupo Pumice Formation (1850 BP) (7 members)	Ngauruhoe Tephra Formation (0-1819 BP)			1500
2000		Mapara Pumice (2270 BP)	Mangatawai Tephra Formation (2500 BP)			2000
2500		Whakaipo Tephra (2800 BP)				2500
3000	Rotokawau Ash (no date)	Waimihia Formation (3150 BP) (2 members)	Papakai Tephra (3420 BP)	Inglewood Tephra (c. 3000 BP)		3000
4000						4000
5000	Whakatane Ash (5180 BP)	Hinemaiaia Tephra Formation (4650 BP) Moutere Tephra Formation (5370 BP)		Korito Tephra (no date)		5000
6000				Oakura Tephra (6900 BP)	Tuhua Tephra (6200 BP)	6000

Years BP				
7000	Mamaku Ash (7050 BP) Rotoma Ash (7330 BP)			
8000		Opepe Tephra (8850 BP)	Stent Ash	
9000		Poronui Tephra (9780 BP)	Okato Tephra (no date) (2 members) Mangamate Tephra (9700 BP) (5 members) Okupata Tephra (9790 BP)	
10000	Waiohau Ash (11 250 BP) Rotorua Ash (13 450 BP) Rerewhakaaitu Ash (14 700 BP)	Karipiti Ash (no date) Puketarata Ash (no date)	Rotoaira Lapilli (13 800 BP)	
15000		Oruanui Formation = Kawakawa Tephra (19 850 BP) (3 members) (20 500 BP)	Saunders Ash (16 000 BP) Carrington Tephras	
20000	Okareka Ash (c.17 000 BP) Te Rere Ash (no date)	Poihipi Tephra Okaia Tephra	Koru Tephra (no date) (2 members)	
30000	Mangaone Lapilli (10 formations) (30 100 BP)		Pukeiti Tephra (no date)	
40000	Rotoehu Ash (41 700 BP) (3 members)		Well Tephra (no date)	
			New Plymouth Ashes (> 70 000 BP)	Hamilton Ash Formation (120–210 000 yrs BP) Ohinewai Tephra (150 000 yrs BP) = Kauroa Ash Formation

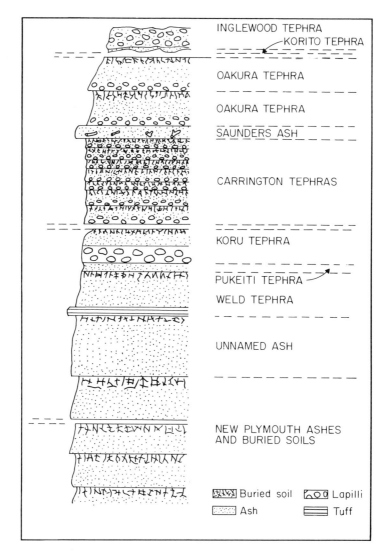

Figure 10. Weathering profile of composite tephra formations at four localities in Taranaki, western North Island (not all seen together), (after Neall, 1972).

Younger Paleosols from Tephra

Paleosols are frequently found in the younger tephra deposits from each of the volcanic centres and the main difference compared with the older paleosols is that they are much less weathered. The stratigraphy and chronology of late Pleistocene (15 000−42 000 yrs BP) rhyolitic ash beds from the Okataina and Maroa-Taupo Volcanic Centres were described by Vucetich and Pullar (1969) who identified and mapped the tephra formations and markers throughout the Central North Island region. They combined lithologic characteristics of individual members with their weathering characteristics, including paleosol formation, to distinguish and map the various tephra. Where the thickness of an ash was less than optimum (60—120 cm), the colour, consistence and texture were used for correlation of paleosols rather than the distinctive lithologic characteristics such as pinkish or greyish colours, massive tuff-like form or presence of pumice blocks, lapilli or chalazoidites. Because of the interlayering of tephra from various sources, careful tracing of paleosols and tephra layers was required to establish the tephra successions.

Paleosols recognized in Okareka Ash Formation were mainly weakly weathered. Likewise the underlying Te Rere Ash Formation had a paleosol recognized only by a darker surface colour as paleostructure was lacking. The Oruanui Formation (Kawakawa Tephra) contained little by way of recognizable paleosols but the underlying Mangaone Lapilli, which was multiple bedded, had most breaks marked by paleosols. This formation was later subdivided by Howorth (1975) and Vucetich and Howorth (1976) who identified numerous paleosols which were predominantly pale coloured, weakly to moderately weathered, and at times weathered and cemented. Textural differences over short distances within ash layers and the paleosols, along with weathering differences due to factors such as degree of wetness, frequently created difficulties in interpretation of the sequences. The oldest unit recognized was the Rotoehu Ash Member (Vucetich and Pullar, 1969; Vucetich and Howorth, 1976), which had a well developed paleosol at its surface. Vucetich and Pullar considered that the whole sequence provided a good record of eruptive events in the region with the paleosols also recording the onset of late Quaternary climatic change. Pre-Rotoehu paleosols with their brown and dark brown weathering features contrast with the weakly weathered paleosols in Rotoehu Ash and Mangaone Lapilli and the very weakly weathered paleosols in Oruanui and Okareka Ashes, while the post-Okareka ashes differ again in having more strongly weathered paleosols (Figure 11).

Vucetich (1968) attributed these weathering changes to a decline in weathering with the onset of Last Glaciation cooling with a subsequent increase in the rate of weathering at the end of the Last Glaciation. Weathering of paleosols in Rotoehu Ash was also studied by Birrell *et al.* (1977) who found that the weathering was distinctly greater further from the source in northern areas. Paleosol differences included finer textures, slightly stronger colours and increased aluminium and

Figure 11. Paleosols in younger (named) rhyolitic tephras from Te Puke region central North Island. Weakly weathered paleosols underlying Kawakawa and Mangaone tephras can be seen.

phosphate retention contents. The differences were thought to be due either to the influence of climate differences on weathering, or to the weathering rates caused by differential burial by the next succeeding ash layer.

Younger paleosols from andesitic tephra from the Egmont Volcanic Centre were identified by Neall (1972) when he defined and mapped tephra formations of the Taranaki district. The ashes vary greatly in texture and are strongly shower bedded (Figure 10). The frequency of eruptions was estimated to be of shorter duration than for the tephra of the Okatina-Taupo Centres, and as most occurred during the Otira Glaciation, the paleosols are for the most part weakly developed.

Holocene Paleosols from Tephra

Paleosols in tephras in the Taupo Region that were derived from the Okataina, Maroa and Taupo Volcanic Centres were discussed by Vucetich and Pullar (1973) and Pullar *et al.* (1973). This sequence of tephras which spans the period of 15 000 yrs BP (Rerewhakaaitu Ash (Figure 11)) to the present (Ngauruhoe Ash) is dominantly rhyolitic and contains many beds some of which are air-fall deposits and others of which are flow-breccias. Again many of the beds were not lithologically distinctive and were therefore defined by the paleosol layers on which they rest (Figure 12). Vucetich and Pullar noted that inevitably each tephra formation thinned away from its source, losing its identity as a discrete layer to become part of an over-thick paleosol. Interfingering of thin andesitic tephras within the Taupo sourced tephra formations has led to the development of paleosols with markedly browner colours because of associated weathering differences.

Because of the frequency of eruptions, most of the paleosols are thin and weakly developed and their colours vary greatly depending on the length of soil formation before burial and the extent of the additions of andesitic ashes. In some of the younger paleosols colours of the upper horizons more closely resemble those of an A horizon, while the presence in places of bleached underlying horizons suggests that the paleosols may have been weakly developed podzols. Further south, however, Froggatt (1981) examined the Hinemaiaia Tephra Formation and found that paleosols separating tephra units were deeper and quite distinctive.

Vucetich and Pullar (1973) considered that more effective use of paleosols as markers would require them to be characterized in terms of their mineralogy, as it was evident that many of the paleosols were time-transgressive and in places were composite soils influenced by the accumulation of andesitic or rhyolitic ashes.

The Holocene paleosols of tephra from the Egmont Volcanic Centre (Neall 1972), have almost complete paleosols where they are formed in some of the youngest tephras, as insufficient time has elapsed for significant paleosol modification to have taken place. A paleosol in the Korito Tephra appears to be one of the most distinctive in the sequence described by Neall, but overall, paleosols from the Egmont tephras are not particularly well expressed. Paleosols in tephra erupted from nine vents in the Tongariro Volcanic Centre (Figure 9;

Figure 12. Thin weakly weathered Holocene paleosols in rhyolitic and andesitic tephra, Quaternary tephric loess and rhyolitic tephra, from east Taupo region, central North Island.

Table 2) were described by Topping (1973). He noted that the andesitic tephras often do not have well developed paleosols, as colour, consistence, and structure differentiation within soils developed from andesitic volcanic ash is often weakly expressed. Where the ash accumulation has been semi-continuous few paleosols are formed, but where eruptive lobes have been separated by a distinct time interval paleosols are better expressed and the features of vertical and horizontal cracking, darker colours, and increased clay contents help to define the paleosols. Interbedded rhyolitic tephras, which frequently occur, help to define the tephra formations as well as the paleosols. As was the case for paleosols of the Egmont Volcanic Centre eruptives the younger paleosols in the more recent tephras are most clearly expressed, often with little pedological alteration, while in the older paleosols (for example, 3–5000 yrs) soil features, including colour and structure, are lost.

PALEOSOL IDENTIFICATION AND PALEOSOL STRATIGRAPHY

It has been shown in the foregoing how paleosols have been used to help construct the New Zealand Quaternary glacial chronology and to establish sequences of Quaternary and Holocene events. Limitations are, however, imposed by lateral changes in paleosols over short distances because of weathering differences in response to parent material, climate, soil moisture or drainage, and time variations. The extent of diagenetic alteration after burial also creates doubt in many instances as to the value of some paleosols as stratigraphic entities. Within a particular sequence breaks or changes may readily be interpreted as paleosols, but the required degree of control for positive lateral recognition and correlation from region to region may often be insufficient, particularly where sections are widely scattered or incomplete. Paleosols can be expected to show lateral variation across a landscape similar to that seen in the modern soils which, as shown by soil mapping, is very considerable. There has been interest in improving the recognition of paleosols, however, and results have been reported of a number of studies concerning aspects of their weathering characteristics, ageing, and correlation.

Weathering and Ageing of Paleosols

Several studies of the mineralogical and micromorphological attributes of paleosols have been carried out and they illustrate how the identification of paleosols can be enhanced by using such techniques. For example, Dalrymple (1967) studied the micromorphology of some paleosols sequences in the South Auckland region and showed that there were distinct differences in the amount and character of colloidal matter, both vertically in profiles and laterally in sections, but the paleo-B horizons could readily be distinguished from one another in each case. Along with the micromorphological identifications mineralogical characterizations were considered to be very useful where separations based on field morphological expressions could not be made. The weathering of paleosols in late Pleistocene to Holocene rhyolitic and andesitic tephras from central North Island was also investigated in a number of studies. Birrell et al. (1971) studied intensely dark coloured horizons of recent paleosols formed in rhyolitic materials and found that they were characterized by high organic carbon contents, very high C/N ratios, a preponderance of humic acids in the organic constitutents, and very low methoxyl-carbon contents. These properties were considered most likely to have been inherited from the previous plant cover which was bracken-fern (*Pteridium aquilinum* var. *esculentum*). Birrell and Pullar (1973) also investigated the rate of soil formation in a number of paleosols from tephras in the central North Island region using chemical methods to evaluate weathering. They found that organic C levels were generally low in paleosols but were somewhat higher where andesitic material was present in the tephra. Clay contents in Holocene tephras were largely allophane while the late Pleistocene tephras had somewhat higher clay contents with a predominance of halloysite and occasionally micaceous clays. Aluminium values, extracted by Tamm's reagent, were found to be a good indicator of weathering status with the higher values indicating exposure to weathering for longer periods of time. Paleosols in the late Pleistocene rhyolitic tephras generally had lower Al values than those of the Holocene tephras.

Another study of the weathering of paleosols was carried out by Childs (1975) who determined the element content (Al, Si, P, K, Ca, Mn, Te, Fe, and Zr) in the soils and paleosols of two loess columns. The paleosols were from somewhat differing weathering environments, one being from the Timaru region in the South Island and the other from near Table Flat (east of Wanganui) in the North Island. Differences in mineral contents occurred between the two regions, but within each column various elements such as P, K, Ca, Mg, and possibly Ti and Fe at Table Flat, appeared to confirm, by their variation down the columns, the existence of paleosols. None, however, appeared to show a difference distinctive enough to be interpreted as an interglacial paleosol. This analytical approach would appear to be very useful in establishing the paleosols and their significance within a column or section but would require strongly diagnostic mineralogical characteristics for lateral recognition of individual paleosols.

A similar study was carried out by Runge *et al.* (1974) who examined four paleosols within a loess sequence at Timaru and characterized them on the basis of various chemical and physical properties. They showed that there were systematic differences in the phosphorus fractions (Figure 13), extractable manganese percentage, and bulk density associated with the paleosol of each loess member, but clay differences were not distinctive and did not accurately identify the paleosols.

Another method of distinguishing paleosols was outlined by Goh (1972) using amino acid levels in the profiles as a paleosol indicator. He examined soil profiles and the paleosols underlying them of three

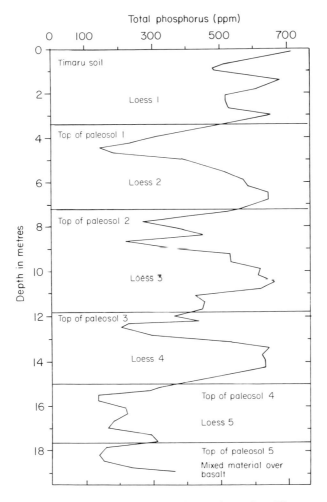

Figure 13. Distribution of total phosphorus in loess columns from Timaru region (after Runge et al., 1974).

widely differing soils; one from old ashes and tephric loess, another from late Last Glaciation quartzo-feldspathic loess, and another from Holocene alluvium. He found that the paleosols had been enriched secondarily with bases, including exchangeable Na, K, and Mg, which had moved down the older profiles by percolating solutions. Pedotranslocation of exchangeable Ca was apparently absent. Goh also found that mineralogical data contributed little to the characterization of the paleosols studied. Amino acid-nitrogen levels decreased with depth in the present day profiles and a decrease also occured in the paleosols. In addition, there was a marked increase in amino acid levels in the top horizon of each paleosol relative to the overlying material. The actual levels of amino acids within the paleosols were found to be quite similar to those present in the subsoils of the present day soils and were considered to depend on factors such as the age of the buried soil, the amount present before burial as well as the rates of degradation, re-synthesis and pedotranslocation processes.

CORRELATION OF PALEOSOLS

There have been a number of contributions to the question of correlation of paleosols and consideration of their spatial relationships and relative significance. Several mineralogical studies, for example, have illustrated how the various beds and their paleosols can be characterized and traced, using their sand and clay mineralogical properties together with their general morphological features, while others have considered the broad stratigraphic relationships of paleosols.

The relationships between thinly bedded tephra and paleosols in a line of comparatively shallow sections which formed the parent material of a soil known as Tirau silt loam were described by Pullar and Birrell (1973). This work highlights the problems of paleosol identification in thinly bedded deposits. The section line over a distance of about 40 km was in the central North Island area where multiple tephra and tephric loess deposits occurred. The tephra and loess beds were identified mainly on lithologic criteria and they showed that these sections contain tephras ranging from the recent Taupo Pumice down to Rotoehu Ash or older (Table 2). Although some paleosols were present, there was little correlation between them. Individual paleosols were not recognized as such on all of the tephras as the accessions of ash had been frequent and thin. As a result, composite rather than discrete soils and paleosols have formed. Clay minerals, clay, organic carbon and Tamm's extractable Fe and Al were used to help distinguish the various contributing tephras while the presence of halloysite was considered to indicate the occurrence of Okareka Ash.

A similar type of investigation in the Hamilton district was carried out by Lowe (1981) who examined the parent material relationships of soils formed from tephric cover beds at a number of reference sites. Underlying the cover beds the Hamilton Ash paleosol was found to be mineralogically and physically distinctive from the cover bed materials above, which were composite distal tephras. Again the presence of individual tephras could be recognized through mineralogical criteria, and while distinctive paleosols in the cover deposits were not always evident due to mixing or homogenizing processes, the presence of weathering breaks was revealed by the mineralogical datums present within the profiles. Another investigation reported was by Hogg and McCraw (1983) who traced paleosols on Rotoehu Ash northward in the Tauranga-Coromandel region and, by using mineralogical characterization of the various younger tephra, showed how progressive mixing of the Rotoehu Ash had occurred in the distal areas One of the younger ashes (from Mayor Island) was especially useful because of its distinctive mineralogy.

Each of these studies mentioned above demonstrates the problems and potential pitfalls associated with paleosol recognition and paleosol stratigraphy. Where the tephra are thinly bedded and the paleosol is not within a distinctive lithological unit, superficial features are not suitable for positive paleosol identification and correlation as they are likely to be time transgressive. The value of many of the paleosols as time stratigraphic units is therefore restricted although they may still be recognized as soil lithostratigraphic units with individual tephra defining separate paleosol facies.

As outlined earlier, the paleosols identified in the loess deposits are more likely to be time stratigraphic units because the deposition of the loess layers and the development of paleosols on them are more clearly linked with Quaternary events such as periods of warming and cooling. Notwithstanding this, however, local tectonic, climatic or geomorphic influences may cause loess to be deposited independently of warming or cooling trends, as is evidenced by some loess deposits which are accumulating at the present time. Furthermore, Ives (1973) considered that pedosphere stripping was an important part of loess accumulation and paleosol formation in the Timaru region, but again this need not be restricted to any particular part of the glacial-interglacial time-scale because the duration of cooling as against warming probably also varied from place to place depending on the local severity of glaciation. The paleosols which occur in loess may therefore also be to some extent time-transgressive. The unclear relationships between some of the Timaru loesses and paleosol sequences (Raeside 1964; Tonkin et al. 1974) illustrate the problems of paleosol correlation in loesses, as even with ^{14}C dates (Goh et al. 1978), correlation between sections is tentative.

An approach towards assisting paleosol identification and correlation was outlined by Leamy and Burke (1973) and Leamy (1975), and they suggested that for each horizon, the comparative intensity of morphological development in a vertical sequence of paleosols could be assessed against the modern soil and assigned a value to give a pedomorphic index. When taken in conjunction with other data such as phosphorus values, mineralogy, and amino acid content, the concept could be very useful for recognition and correlation of paleosols where there is evidence that the processes operating in modern soils are similar to those in the paleosols, and also where lateral variation in the paleosol is slight.

Stratigraphic relationships of paleosols were discussed by Leamy *et al.* (1973) who suggested that consistently recognizable paleosols could be defined as soil stratigraphic units which would demarcate the boundaries of periods of warming and succeeding warmth. They defined a soil stratigraphic unit as a paleosol which, by its intrinsic pedological properties and stratigraphic position, could be consistently recognized and mapped. Another criterion suggested was that it should represent only one period of time (i.e. the weathering of a single interglacial of interstadial period); buried relict soils for example would be unsatisfactory. A number of paleosols were formally named and it was proposed that differing facies (for example, as a result of lateral variation in soil forming factors) be termed variants. It was also suggested that the paleosol be interpreted as a soil stratigraphic unit formed in association with a particular warm climatic period. In Central Otago for example, the Pisa Paleosol is an 'Oturian soil'.

It is likely that with improved dating and mineralogical characterization or with other forms of identification, the paleosols of loess deposits will be found, like those of the North Island tephras, to be much more complex than is presently thought. Changes in the naming of the time unit should not, however, affect the naming of the paleosol.

SUMMARY

In this review, the principal kinds of paleosols which have been described in New Zealand have been briefly summarized. In the South Island they are found chiefly within loess and periglacial slope deposits and are believed to be closely related to late Quaternary cooling and warming cycles. They are not highly distinctive, nor are they well dated and for the most part they are only tentatively correlated from region to region as clear stratigraphic control is lacking. In the North Island on the other hand, most of the paleosols are in tephra units that are much more distinctive stratigraphically and the paleosols are very useful as they aid the recognition of the tephra units. The considerable number of tephra erupted through the late Quaternary provide a remarkable sequence of well-dated paleosols. Because many of the time periods of weathering are short, the paleosols in tephra are less useful for Quaternary subdivision. In distal areas where tephra are thin, paleosol recognition is more difficult and they are likely to be time-transgressive. Mineralogical studies are showing their soil-lithologic and stratigraphic relationships. Improved recognition using analytical techniques and better dating and other forms of signature recognition, will enhance the usefulness of paleosols as stratigraphic units and improve our knowledge of the Quaternary.

REFERENCES

Birrell, K.S. and Pullar, W.A. 1973. Weathering of paleosols of Holocene and Late Pleistocene tephras in central North Island, New Zealand. *N. Z. J. Geol. Geophys.* **16**, 687–702.

Birrell, K.S. Pullar, W.A. and Heine, J.C. 1971. Pedological chemical and physical properties of the organic horizons of paleosols underlying Tarawera Formation. *N. Z. J. Sci.* **14**, 187–218.

Birrell, K.S., Pullar, W.A. and Searle, P.L. 1977. Weathering of Rotoehu Ash in the Bay of Plenty district. *N. Z. J. Sci.* **20**, 303–10.

Bruce, J.G. 1973a. Loessial deposits in southern South Island, with a definition of Stewarts Claim Formation. *N. Z. J. Geol. Geophys.* **16**, 533–48.

Bruce, J.G. 1973b. A time-stratigraphic sequence of loess deposits on near coastal surfaces in the Balclutha district. *N. Z. J. Geol. Geophys.* **16**, 549–56.

Bruce, J.G. Ives, D.W. and Leamy M.L. 1973. Maps and sections showing the distribution and stratigraphy of South Island loess deposits, New Zealand 1:1 000 000 *N.Z. Soil Survey Report* 7.

Campbell, I.B. 1979. Occurrence of Kawakawa Tephra near Nelson—Note. *N. Z. J. Sci.* **22**, 133–6.

Childs, C.W. 1975. Distribution of elements in two New Zealand Quaternary loess columns, in (R.P. Suggate, M.M. Cresswell), *R. Soc. N. Z. Bull.* (ed.) *Quaternary Studies* (INQUA), **13**, 35–44.

Cowie, J.D. and Milne, J.D.G. 1973. Maps and sections showing the distribution and stratigraphy of North Island loess and associated deposits, New Zealand. *1:1 000 000 N. Z. Soil Survey Report* 6.

Dalrymple, J.B. 1967. A study of paleosols in volcanic ash-fall deposits from northern North Island, New Zealand, and the evaluation of soil micromorphology for establishing their stratigraphic correlation. *Quaternary Soils, Proceedings of the 7th International Congress on Quaternary Research* **9**, 104–22.

Froggatt, P.C. 1981. Motutere Tephra Formation and redefinition of Hinemaiaia Tephra Formation, Taupo Volcanic Centre, New Zealand. *N. Z. J. Geol. Geophys.* **24**, 99–106.

Gibbs, H.S. 1971. Nature of paleosols in New Zealand and their classification, in *Paleopedology*, Dan H. Yaalon (ed.). Israel Universities Press, Jerusalem. 229–44.

Grange, L.I. 1931. Volcanic-ash showers. *N. Z. J. Sci. Technol.* **12**, 228–40.

Grange, L.I., Taylor, N.H., Sutherland, C.F., Dixon, J.K., Hodgson, L. and Seelye, F.T. 1939. Soils and agriculture of part of Waipa County. *N. Z. Dep. Sci. Indust. Res. Bull.* **76**, 32–63.

Griffiths, E. 1973. Loess of Banks Peninsula. *N. Z. J. Geol. Geophys.* **16**, 611–22.

Goh, K.M. 1972. Amino acid levels as indicators of paleosols in New Zealand soil profiles. *Geoderma* **7**, 33–47.

Goh, K.M. Tonkin, P.J. and Rafter, A.T. 1978. Implications of improved radiocarbon dates of Timaru peats on Quaternary loess stratigraphy. *N. Z. J. Geol. Geophys.* **21**, 463–6.

Hogg, A.G. and McCraw, J.D. 1983. Late Quaternary tephras of Coromandel Peninsula, North Island, New Zealand. A mixed peralkaline and calcalkaline tephra sequence. *N. Z. J. Geol. Geophys.* **26**, 163–87.

Howorth, R. 1975. New formations of late Pleistocene tephras from the Okataina Volcanic Centre, New Zaeland. *N. Z. J. Geol. Geophys.* **18**, 683–713.

Ives, D.W. 1973. Nature and distribution of loess in Canterbury. *N. Z. J. Geol. Geophys.* **16**, 587–610.

Laffan, M.D. and Cutler, E.J.B. 1977. Landscapes, soils and erosion of a small catchment in the Wither Hills, Marlborough. 1. Landscape periodicity slope deposits and soil pattern. *N. Z. J. Sci.* **20**, 37–48.

Leamy, M.L. 1964. Some fossil soils in Pleistocene and Holocene deposits near Wellington. *N. Z. Soil Bureau Report*, 5/1964.

Leamy, M.L. 1975. Paleosol identification and soil stratigraphy in South Island, New Zealand. *Geoderma* **13**, 53–60.

Leamy, M.L. and Burke, A.S. 1973. Identification and significance of paleosols in cover deposits in Central Otago. *N. Z. J. Geol. Geophys.* **16**, 611–22.

Leamy, M.L. Milne, J.D.G. Pullar, W.A. and Bruce, J.G. 1973. Paleopedology and soil stratigraphy in the New Zealand Quaternary succession. *N. Z. J. Geol. Geophys.* **16**, 723–44.

Leslie, D.M. 1973. Relationship between soils and regolith in a volcanic landscape on Otago Peninsula. *N. Z. J. Geol. Geophys.* **16**, 567–74.

Lowe, D.J. 1981. *Origin and Composite Nature of Late Quaternary Air-Fall Deposits, Hamilton Basin, New Zealand.* Unpublished M.Sc. thesis, University of Waikato, Hamilton, New Zealand.

Milne, J.D.G. 1973a. Maps and sections of river terraces in the Rangitikei Basin, North Island, New Zealand. *N. Z. Soil Survey Report*, **4**.

Milne, J.D.G. 1973b. *Upper Quaternary Geology of Rangitikei Valley.* Unpublished Ph.D. thesis, Victoria University of Wellington, New Zealand.

Neall, V.E. 1972. Tephrochronology and tephrostratigraphy of western Taranaki, New Zealand. *N. Z. J. Geol. Geophys.* **15**, 507–57.

Pain, C.F. 1975. Some tephra deposits in the south-west Waikato area, North Island, New Zealand. *N. Z. J. Geol. Geophys.* **18**, 541–50.

Pearce, A.J. Phillips, C.J. and Campbell, I.B. 1983. Regolith profiles on slopes underlain by Moutere Gravel Formation, Big Bush State Forest. Hydrologic and geomorphic implications. *N. Z. J. Geol. Geophys.* **26**, 57–70.

Pohlen, I.J., Harris, C.S., Gibbs, H.S. and Raeside, J.D. 1947. Soils and some related agricultural aspects of mid Hawke's Bay. *N. Z. Dep. Sci. Ind. Res. Bull.* **94**.

Pullar, W.A. and Birrell, K.S. 1973. Parent materials of Tirau silt loam. *N.Z. J. Geol. Geophys.* **16**, 677–86.

Pullar, W.A., Birrell, K.A. and Heine, J.C. 1973. Named tephras and tephra formations occurring in the central North Island, with notes on soils and buried paleosols. *N. Z. J. Geol. Geophys.* **16**, 497–518.

Raeside, J.D. 1964. Loess deposits of the South Island, New Zealand and the soils formed on them. *N. Z. J. Geol. Geophys.* **7**, 811–38.

Runge, E.C.A., Walker, T.W. and Howarth, D.T. 1974. A study of late Pleistocene loess deposits, South Canterbury, New Zealand. Part 1. Forms and amounts of phosphorus compared with other techniques for identifying paleosols. *Quaternary Research* **4**, 76–84.

Stevens, G.R. 1957. Solifluxion phenomena in the Lower Hutt area. *N. Z. J. Sci. Technol.* **38**, 279–96.

Suggate, R.P. 1965. Late Pleistocene geology of the northern part of the South Island, New Zealand. *N. Z. Geol. Surv. Bull.* **77**.

Suggate, R.P. 1978. (The Late Mobile phase: Quaternary) Introduction, in *The Geology of New Zealand.* Vol. 2, Suggate, R.P., Stevens, G.R. and Te Punga, M.T. (eds). Government Printer, Wellington, New Zealand. 542–4.

Taylor, N.H. 1933. Soil processes in volcanic ash-beds. *N. Z. J. Sci. Technol.* **14**, 193–202.

Tonkin, P.J. Runge, E.C.A. and Ives, D. 1974. A study of late Pleistocene loess deposits, South Canterbury, New Zealand. Part II. Paleosols and their stratigraphic implications. *Quaternary Research* **4**, 217–31.

Topping, W.W. 1973. Tephrostratigraphy and chronology of the late Quaternary eruptives from Tongariro Centre, New Zealand. *N. Z. J. Geol. Geophys.* **16**, 397–423.

Vucetich, C.G. 1968. Soil-age relationships for New Zealand based on tephrachronology. *Transactions of the 9th International Congress of Soil Science, Adelaide* III, 121–30.

Vucetich, C.G. Birrell, K.S. and Pullar, W.A. 1978. Ohinewai Tephra Formation, a c. 150 000 year old tephra marker in New Zealand. *N. Z. J. Geol. Geophys.* **21**, 50–71.

Vucetich, C.G. and Howorth, R. 1976. Proposed definition of the Kawakawa Tephra, the c. 20 000 years B.P. marker horizon in the New Zealand region *N. Z. J. Geol. Geophys.* **19**, 43–50.

Vucetich, C.G. and Howorth, R. 1976. Late Pleistocene tephrostratigraphy in the Taupo district, New Zealand. *N. Z. J. Geol. Geophys.* **19**, 51–69.

Vucetich, C.G. and Kohn, B.P. 1973. The stratigraphic significance of a dated late Pleistocene ash-bed, near Amberley, South Island, New Zealand. *Abstracts for 9th Congress of the International Union for Quaternary Research, Christchurch, New Zealand*, 390.

Vucetich, C.G. and Pullar, W.A. 1969. Stratigraphy and chronology of late Pleistocene volcanic ash beds in central North Island, New Zealand. *N. Z. J. Geol. Geophys.* **12**, 784–837.

Vucetich, C.G. and Pullar, W.A. 1973. Holocene tephra formations erupted in the Taupo area, and interbedded tephras from other sources. *N. Z. J. Geol. Geophys.* **16**, 745–80.

Ward, W.T. 1967. Volcanic ash beds of the lower Waikato Basin, North Island, New Zealand. *N. Z. J. Geol. Geophys.* **10**, 1109–35.

Wilde, R.H. 1972. An age sequence of parent materials and the soils formed from them in Central Otago, New Zealand. *N. Z. J. Sci.* **15**, 637–64.

Young, D.J. 1964. Stratigraphy and Petrography of North East Otago Loess. *N. Z. J. Geol. Geophys.* **7**, 839–63.

Young, D.J. 1967. Loess deposits of the West Coast of South Island, New Zealand. *N. Z. J. Geol. Geophys.* **10**, 647–58.

Chapter 8

PRE-FLANDRIAN QUATERNARY SOILS AND PEDOGENIC PROCESSES IN BRITAIN

R. A. KEMP*
Department of Geography, Birkbeck College, London, Great Britain
*now at, Department of Soil Science, Lincoln College, Canterbury, New Zealand

INTRODUCTION

The stratigraphical and palaeoenvironmental contributions that soils are capable of making to Quaternary studies have only been fully appreciated in Britain during the last decade. Prior to this period there were few guidelines available to non-pedologists as to how a buried soil might be recognized in the field. Even in the rare situations where one was identified, too many workers were content to derive vague stratigraphical and palaeoenvironmental conclusions from poorly detailed descriptions. A major advance in Britain has been the relatively recent appreciation that buried soils can only be applied correctly, and to their fullest extent, if they are properly identified and interpreted in terms of not only pedological features but also pedogenic processes.

In contrast to North America and the rest of Europe, it appears that very few interglacial or interstadial soils have been buried and remained intact in Britain (Catt, 1979). Consequently, relatively little is known about the types of soils, features and processes occurring during particular stages and sub-stages. In fact, most of the information available has been derived from non-buried soils which contain 'relict' pedological features formed prior to the Flandrian.

After briefly outlining the British Quaternary stratigraphy, this review restricts itself to a consideration of Quaternary buried soils and relict features in non-buried soils formed prior to the Flandrian stage in Britain. This includes details of pre-Flandrian pedogenic processes active in Britain during the Quaternary. An overall assessment of these processes is summarized within the conclusions.

QUATERNARY STRATIGRAPHY IN BRITAIN

The presently accepted stratigraphical table of the British Quaternary is summarized in Table 1. Eight major temperate and eight major cold stages are recognized in addition to a number of minor cool or brief warm phases (interstadials), of which all but one occur within the Devensian.

The stratigraphy is based upon climatic inferences derived largely from pollen assemblages and sediment properties. The presence of glacial ice over parts of Britain during each of the last three cold stages (Anglian, Wolstonian and Devensian) is inferred from the extensive till deposits that are frequently associated with glacifluvial and aeolian sediments. Pre-Anglian glaciations have been inferred less directly from deeply weathered till-like deposits (Catt, 1979) and from erratics in fluvial gravels (Hey, 1980; Green et al., 1982), although glacial or interglacial status is only normally assigned to cold and temperate stages younger than the Cromerian (West, 1977). However, the term 'interglacial' is often used synonymously with 'temperate', and some undatable buried soils and pedological features have been considered to be interglacial in age even though they may have originated prior to the Anglian (Bullock and Murphy, 1979; Clarke and Fisher, 1983).

All of the unconsolidated glacial, glacifluvial, fluvial, aeolian or gelifluction sediments deposited during the cold stages provided parent material from which soils could develop during more stable phases, particularly the intervening temperate stages. These soils were sometimes buried by subsequent deposits, although extensive erosion during the Devensian and earlier cold stages, particularly in the northern parts of the country, resulted in the removal of substantial thicknesses of not only earlier glacial sediments, but also the soils and deposits dating from the temperate stages. Consequently, very few buried pre-Flandrian soils exist within areas glaciated during the Devensian. Even beyond the ice limits gelifluction processes probably removed many soils, and others were often not buried because loess and other extensive cover sediments were infrequently deposited (Catt, 1979).

Buried soils are normally dated relative to the ages of parent materials and overlying sediments, and similarly the maximum age of a relict feature within a non-buried soil is indicated by the youngest dated parent material of the soil. Many Quaternary sediments cannot be dated absolutely, primarily because their ages lie outside the range afforded to radiocarbon dating. Although new techniques such as thermoluminscence and amino acid racemization may in the future assist in the dating of deposits and soils, most sediments can still only

Table 1. British Quaternary stages (after Mitchell et al. (1973) with modifications from Bryant et al. (1983), Catt (1979), Funnell et al. (1979) and Boardman (1985b)).

Stages	Sub-division and dating	Environment
Flandrian	Begins at 10 000 BP	Temperate
Devensian	Late†: 26 000–10 000 BP	Cold
	Windermere interstadial	Cool
		Cold
	Middle†: 50 000–26 000 BP	Cold
	Upton Warren interstadial complex	Cool
		Cold
	Early†: prior to 50 000 BP	
	Brimpton interstadial	Cool
		Cold
	Chelford interstadial	Cool
		Cold
	Wretton interstadial	Cool
		Cold
Ipswichian	128 000 to 118 000 BP (?)	Temperate
††		
Wolstonian		Cold
Hoxnian		Temperate
Anglian†		Cold
	Corton interstadial	Cool
		Cold
Cromerian		Temperate
Beestonian		Cold
Pastonian		Temperate
Pre-Pastonian		Cold
	——— Hiatus ? ———	
Bramertonian		Temperate
	——— Hiatus ? ———	
Baventian		Cold
Antian		Temperate
Thurnian		Cool
Ludhamian		Temperate
Pre-Ludhamian	Prior to 2 000 000 BP	Cool

† Anglian, Late Devensian, Middle Devensian and Early Devensian were predominantly cold, with brief intermittent cool or even temperate phases (interstadials).

†† There is growing evidence to suggest that an additional temperate stage occurred between the Ipswichian and the Hoxnian (Bowen, 1978; Green et al., 1984; Shotton, 1985), though no formal name has been proposed for it or the associated intervening cold stage.

be dated relative to others using lateral correlations based on lithological criteria and fossil assemblages determined by environmental conditions. These approaches, however, are liable to introduce errors in correlation, which could lead to incorrect dating of soils (Catt, 1979). Particular doubts have been expressed over the differentiation of tills of successive cold stages and thus of the position of the main Anglian, Wolstonian and Devensian ice limits. The chalky tills in East Anglia were originally ascribed to one glaciation by Harmer (1904). Later work suggested that two glaciations, the Lowestoft (Anglian) and Gipping (Wolstonian), were responsible for the tills in both the east Midlands and East Anglia (Baden-Powell, 1948; West and Donner, 1956). More recent views favour a reversion to the ideas of Harmer in that the chalky tills in East Anglia are thought to represent different facies of the one Lowestoft Formation (Perrin et al., 1979), and the consensus of opinion appears to favour an Anglian age for all the tills in this region (Shotton et al., 1977). In the east Midlands, however, several sites have been described which appear to demonstrate a post-Hoxnian till (Shotton et al., 1977). Satisfactory correlations between the two regions are difficult to establish and are still a source of considerable disagreement (Shotton et al., 1977; Perrin et al., 1979; Straw, 1983). The Anglian till in East Anglia, however, is very important as it provides a marker horizon by which soils and other deposits may be relatively dated.

The difficulties in dating sediments and thus buried soils or relict features are compounded by the undoubted gaps in the British Quaternary succession (Table 1). For example, Bowen (1978) summarized evidence to indicate an interglacial between the Ipswichian and the Hoxnian. This suggestion was discussed in more detail by Shotton et al. (1983), and further evidence presented by Green et al. (1984) and Shotton (1985). However, no formal stage names have so far been proposed or accepted. Other stages may yet be identified so as to allow direct correlation with the rest of Europe, and to put terrestial evidence more in line with information derived from deep sea ocean cores. However, in the absence of an acceptable revised version, the stratigraphical table outlined in Table 1 will be used throughout this paper.

BURIED PRE-FLANDRIAN SOILS

Most buried pre-Flandrian soils in Britain can be broadly related to either stadial, interstadial or temperate environments—a division that forms the basis for the structure of the following review. As the dates assigned to many soils are unreliable, the order in which the soils are considered within these three divisions is largely determined by geographical rather than chronological criteria (Table 2).

One of the first recorded descriptions of a buried soil in Britain was by Reid and Reid (1904) from Prah Sands in West Cornwall. This grey sandy loam, containing vertical root casts, is overlain by a thin black loam with included charcoal fragments and Palaeolithic quartz implements. This landsurface was buried beneath 12 m of head deposits. French (1983) suggested the soil is early Devensian in age, although whether it resulted from stadial or interstadial pedogenesis is uncertain.

Soils akin to present-day Arctic and Sub-Arctic soils undoubtedly formed on stable landscapes in pre-glacial cold stages and in regions outside glacial limits during glacial stages. Their preservation and burial, however, has been rarely reported. One exception is the possible Anglian humic gleysol, superimposed upon the temperate Valley Farm Soil and buried beneath Anglian glacifluvial and glacial sediments, at Broomfield in Essex (Rose *et al.*, 1978). In addition, Valentine and Dalrymple (1975) identified a weakly developed buried podzol at West Runton in Norfolk which they related to pedogenesis during the cold environment of the late Beestonian. This chronology differed from that of West and Wilson (1966), however, who had earlier suggested an early Cromerian age for purportedly the same soil.

The most common buried soils associated with cold stages are the arctic structure soils characterized by ice wedge casts, sand wedges, involutions and other cryoturbation structures. Catt (1979) and Rose *et al.* (1985a) discussed a number of these soils from diverse parts of Britain which developed during the Devensian. Examples from earlier stages have also been reported, either as single units (West and Wilson, 1966; Gruhn and Bryan, 1969; Gladfelter, 1972; Gruhn *et al.*, 1974) or superimposed on temperate soils (Rose *et al.*, 1976; 1985b; Rose and Allen, 1977). The early Anglian Barham Soil in East Anglia is either developed in Cromerian freshwater sediments or superimposed upon the 'temperate' Valley Farm Soil (Rose *et al.*, 1985b). In addition to the large-scale involutions, sand wedges and ice-wedge casts, it contains small-scale soil features such as silty clay cappings on grains, fragmented clay coatings, fractured grains and banded fabrics, which are all indicative of extensive freeze-thaw processes.

Buried soils assigned to Late Devensian interstadials on the basis of radiocarbon dates or associated molluscan assemblages are normally weakly developed, occurring either in the form of raw humus layers or as rendzinas (Catt, 1979). A rendzina developed in the Windermere interstadial and formally designated as the Pitstone Soil (Rose *et al.*, 1985a) has been described from a number of locations in chalk dry valleys in Buckinghamshire (Evans, 1966), Kent (Kerney, 1965) and Berkshire (Paterson, 1971).

Table 2. Regional distribution and ages of buried soils discussed in this chapter.

Buried stadial and interstadial soils

(1) West Cornwall: Devensian (Reid and Reid, 1904; French, 1983)
(2) East Anglia: (a) West Runton: Beestonian (Valentine and Dalrymple, 1975)
                (b) Barham Soil: Anglian (Rose *et al.*, 1976; 1985b)
                (c) Norfolk: Devensian interstadial (Straw, 1980)
                (d) SE Essex: Middle Devensian interstadial (Gruhn *et al.*, 1974)
(3) Sussex: Wolstonian interstadial(?) transported soil (Dalrymple, 1957)
(4) Buckinghamshire: Windermere interstadial (Evans, 1966; Rose *et al.*, 1985a)
(5) Berkshire: Windermere interstadial (Paterson, 1971)
(6) Kent: Windermere interstadial (Kerney, 1965)

Buried temperate soils

(1) Scotland: (a) Teindland: Ipswichian(?) (Fitzpatrick, 1965; Edwards *et al.*, 1976; Romans, 1977)
             (b) Kirkhill: Ipswichian and Hoxnian(?) (Connell *et al.*, 1982)
(2) Cumbria: (a) Troutbeck: Ipswichian(?) (Boardman, 1979; 1981; 1983; 1985a)
            (b) Laddray Wood, Keswick: Flandrian (Boardman, 1981, 1983)
(3) SW Britain   (a) Llansantffraid, N. Wales: Ipswichian(?) (Mitchell, 1960; Rudeforth, 1970)
              (b) Devon and Cornwall: Ipswichian(?) (Stephens, 1970)
              (c) SW Wales: Ipswichian(?) transported soil (Ball, 1960; Bowen, 1966)
(4) Hampshire: Pre-Devensian (Clarke and Fisher, 1983)
(5) Sussex: Hoxnian(?) transported soil material (Dalrymple, 1957)
(6) Kent: (a) Northfleet: Ipswichian (Zeuner, 1955; Dalrymple, 1958)
         (b) Sittingbourne: Ipswichian(?) (Tilley, 1964)
         (c) Swanscombe: Hoxnian(?) (Conway and Waechter, 1977; Kemp, 1985d)
(7) East Anglia   (a) Ipswich: Pre-Anglian (MacClintock, 1933)
              (b) Cambridge: Pre-Devensian (MacClintock, 1933)
              (c) Valley Farm Soil: Cromerian plus Beestonian and Pastonian(?) (Rose *et al.*, 1976; 1978; Rose and Allen, 1977; Rose, 1983; Kemp, 1983; 1985a)
(8) Hertfordshire (a) Epping Green: Pre-Anglian (Moffat and Catt, 1982)
                 (b) Gaddesden Row: Ipswichian and/or Hoxnian(?) (Avery *et al.*, 1982)

Straw (1980) identified a buried podzol within fluvial gravels in north-west Norfolk. His interpretation of the radiocarbon dates from the soil led to the assignment of a Middle Devensian age to this interstadial soil. A similar minimum age was intimated by radiocarbon dating for a buried podzol beneath glacial deposits at Teindland in Morayshire (FitzPatrick, 1965). However, FitzPatrick (1965) preferred to ascribe an Ipswichian interglacial age to this soil because it is more strongly developed than the overlying Flandrian soil. Palynological investigations by Edwards *et al.* (1976) substantiated this interpretation, although Romans (1977) considered that the soil was buried by solifluction rather than glacial sediments.

At Kirkhill Quarry, approximately 50 km to the east of the Teindland site, Connell *et al.* (1982) reported two buried pre-Flandrian soils within a single sequence. The older podzolic soil, developed in fluvial or glaciofluvial sands and gravels, contains altered felsite clasts and evidence for augite, hornblende and biotite weathering. It is disturbed and buried beneath solifluction and till deposits—the latter providing the parent material for the younger soil, which is weathered, gleyed (but not rubified) and contains void argillans that were fragmented prior to truncation and burial by later tills. Both soils were thought to represent interglacial pedogenesis and tentatively assigned to the Hoxnian and Ipswichian stages respectively (Connell *et al.*, 1982).

The preservation of the two soils beneath till at Kirkhill is likely to be due to the protection given by the irregular bedrock topography (Connell *et al.*, 1982), an explanation also favoured by Boardman (1981) to account for the similar occurrence of the Troutbeck Soil beneath Devensian till in the Vale of Threlkeld, Cumbria. This truncated soil was shown to be *in situ* by the systematic decrease in intensity of rock fragment weathering down the profile to the unaltered till parent material. Micromorphological evidence of biological activity, gleying and silty clay illuviation confirms the pedogenic origin of the weathering profile (Boardman, 1983). The buried soil, which can be traced over an area of 8 km^2, developed during a temperate phase that must have included the Ipswichian (Boardman, 1983, 1985a). In contrast with the pre-Devensian buried soils reported further south in the country, the Troutbeck Soil is neither rubified nor contains any evidence for extensive periods of clay illuviation.

The Laddray Wood Soil near Keswick, Cumbria was originally assumed to be Ipswichian in age largely on the basis of the inferred climatic significance of the red coatings around voids and pebbles (Boardman, 1979). Contradictory stratigraphical evidence, however, persuaded Boardman (1981, 1983) to revise the age to Flandrian. No hematite could be detected in the soil despite the coatings having Munsell colour hues redder than 2.5YR, the major iron oxide present being lepidocrocite (Boardman, 1981).

A reddish-brown loamy soil beneath glacial gravels at Llansantffraid in Cardiganshire was assigned to the Ipswichian interglacial by Mitchell (1960). Although Rudeforth (1970) suggested that there was no reason why it could not have formed during a Devensian interstadial, the soil has been correlated with a buried soil of assumed Ipswichian age in south-west England by Stephens (1970). Detailed information on both soils is sparse, although the latter, formed in Fremington Till and buried by head, is red, decalcified and weathered (Stephens, 1970). Other buried weathered materials were described by Stephens (1970) in this part of the country, notably head containing disintegrated shales and slates beneath unweathered solifluction deposits near Hoyle in Cornwall. In many cases, however, it is difficult to ascertain whether these are *in situ* soils or deposits containing transported soil material.

A red sandy clay loam soil, resting on a Patella raised beach and beneath over 5 m of sediment, was reported from south-west Wales by Ball (1960). Bowen (1966) later showed that this terra fusca soil is widespread in this part of Wales, forming in not only limestone but also head deposits. From stratigraphic evidence it was assigned to the Ipswichian interglacial and correlated with the soil at Llansantffraid (Bowen, 1966). However, Catt (1979) suggested that this dating may not necessarily be correct as the soil is probably not *in situ* having been redeposited by solifluction or sheet wash processes.

A buried interglacial soil has been reported near Aldershot in southern England (Clarke and Fisher, 1983). The soil, developed in Caesar's Camp Gravels—an early Pleistocene fluvial deposit, has red mottles (10R 4/6), a columnar to platy structure and contains well oriented free grain argillans, fragmented argillans (papules) and silt droplets (Romans and Robertson, 1974). As Clarke and Fisher (1983) suggested that the overlying silty clay loam may represent reworked Devensian loess, the soil could have developed over a number of warm and cold stages during most of the Pleistocene.

Dalrymple (1957) distinguished Tertiary and Quaternary transported soil lenses in head overlying the assumed Hoxnian raised beach at Slindon, Sussex by their respective micromorphological fabrics. Rotlehm and parabraunerde fabrics were ascribed to Tertiary and Quaternary pedogenesis respectively. Prefixes 'rot' and 'braun' refer to a simplistic red/brown colour differentiation. A lehm fabric has a continuous colloidal groundmass characterized by flow structures identifiable under crossed nicols by adjacent birefringent streaks and attributable to either clay illuviation or stress reorganization. In contrast, an erde fabric is stable with flocculated aggregates showing little or no alteration under crossed nicols. A parabraunerde is a braunerde fabric containing illuvial clay in voids (Dalrymple, 1957; Kubiena, 1970). Assumed Hoxnian orange-brown clay-rich soil material and Wolstonian interstadial brown silty loam lenses, formed in loess and incorporated into the head, both had parabraunerde fabrics but were differentiated on the basis of colour, texture and stratigraphical position.

An Ipswichian soil formed in Wolstonian loess and buried beneath late Ipswichian shelly fluvial deposits at Northfleet in Kent (Kerney and Sieveking, 1977) was described by Zeuner (1955) and Dalrymple (1958). The soil was recognized and interpreted on the basis of its red colour, decalcified surface horizons and parabraunerde fabric. Similar colours combined with the development of columnar structure in loess beneath 'greyish wash' near Sittingbourne (Kent) were quoted as evidence for a buried (Ipswichian?) interglacial soil by Tilley (1964).

Considerable work has been done during the last 60 years on the stratigraphical and archaeological aspects of the Thames sand, gravel and loam sequence at Swanscombe in Kent. The Lower Loam, which occurs beneath a series of sand and gravel units, is mottled throughout and is decalcified in its upper part with carbonate nodules present lower down. This decalcification has been related to both diagenetic groundwater leaching (Kerney, 1971) and pedogenic chemical weathering (Zeuner, 1959; Conway, 1972; Conway and Waechter, 1977). Kemp (1985d) has recently provided micromorphological evidence to substantiate the latter interpretation, and proposed a development sequence for the buried soil. It began with decalcification of the upper horizons and precipitation of secondary carbonate lower down in the profile. This was succeeded by a phase of gleying associated with a rise in groundwater levels, prior to soil truncation and burial beneath sands and gravels. The age of the soil is uncertain, although according to the presently accepted stratigraphy at Swanscombe, it probably developed during a temperate phase of either a part of the Hoxnian interglacial or a Hoxnian complex (Kemp, 1985d). Gruhn et al. (1974) reported a parabraunerde fabric from a buried soil within loess deposits in southeast Essex. Their assertion that the soil reflected Middle Devensian interstadial pedogenesis, however, is in conflict with the view of Catt (1983) who suggested there is little or no evidence for clay illuviation in Britain during the Devensian. Gruhn et al. (1974) also noted an Ipswichian soil in the same area, formed in fluvial gravels and buried beneath Devensian loess. Interglacial status was given to the soil on the basis of a 'parabraunerde verging to braunlehm fabric', interpreted as indicative of 'a warmer humid climate than an ordinary parabraunerde' (Gruhn et al., 1974, p. 67).

A dark-buff silty clay B horizon of a soil formed in 'brickearth' above calcareous gravels near Ipswich, Suffolk was reported by MacClintock (1933). Buried beneath (Anglian?) chalky till the soil had 'a tendency to columnar jointing' (MacClintock, 1933, p. 1049) and was presumed to be formed by weathering of older glacial material. MacClintock (1933, p. 1048) also referred to a buried interglacial soil of unknown age within sand and gravel deposits near Cambridge. He described it as being 'silty' and 'rusty' in its upper part with weathering channels of 'dark brown, highly oxidized, rotten silty clayey gravel extending downwards'.

The recognition of the Valley Farm Soil beneath Anglian glacifluvial, aeolian and glacial deposits in southern East Anglia led to a revision of the Middle Pleistocene stratigraphy for the region (Rose *et al.*, 1976; Rose and Allen, 1977; Kemp, 1983; 1985a). This soil, which developed in the pre-Anglian (proto-Thames) Kesgrave Sands and Gravels on low-relief terrace surfaces, is red, reddish brown or reddish yellow (10R to 7.5YR hues) with common grey mottles and high illuvial clay contents (Kemp, 1985a). The reddish colours have been directly correlated with hematite contents of the soil (Kemp, 1985c). In addition to rubification (pedogenic formation of hematite: Kemp, 1985b), clay illuviation and surface water gleying there is limited evidence of *in situ* weathering of glauconite and feldspar grains, although not sufficient to produce the large amounts of iron oxides and clay present within the Valley Farm Soil. Consequently, Kemp (1985a) suggested that these components originated from fine-textured eluvial horizons, developed in overbank or loess-enriched sediments, which were removed during the superimposition of the periglacial Barham Soil and burial beneath Anglian sediments.

A Cromerian age for the Valley Farm Soil was initially suggested on the basis of its probable development under a temperate environment and its stratigraphic position between the assumed Beestonian Kesgrave Formation and the overlying Anglian sediments. This age was later extended to cover earlier stages (Rose, 1983) following the establishment of a Pre-Pastonian age for topographically higher units of the Kesgrave Formation (Hey, 1980). Preliminary studies of the Valley Farm Soil developed on both high and low levels reveal considerable differences in colour and micromorphology, which may reflect a longer soil-forming interval on the upper surface (Kemp, 1985a). However, until the significance of these results has been confirmed, the Valley Farm Soil is considered as a single (albeit complex) soil stratigraphic unit having a Cromerian age as a minimum, and a maximum age range equivalent to at least the Cromerian, Beestonian and Pastonian stages (Kemp, 1985a).

A detailed micromorphological study of a composite soil comprising the Valley Farm and Barham Soils at Ipswich Airport in Suffolk revealed a complex development history (Kemp, 1984). The proposed pedogenic and environmental reconstruction indicated that there was a gradual change from fine clay to coarse clay and silt translocation as the environment deteriorated from temperate to periglacial. This environmental change, which was further substantiated by the associated evidence of small and large scale cryogenic features (papules, banded fabrics, frost-shattered clasts and frost cracks), was dated to the transitional phase between the Cromerian and Anglian stages (Kemp, 1984). This is one of the few studies where gradual, rather than sharp, changes in type of pedogenic processes have been recognized in soils developed over a time period encompassing both temperate and cold stages.

Despite initial objections to its existence and palaeoenvironmental significance (Lake et al., 1977; Baker and Jones, 1980; Wilson and Lake, 1983) the Valley Farm Soil has been identified and traced over large parts of East Anglia, either buried or in a relict form at the present surface (Kemp, 1985a). Moffat and Catt (1982) also suggested that the Valley Farm Soil may be laterally equivalent to part of the Pebbly Clay Drift in south-east Hertfordshire. The latter was originally mapped by Thomasson (1961) as a weathered pre-Anglian till. At Ashendene Farm, Epping Green, a bore-hole revealed Pebbly Clay Drift beneath Anglian coversand and till and above Pebble Gravel (Moffat and Catt, 1982). The similarity in particle size, stone content and fine sand mineralogy between the drift and the gravel suggests a common origin. Thin sections from the former contain considerable amounts of illuvial clay mainly in the form of fragmented argillans, loose or locally incorporated into a stress-oriented fine matrix. These observations led Moffat and Catt (1982) to propose that the Pebbly Clay Drift at this site is a temperate soil modified by cryogenic processes and buried beneath Anglian aeolian and glacial sediments. The deduced Pleistocene succession was therefore thought to be analogous to the one established in East Anglia by Rose et al. (1976).

Avery et al. (1982) reported a brown soil with yellowish red and grey mottles developed in brickearth and buried beneath solifluction deposits at Gaddesden Row in Hertfordshire. Thin sections contain silty aggregates enveloped by thick coats of well oriented clay which are partly fractured in places. These coats were interpreted as due to the deposition of suspended clay down large voids which possibly have been created by earlier ground ice segregation. Avery et al. (1982) could not date the soil accurately but it probably represents Hoxnian or Ipswichian pedogenesis.

PRE-FLANDRIAN RELICT FEATURES IN NON-BURIED SOILS

The contribution made by pre-Devensian processes to many of the soils in Britain, particularly south of the Devensian glacial limits, was formally recognized with the introduction of the paleo-argillic horizon into the classification system of the Soil Survey of England and Wales

(Avery, 1973, 1980; Bullock, 1974). This is defined as 'an argillic B horizon with additional characteristics attributable to pedogenesis in the Ipswichian interglacial period or earlier' (Avery, 1980, p. 30). An argillic horizon is a horizon containing translocated silicate clay (normally indicated by the presence of undisturbed or disrupted, strongly oriented clay coatings or infillings), which is presently, or was originally, overlain by an eluvial horizon from which the clay was translocated (Avery, 1980). The main distinguishing field characteristics of paleo-argillic horizons are the reddish mottles or matrix colours (hues of at least 7.5YR or 5YR depending on texture and size or quantity of mottles) which have not been directly inherited from pre-Quaternary rocks (Avery, 1980). Bullock (1974) suggested four micromorphological criteria by which argillic and paleo-argillic horizons may be differentiated. Nodules in the former have hues of 5YR or yellower, whilst the s-matrix has only been weakly to moderately reorganized resulting in asepic to moderately sepic plasmic fabrics. In contrast, paleo-argillic horizons contain nodules with hues redder than 5YR and have strongly sepic plasmic fabrics in clayey materials reflecting strong pedological reorganization of both illuvial and non-illuvial clay. Additionally, the recognizable egg yellow clay coatings (argillans) are highly disrupted, contrasting strongly with the relatively undisturbed yellowish brown coatings in normal argillic horizons (Bullock, 1974). Total amounts of illuvial clay are generally greater in paleo-argillic horizons.

The rationale behind this high level distinction between normal argillic and paleo-argillic horizons lies primarily in the assumed palaeoenvironmental significance of the intrinsic properties of the latter. The reddish colours, which are thought to reflect rubification under warmer conditions than occurred in Britain during any phase of the Devensian or Flandrian, are assumed to be an interglacial relict characteristic (Catt, 1979). Similarly, the differences in amounts of illuvial clay and the contrast in sepic plasmic fabrics are interpreted in terms of more extensive illuviation and the increased stress related to repeated shrink-swell cycles, possibly brought about by more pronounced wet and dry seasons during interglacials (Catt, 1983). However, the possible longer duration of interglacials compared to the Flandrian (Catt, 1979) and the cumulative effect of more than one phase of interglacial, interstadial or even stadial pedogenesis are important factors to consider when attempting to account for these gross differences. Boardman (1985b) suggested that differences in climate between interglacials are probably overemphasized, and that there is little evidence to substantiate the assumption that pre-Devensian interglacials were warmer than the Flandrian. Consequently, he placed greater emphasis on the time factor in accounting for differences between soils formed solely during the Flandrian and those developed over a time period encompassing the Flandrian and earlier stages.

The scale and extent of clay coating disruption in paleo-argillic horizons is too great to explain by normal pedoturbation processes and is attributed primarily to cryoturbation or transport by gelifluction during cold stages (Bullock, 1974). The coincidental occurrence within paleo-argillic horizons of ice wedge casts and festooned or vertically-oriented stones (Avery, 1980) tends to corroborate the fact that periglacial processes have made significant contributions to the development of the soil horizons. These and other features such as involutions and fragipans have also been observed in soils from all over Britain that have formed in non-paleo-argillic Devensian materials (Dimbleby, 1952; FitzPatrick, 1956; 1969; 1974; Matthews, 1976; Catt, 1979; Payton, 1980). Relict features resulting from podzolization, gleying or clay illuviation processes during the Devensian or earlier cold stages have not been reported. If present, they have probably been masked by more intense soil development during the Flandrian and earlier temperate stages in Britain (Catt, 1979).

Paleo-argillic brown earths, consisting of paleo-argillic horizons overlain by horizons (< 1 m thick) frequently derived in part from incorporated Devensian or older loess, have been mapped in many parts of Britain by the Soil Survey of England and Wales. They have developed not only in Quaternary deposits but also on older rocks such as the Lower Greensand in Kent, the Carboniferous Limestone in South Wales, Derbyshire and Staffordshire and the Jurassic rocks in South Wales and north-east Yorkshire (Catt, 1979). The description of the paleo-argillic characteristics in soils on Jurassic estuarine clays in Yorkshire and Carboniferous shales in Northumberland by Bullock (1972) and Bullock et al. (1973) was particularly important on a number of accounts. Not only did they discuss the palaeoenvironmental implications of the soils, but they also related their red colours to the presence of hematite. Despite the correlation frequently made in the literature between red colours, hematite and the rubification process, this is one of the few studies in Britain in which hematite has actually been identified. Kemp (1985c) provided some substantiation for the general assumptions by showing that the reddish colours of some soils in eastern England were strongly correlated with hematite content. However, he also quoted other examples where non-inherited reddish colours did not appear to reflect hematite content, or pre-Devensian pedogenesis, and thus suggested that further work is necessary to establish iron oxide/colour/environment/age inter-relationships.

Most of the groundwork for the paleo-argillic concept was done on soils formed in the drift deposits on chalk plateaux in southern England (Avery et al., 1959). Some of the drift deposits themselves, particularly the Clay-with-flints, are thought to result from pedological reorganization of primarily pre-Quaternary parent materials (Catt, 1983). The processes responsible for the formation of the Clay-with-flints *sensu stricto* include the accumulation of illuvial clay in voids produced by the slow dissolution of chalk or limestone beneath a thin layer of denuded Reading Beds. Intermittent phases of slumping and cryoturbation have resulted in comprehensive mixing of the residual flints, illuvial clay and the remnants of weathered Reading Beds (Catt, 1983).

As these paleo-argillic horizons and drift deposits are apparently derived from pre-Quaternary parent materials, it is possible that their relict pedological characteristics could have developed at any time during the Quaternary and possibly even the Tertiary (Catt, 1979). However, in south-east England at least, Catt (1983) suggested that all such features have resulted from Quaternary pedogenesis.

Recent studies of paleo-argillic horizons in this country have been generally restricted to those developed in parent material of assumed Quaternary age. Bullock and Murphy (1979), for instance, deduced a complex history of development in a paleo-argillic brown earth formed in Plateau Drift in Oxfordshire. The major micromorphological features were interpreted in terms of pedogenic or sedimentary processes and dated relative to each other by using standard laws of superposition. Certain features such as partially reddened embedded grain argillans, large yellowish brown papules and fossil aggregates were thought to be pedorelicts, derived from pre-existing soils that had been incorporated within materials transported and deposited as the Plateau Drift. The age of this deposit is uncertain although it probably dates from before the Anglian (Bullock and Murphy, 1979). Later irregular or linear clay concentrations, red segregations and red and/or egg yellow papules were considered to represent four major pre-Devensian processes (clay translocation, gleying, rubification and cryoturbation) occurring over a time period encompassing at least one glacial and two interglacials. Rubification was intimated to have occurred over both these interglacial stages. Following truncation, cryoturbation and addition of loess during the Devensian, further Flandrian illuviation, gleying and podzolization features were superimposed on the upper part of the profile.

Similar complex microfabrics were reported from paleo-argillic soils formed in brickearths on the Chilterns (Avery et al., 1982). At Cholesbury mineralogical dating of the loess components in the parent material indicated that the *in situ* disrupted red and yellow illuvial features originated during at least one interglacial between the Anglian and the Devensian. A major implication of this study is that rubification must have been an active process during some period since the Anglian. This confirms the conclusions of Sturdy *et al.* (1979), who studied three stagnogleyic paleo-argillic brown earths developed in Anglian chalky boulder clay (till) near Chelmsford in Essex. The major pre-Devensian processes appear to have been decalcification (down to 1.9 m), clay illuviation, gleying, cryoturbation and localized rubification. The illuvial clay, undisturbed or consisting of fragmented egg yellow and red coatings, contributes up to 30% of the areal cover of thin sections from two of the profiles. This contrasts with values of less than 10% observed in Flandrian soils (Sturdy *et al.*, 1979). Hematite was identified in one profile and extensive weathering indicated from the clay, silt and sand mineral suites. In particular, it appears that there has been partial or complete alteration of micas, chlorite, glauconite, collophane, apatite, pyroxenes and amphiboles, and some clay-size micas had apparently been weathered to interstratified mica-smectite (Sturdy *et al.*, 1979).

Weathering trends in most paleo-argillic horizons are surprisingly weakly developed in view of the assumed duration and intensity of soil development (Catt, 1979). One possible reason may be that some of the soils have been repeatedly rejuvenated by loess additions and remixed by cryoturbation, thus making trends difficult to detect (Catt, 1979).

Chartres (1980) contrasted the soils on four topographically distinct terraces of the River Kennet. The paleo-argillic brown earths on the higher levels, which have Devensian loess incorporated in their upper horizons, and the argillic Flandrian soil on the lowest level can all be differentiated on micromorphological criteria. By applying similar concepts to those used by Bullock and Murphy (1979), he identified three phases of clay illuviation. On the upper two terrace levels, disrupted red ferriargillans appear to predate egg yellow coatings which in turn were formed and fragmented (by cryoturbation) prior to the (presumed) Flandrian undisturbed reddish orange to yellowish brown coatings. The occurrence of egg yellow and reddish orange/yellowish brown papules and ferriargillans in soils on a lower terrace and only the latter features in the lowest terrace soils provides the basis for an important pedological chronology. By employing a 'counting back' technique and using the extremely limited dating evidence available, Chartres (1980) tentatively assigned ages to the formation of each terrace and pedogenic phase. The correlation of egg yellow and red coatings to Ipswichian and Hoxnian pedogenesis respectively was the first attempt in this country to relate specific pedological features to particular Quaternary stages. However, the widespread applicability of this correlation, and use of these features as intrinsic dating tools, are doubtful, particularly in view of the possibility of the introduction of further stages into the British Quaternary system.

CONCLUSIONS

Evidence from buried and non-buried pre-Flandrian soils indicates that the most important soil processes active during temperate stages in southern Britain were decalcification, clay translocation, rubification, stress reorganization of finer materials, gleying and weathering of mineral or rock components. Podzolization was probably a significant additional process further north, particularly in Scotland. All of these processes, with the exception of matrix rubification, seem to have occurred during the Flandrian, though to lesser extents, reflecting a shorter soil-forming interval and/or milder environment (Catt, 1979; Boardman, 1985b). Intervening cold stages were characterized by more disruptive processes such as frost cracking or ice/sand wedge formation and cryoturbation with production of large- and small-scale freeze-thaw features. During these unstable phases profiles were frequently truncated and had loess incorporated into the upper parts of remnant horizons. Evidence of other soil processes during cold stages is limited to a few examples of interstadial decalcification, organic matter accumulation or weak podzolization and clay illuviation. This contrasts with some other parts of Europe (e.g. Belgium) where intensive clay illuviation is thought to have occurred during the Weischelian (Devensian) (Langohr and Sanders, 1985).

The correlation of specific features and processes with particular temperate stages is beset with numerous problems, not the least being the difficulties in dating the parent materials. The till extending over large parts of eastern England appears to offer the most likely marker deposit by which soils and pedological features may be relatively dated. Although correlations with the Midland tills are still uncertain, it is generally accepted that this deposit is Anglian in age. Consequently, the identification of rubified soils buried beneath and paleo-argillic horizons developed within this till (Rose *et al.*, 1978) is evidence that rubification occurred both before and after the Anglian glaciation. As no soil horizons have been identified in dated sediments prior to the Pre-Pastonian, the early phase is tentatively related to the Cromerian and/or Pastonian stages (Catt, 1983; Kemp, 1985a). Rubification cannot be confidently ascribed to any one stage after the Anglian as most information is derived from relict features in non-buried soils having no upper age limit. Indeed, where no chronology based upon argillans of different colours is apparent, the soil colours may just reflect the cumulative effect of the processes occurring during more than one interglacial.

Catt (1979) noted that 10R and 2.5YR mottles appear to be more common in Cromerian than Hoxnian or Ipswichian (generally 5YR) soils. He qualified this statement, however, by emphasizing the dangers in making such generalizations in view of the relative lack of examples to base them on. In particular, local variations in texture, mineralogy or hydrological conditions may be at least partly responsible for these observations. These factors must also be taken into account when considering regional implications of argillan stratigraphy derived from a number of paleo-argillic horizons in southern Britain. The possible introduction of further stages into the British Quaternary system would further weaken the present tentative association of egg yellow argillans with the Ipswichian and red (rubified) ferriargillans with the Hoxnian stages (Chartres, 1980).

ACKNOWLEDGEMENTS

The author thanks Dr J.A. Catt, Mr J. Rose, Dr P. Bullock and Dr J. Boardman for their helpful criticism of an earlier draft of this paper. Receipt of a NERC CASE Studentship is also gratefully acknowledged.

REFERENCES

Avery, B.W. 1973. Soil classification in the Soil Survey of England and Wales. *J. Soil Sci.* **24**, 324–8.

Avery, B.W. 1980. *Soil Classification for England and Wales (higher categories)*. Soil Survey Technical Monograph 14, 67 pp.

Avery, B.W., Stephen, I., Brown, G. and Yaalon, D.H. 1959. The origin and development of brown earths on clay-with-flints and Coombe deposits. *J. Soil Sci.* **10**, 177–95.

Avery, B.W., Bullock, P., Catt, J.A., Rayner, J.H. and Weir, A.H. 1982. Composition and origin of some brickearths on the Chiltern Hills, England. *Catena* **9**, 153–74.

Baden-Powell, D.F.W. 1948. The Chalky boulder clays of Norfolk and Suffolk. *Geol. Mag.* **85**, 279–96.

Baker, C.A. and Jones, D.K.C. 1980. Glaciation of the London Basin and its influence on the drainage pattern: a review and appraisal, in *The Shaping of Southern England*, D.K.C. Jones (ed.). Academic Press, London. 131–75.

Ball, D.F. 1960. Relic-soil on limestone in South Wales. *Nature* **187**, 497–8.

Boardman, J. 1979. Pre-Devensian weathered tills near Threlkeld Common, Keswick, Cumbria. *Proc. Cumberland Geol. Soc.* **4**, 33–44.

Boardman, J. 1981. *'Quaternary Geomorphology of the Northeastern Lake District'*. Ph.D. Thesis, University of London.

Boardman, J. 1983. The role of micromorphological analysis in an investigation of the Troutbeck Paleosol, Cumbria, England, in *Soil Micromorphology*, P. Bullock and C.P. Murphy (eds.). AB Academic, Berkhamsted. 281–8.

Boardman, J. 1985a. The Troutbeck Paleosol, Cumbria, England, in *Soils and Quaternary Landscape Evolution*, J. Boardman (ed.). Wiley, Chichester. 231–60.

Boardman, J. 1985b. Comparison of soils in Midwestern United States and Western Europe with the interglacial record. *Quaternary Research* **23**, 62–75.

Bowen, D.Q. 1966. Dating Pleistocene events in south-west Wales. *Nature* **211**, 475–6.

Bowen, D.Q. 1978. *Quaternary Geology*. Pergamon, Oxford. 221 pp.

Bryant, I.D. Holyoak, D.T. and Moseley, K.A. 1983. Late Pleistocene deposits at Brimpton, England. *Proc. Geol. Assoc.* **94**, 321–44.

Bullock, P. 1972. Paleosol features in soils of the North York Moors. *Proc. N. Engl. Soil Discuss. Group* **9**, 20–4.

Bullock, P. 1974. The use of micromorphology in the new system of soil classification for England and Wales, in *Soil Microscopy*, G.K. Rutherford (ed.). Limestone Press, Kingston. 607–31.

Bullock, P. and Murphy, C.P. 1979. Evolution of a paleo-argillic brown earth (Paleudalf) from Oxfordshire, England. *Geoderma* **22**, 225–53.

Bullock, P., Carrol, D.M. and Jarvis, R.A. 1973. Paleosol features in Northern England. *Nature* **242**, 53–4.

Catt, J.A. 1979. Soils and Quaternary geology in Britain. *J. Soil Sci.* **30**, 607–42.

Catt, J.A. 1983. Cenozoic pedogenesis and landform development in southeast England, in *Residual Deposits: Surface Related Weathering Processes and Materials*, R.C.L. Wilson (ed.). Published for The Geological Society of London by Blackwell Scientific Publications, Oxford. 251–8.

Chartres, C.J. 1980. A Quaternary soil sequence in the Kennet Valley, Central southern England. *Geoderma* **23**, 125–46.

Clarke, M.R. and Fisher, P.E. 1983. The Caesar's Camp Gravel—an early Pleistocene fluvial periglacial deposit, southern England. *Proc. Geol. Ass.* **94**, 345–57.

Connell, E.R., Edwards, K.J. and Hall, A.M. 1982. Evidence for two pre-Flandrian paleosols in Buchan, northeastern Scotland. *Nature* **297**, 570–2.

Conway, B.W. 1972. Geological investigation of Boyn Hill Terrace deposits at Barnfield Pit, Swanscombe, Kent during 1971. *Proc. R. Anthrop. Inst. G.B. and Ireland 1971*, 80–5.

Conway, B.W. and Waechter, J. de A. 1977. Lower Thames and Medway Valleys—Barnfield Pit, Swanscombe, in *Southeast England and the Thames Valley: INQUA Guidebook*, E.R. Shephard-Thorn and J.J. Wymer (eds.). Geo Abstracts, Norwich. 38–44.

Dalrymple, J.B. 1957. The Pleistocene deposits of Penfolds Pit, Slindon, Sussex and their chronology. *Proc. Geol. Ass.* **68**, 294–303.

Dalrymple, J.B. 1958. The application of soil micromorphology to fossil soils and other deposits from archaeological sites. *J. Soil Sci.* **9**, 199–209.

Dimbleby, G.W. 1952. Pleistocene ice wedges in North-east Yorkshire. *J. Soil Sci.* **3**, 1–19.

Edwards, K.J., Castledine, C.J. and Chester, D.K. 1976. Possible interstadial and interglacial pollen floras from Teindland, Scotland. *Nature* **264**, 742–4.

Evans, J.G. 1966. Late-glacial and Post-glacial subaerial deposits at Pitstone, Buckinghamshire. *Proc. Geol. Ass.* **77**, 347–64.

FitzPatrick, E.A. 1956. An indurated soil horizon formed by permafrost. *J. Soil Sci.* **7**, 248–54.

FitzPatrick, E.A. 1965. An interglacial soil at Teindland, Morayshire. *Nature* **207**, 621–2.

FitzPatrick, E.A. 1969. Some aspects of soil evolution in northeast Scotland. *Soil Sci.* **107**, 403–8.

FitzPatrick, E.A. 1974. Cryons and isons. *Proc. N. Engl. Soils Discuss. Group* **11**, 31–43.

French, C. 1983. The Prah Sands loam. *Quat. News.* **40**, 14–23.

Funnell, B.M., Norton, P.E.P. and West, R.G. 1979. The Crag at Bramerton, near Norfolk. *Phil. Trans. R. Soc. Lond. B.* **287**, 490–534.

Gladfelter, B.G. 1972. Cold climate features in the vicinity of Clacton-on-sea, Essex (England). *Quaternaria* **14**, 121–66.

Green, C.P., McGregor, D.F.M. and Evans, A.H. 1982. Development of the Thames drainage system in Early and Middle Pleistocene times. *Geol. Mag.* **119**, 281–90.

Green, C.P., Coope, G.R., Currant, A.P., Holyoak, D.T., Ivanovich, M., Jones, R.L., Keen, D.H., McGregor, D.F.M. and Robinson, J.E. 1984. Evidence of two temperate episodes in late Pleistocene deposits at Marsworth, U.K. *Nature* **309**, 778–81.

Gruhn, R. and Bryan, A.L. 1969. Fossil ice wedge polygons in southeast Essex, England, in *The Periglacial Environment: Past and Present*, T.L. Pewe (ed.). McGill-Queen's University Press, Montreal. 351–63.

Gruhn, R., Bryan, A.L. and Moss, A.J. 1974. A contribution to the Pleistocene chronology in southeast Essex, England. *Quaternary Research* **4**, 53–71.

Harmer, F.W. 1904. The Great Eastern Glacier. *Geol. Mag.* **51**, 509–10.

Hey, R.W. 1980. Equivalents of the Westland Green Gravels in Essex and East Anglia. *Proc. Geol. Ass.* **91**, 279–90.

Kemp, R.A. 1983. Stebbing: the Valley Farm Paleosols layer, in *Diversions of the Thames: Q.R.A. Field Guide*, J. Rose (ed.). Quaternary Research Association, Cambridge. 154–8.

Kemp, R.A. 1984. *Quaternary Soils in Southern East Anglia and the Lower Thames Basin*. Unpublished Ph.D. Thesis, University of London.

Kemp, R.A. 1985a. The Valley Farm Soil in southern East Anglia, in *Soils and Quaternary Landscape Evolution*, J. Boardman (ed.). Wiley, Chichester. 179–96.

Kemp, R.A. 1985b. A consideration of the use of the terms 'paleosol' and 'rubification'. *Quat. News.* **45**, 6–11.

Kemp, R.A. 1985c. The cause of redness in some buried and non-buried soils in Eastern England. *J. Soil Sci.* **36**, 329–34.

Kemp, R.A. 1985d. The decalcified Lower Loam at Swanscombe, Kent: a buried Quaternary soil. *Proc. Geol. Ass.* **96**, 343–54.

Kerney, M.P. 1965. Weichselian deposits in the Isle of Thanet, east Kent. *Proc. Geol. Ass.* **76**, 269–74.

Kerney, M.P. 1971. Interglacial deposits in Barnfield pit, Swanscombe and their molluscan fauna. *J. geol. Soc. Lond.* **127**, 69–93.

Kerney, M.P. and Sieveking, G. de G. 1977. Lower Thames and Medway Valleys—Northfleet, in *Southeast England and the Thames Valley: INQUA Guidebook*, E.R. Shephard-Thorn and J.J. Wymer (eds). Geo Abstracts, Norwich. 44–9.

Kubiena, W.L. 1970. *Micromorphological Features of Soil Geography*. Rutgers University Press, New Brunswick and New Jersey. 249 pp.

Lake, R.D., Ellison, R.A. and Moorlock, B.S.P. 1977. Middle Pleistocene stratigraphy in southern East Anglia. *Nature* **265**, 663.

Langohr, R. and Sanders, J. 1985. The Belgian loess belt in the last 20 000 years. Evolution of soils and relief in the Zonien Forest, in *Soils and Quaternary Landscape Evolution*, J. Boardman (ed.). Wiley, Chichester. 359–71.

MacClintock, P. 1933. *Interglacial Soils and the Drift Sheets of Eastern England*. Rept. XVI Int. Geol. Congr., Washington.

Matthews, B. 1976. Soils with discontinuous induration in the Penrith area of Cumbria. *Proc. N. Engl. Soils Discuss. Group* **13**, 11–19.

Mitchell, G.F. 1960. The Pleistocene History of the Irish Sea. *Adv. Sci.* **17**, 313–25.

Mitchell, G.F., Penny, L.F., Shotton, F.W. and West, R.G. 1973. *A Correlation of Quaternary Deposits in the British Isles*. Geol. Soc. Lond. Spec. Rep.

Moffat, A.J. and Catt, J.A. 1982. The nature of the Pebbly Clay Drift at Epping Green, south-east Hertfordshire. *Herts. Nat. Hist. Soc.* **28**, 16–24.

Paterson, K. 1971. Weichselian deposits and fossil periglacial structures in north Berkshire. *Proc. Geol. Ass.* **82**, 455–68.

Payton, R.W. 1980. Pedogenic compaction: the character and formation of compact soil horizons. *Proc. N. Engl. Soil Discuss Group* **16**, 103–25.

Perrin, R.M.S., Rose, J. and Davies, H. 1979. The distribution, variation and origins of pre-Devensian tills in Eastern England. *Phil. Trans. R. Soc. Lond. B.* **287**, 536–70.

Reid, C. and Reid, E.M. 1904. On a probable Palaeolithic floor at Prah Sands (Cornwall). *J. geol. Soc. Lond.* **60**, 106–12.

Romans, J.C.C. 1977. Stratigraphy of buried soil at Teindland Forest, Scotland. *Nature* **268**, 622–3.

Romans, J.C.C. and Robertson, L. 1974. Some aspects of alpine and upland soils in the British Isles, in *Soil Microscopy*, G.K. Rutherford (ed.). Limestone Press, Kingston. 498–510.

Rose, J. 1983. Early Middle Pleistocene sediments and palaeosols in west and central Essex, in *Diversions of the Thames: Q.R.A. Field Guide*, J. Rose (ed.). Quaternary Research Association, Cambridge. 135–9.

Rose, J. and Allen, P. 1977. Middle Pleistocene stratigraphy in southeast Suffolk. *J. geol. Soc. Lond.* **133**, 83–102.

Rose, J., Allen, P. and Hey, R.W. 1976. Middle Pleistocene stratigraphy in southern East Anglia. *Nature* **263**, 492–4.

Rose, J., Boardman, J., Kemp, R.A. and Whiteman, C.A. 1985a. Palaeosols and the interpretation of the British Quaternary stratigraphy, in *Geomorphology and Soils*, K. Richards, R. Arnett and S. Ellis (eds). George Allen and Unwin, Hemel Hempstead. 348–75.

Rose, J., Sturdy, R.G., Allen, P. and Whiteman, C.A. 1978. Middle Pleistocene sediments and paleosols near Chelmsford, Essex. *Proc. Geol. Ass.* **89**, 91–6.

Rose, J., Allen, P., Kemp, R.A., Whiteman, C.A. and Owen, N. 1985b. The early Anglian Barham Soil of Eastern England, in *Soils and Quaternary Landscape Evolution*, J. Boardman (ed.). Wiley, Chichester. 197–230.

Rudeforth, C.C. 1970. *Soils of North Cardiganshire (sheets 163 and 178)*. Memoirs Soil Survey G.B. 1–153.

Shotton, F.W. 1985. IGCP 24 Quaternary glaciations in the northern hemisphere. *Quat. News* **45**, 28–36.

Shotton, F.W., Banham, P.H. and Bishop, W.W. 1977. Glacial-interglacial stratigraphy of the Quaternary in Midland and Eastern England, in *British Quaternary Studies: Recent Advances*, F.W. Shotton (ed.). Clarendon Press, Oxford. 267–82.

Shotton, F.W., Sutcliffe, A.J., Bowen, D.Q., Currant, A.P., Coope, G.R., Harmon, R., Shackleton, N.J., Stringer, C.B., Turner, C., West, R.G. and Wymer, J. 1983. United Kingdom contribution to the International Geological Correlation Programme; Project 24, Quaternary Glaciations of the Northern Hemisphere. Interglacials after the Hoxnian in Britain. *Quat. News.* **39**, 19–24.

Stephens, N. 1970. The west country and southern Ireland, in *The Glaciations of Wales and Adjoining Regions*, C.A. Lewis (ed.). Longman, London. 267–314.

Straw, A. 1980. The age and geomorphological context of a Norfolk paleosol, in *Timescales in Geomorphology*, R.A. Cullingford, D.A. Davidson and J. Lewin (eds). Wiley, London. 305–15.

Straw, A. 1983. Pre-Devensian glaciation of Lincolnshire (eastern England) and adjacent areas. *Quat. Sci. Rev.* **2**, 239–60.

Sturdy, R.G., Allen, R.H., Bullock, P., Catt, J.A. and Greenfield, S. 1979. Paleosols developed on Chalky Boulder Clay in Essex. *J. Soil Sci.* **80**, 117–37.

Tilley, P.D. 1964. The significance of loess in southeast England. *Rep. VIth INQUA*

Congr., Warsaw, 1961 **4**, 591–6.

Thomasson, A.J. 1961. Some aspects of the drift deposits and geomorphology of southeast Hertfordshire. *Proc. Geol. Ass.* **72**, 287–302.

Valentine, K.W.G. and Dalrymple, J.B. 1975. The identification, lateral variation, and chronology of two buried paleocatenas at Woodhall Spa and West Runton, England. *Quaternary Research* **5**, 551–90.

West, R.G. 1977. *Pleistocene Geology and Biology*, 2nd ed. Longman, London and New York. 440 pp.

West, R.G. and Donner, J.J. 1956. The glaciations of East Anglia and the east midlands: a differentiation based on stone-orientation measurements of the tills. *J. geol. Soc. Lond.* **112**, 69–91.

West, R.G. and Wilson, D.G. 1966. Cromer Forest Bed Series. *Nature* **209**, 497–8.

Wilson, D. and Lake, R.D. 1983. Field meeting to north Essex and west Suffolk, 20–22 June 1980. *Proc. Geol. Ass.* **94**, 75–9.

Zeuner, F.E. 1955. Loess and Palaeolithic chronology. *Proc. Prehist. Soc.* **21**, 51–64.

Zeuner, F.E. 1959. *The Pleistocene Period. Its Climate, Chronology and Faunal Successions.* Hutchinson, London, 447 pp.

Chapter 9

PALEOSOLS IN ARCHAEOLOGY: THEIR ROLE IN UNDERSTANDING FLANDRIAN PEDOGENESIS

RICHARD I. MACPHAIL
Institute of Archaeology, University of London, Great Britain

INTRODUCTION

Soils buried by archaeological monuments and from archaeologically and environmentally known contexts provide a wealth of information for the study of Flandrian pedogenesis, firstly because such dateable soil profile features allow the investigation of natural pedogenesis through time, and secondly because man-induced soil changes are also readily identified. The latter is of more interest to archaeologists who use other environmental data preserved by burial to interpret conditions in the human landscape prevalent up to the period of monument construction.

The study of archaeological paleosols is a new subject with most investigations relating to Flandrian pedogenesis in Western Temperate areas (Table 1). Nevertheless, the identification of a few sites from other climatic zones (Goudie, 1977) has allowed possible future lines of research to be recognized from parts of the world featuring markedly different environments such as the New World, Middle East and Central Asia. Some examples are included below.

HISTORY

Cornwall (1953) was probably the first British soil scientist to start drawing inferences about Flandrian pedogenesis by comparing present day soils with soils buried beneath archaeological monuments. The value of such soils buried beneath, for example, Neolithic, Bronze Age, Iron Age or later monuments, lay in the fact that even before regular ^{14}C dating, archaeological burial gave them a *terminus antiquem* and thus they could be confidently placed in the Flandrian soil succession. Soils and individual horizons buried by natural processes such as colluvium, alluvium or wind-blown deposits could also be dated by the artefacts they contained.

Preliminary studies in Europe centred on pedogenesis on freely-draining substrates and soils buried beneath prehistoric monuments and unburied soil profiles were compared. In Holland brown soils were found to have developed into podzols by the Neolithic (Waterbolk, 1957), whereas in England this was mainly a Bronze Age event (Cornwall, 1958; Dimbleby, 1962). Correlatory soil pollen studies provided, for the first time, strong evidence that man had played a significant role in this so-called soil 'degradation sequence' by his interference with the natural vegetation cover, encouraging broad-leaved woodlands to be replaced by heathlands and moorlands (Dimbleby, 1962). Similar influences were reported by workers investigating unburied soil 'developmental sequences' in England (Mackney, 1961) and in France (Duchaufour, 1965). In comparison, progressive glacial lake retreats over the last 10 000 years in northern Michigan gave rise to a natural podzol chronosequence where profile maturity was related to age and broad-leaved woodland succession (Franzmeier and Whiteside, 1963).

Climatic change has also been inferred to account for Flandrian soil changes, for example in explaining massive prehistoric erosion in the Mediterranean region (Vita-Finzi, 1969). The effects on soil formation of climatic fluctuations, i.e. wetter and drier phases, have also been reported from the Jordan Valley, Israel (Goldberg, 1983). In northwest India, archaeological paleosols dating to 2500 BC, which contain evidence of once being very fertile, occur in areas now suffering desertification and salinization, and thus indicate climatic alteration in this region (Courty and Federoff, in press). In contrast, the reported progressive soil hydromorphism which occurred during the sub-Atlantic period in western temperate areas of Europe in the Iron Age (Ball, 1975; Pennington, 1975) may be over-emphasized. Investigations of buried soils suggest that man was probably responsible for the development of stagnopodzols and peats by the Neolithic in Northern Ireland (Proudfoot, 1958) and by the Bronze Age in south-west England (Keeley and Macphail, in Smith *et al.*, 1981).

Paleosols at archaeological sites have also been studied in the New World. For example, variations in paleosols were noted in Indian soil mounds of differing ages from Iowa (Parsons *et al.*, 1962). More recently, anthrosols, such as *terra preta* (Eden *et al.*, 1984; Eidt, in press) and large areas of terraced soils (Keeley, in press) from central and southern America have come under scrutiny. Surveying of natural, pre-Inca and Inca terrace soils from Peru for instance, is beginning to

Table 1. Classic view of north-west European pedogenic trends in Flandrian (a more recent appraisal is given in the text).

Period	Climate	Pedogenic process	Archaeology
Sub-Atlantic	Cool, wet oceanic	Hydromorphism (peat formation)	Medieval Saxon Romano-British
			Iron Age
2500			
Sub-Boreal	Warm, dry,	Podzolization	Bronze Age
			Neolithic (Copper Age)
4500			
Atlantic	Warm, wet 'climatic optimun'	Lessivage	
			Mesolithic
7500			
Boreal	Relatively warm, dry	Decalcification	
9000			
Pre-Boreal	Sub-Arctic	Raw Soils	
10 000	(cool temperate Alleröd Interstadial)		Upper Palaeolithic
Late Glacial	Sub-Arctic		
−13 000 yrs BP			

show how large areas of artificial soils can be produced and agriculturally utilized in montane (3−4000 m) areas naturally unsuitable for cultivation, thus markedly affecting the upland pedogenic trends of such areas (Keeley, in press).

The success of interdisciplinary approaches to the study of archaeological paleosols (Dimbleby, 1962), encouraged Evans (1971, 1972) to link morphological studies of base-rich soil profiles beneath Neolithic monuments with buried molluscan sequences which indicated various clearance, tillage and grassland phases. Such teamwork, with soil investigations and associated palaeoenvironmental studies (Jorda and Vaudour, 1980; Simmons and Tooley, 1981) of soil (Scaife and Macphail, 1983), alluvium (Burrin and Scaife, 1984), peat (Smith et al., 1981) or dry valley sediments (Bell, 1983) has allowed strong corroboration of interpretative statements on Flandrian pedogenesis based on the paleosols themselves, and which are relevant to whole landscape development.

The second part of this chapter deals with methodology, detailing techniques employed, the types of information available, and problems which may be encountered in these types of studies. The following sections concentrate on brown soil development (decalcification and lessivage) and the soil deterioration processes of podzolization and upland hydromorphism. Much detail is available on such material (Macphail, in press), but examples and themes have had to be selected for both brevity and to highlight the course of Flandrian pedogenesis. In addition, analogues hopefully useful to workers studying earlier paleosols are included. Some of the findings are recent or not yet published, but these are used because they show the quality of data which can be determined if up to date techniques, such as soil micromorphology and interdisciplinary approaches, are applied.

METHODOLOGY

TECHNIQUES

In early studies, soils from archaeological sites were preliminarily identified on their profile morphology, with soil ignition establishing the degree of leaching in podzols (Dimbleby, 1962). The more sophisticated techniques of alkali soluble humus and ferric iron analysis were carried out by Cornwall (1958), who also sought to identify horizons influenced by human occupation by phosphate determinations—a method still common to most archaeological investigations (Proudfoot, 1976; Eidt, in press). Cornwall (1953) also carried out detailed grain-size analyses of Bronze Age ditch fills on the erroneous basis (Limbrey, 1975) that these would reflect the dry conditions of the Sub-Boreal.

Archaeologists are unfamiliar with soil science and so a number of basic pedological techniques have been suggested for them to use (Limbrey, 1975; Keeley and Macphail, 1981). In addition, non-specialist soil scientists unused to archaeology, when faced with archaeological paleosols have applied standard techniques (Avery and Bascomb, 1974; Soil Survey Staff, 1975; Bonneau and Souchier, 1982), as if studying normal unburied soils. This, as shown below, can lead to a range of interpretive problems. Other methods can benefit from supportive micromorphological studies, such as the chemical characterization and ^{14}C dating of organic horizons (Guillet, in Bonneau and Souchier, 1982), or palaeomagnetic susceptibility enhancement measurements used to establish the source and nature of soils and sediments (Oldfield et al. 1978). The use of micromorphology in archaeology (Fisher and Macphail, in press) was pioneered by Cornwall (1958), and other workers were encouraged to apply this method to paleosols in general (Dalrymple, 1958). The benefits of using micromorphology, now that there is an universal descriptive scheme (Bullock et al., 1984), are that pre-burial features, burial features and post-burial effects can be readily differentiated (Courty et al., in prep.).

Sources of Information

It is not only soil profiles buried beneath archaeological monuments which are of interest, but sometimes the monument itself may provide valuable information. Barrow mounds made of soil may contain turves while even stone monuments may contain the odd soil fragment. Investigation of these can be worthwhile, especially when the buried soil itself may be truncated. Local soils may also be compared with their buried counterparts. Even unburied soils may reveal an anthropogenic impact if they contain polycyclic microfabrics which can be interpreted on an hierarchical basis, i.e. by the differentiation of superimposed pedofeatures. Soils are not only found buried beneath Neolithic long cairns, Bronze Age round barrows, Iron Age hillfort ramparts, Roman walls or Medieval castles, (Macphail, in press); a whole range of natural events can also bury soils, such as dune sand accumulations in India (Courty and Federoff, in press), beach sand deposits in Denmark (Courty and Nørnberg, in press) and hillwash colluvium in the Belgian Ardennes (Kwaad and Mücher, 1977, 1979). Accumulations of the latter can be identified in thin section (Mücher, 1974), and the paleosol horizons can be dated and characterized by palaeo-magnetic techniques (Oldfield, et al., 1978; Allen, 1983). Their environmental significance in landscape studies may also be deduced where the deposit is alkaline, by using molluscs (Bell, 1983), and where the accumulation is acid, through pollen analyses (Kwaad and Mücher, 1977).

Both buried and unburied podzols may reflect a supposed argillic brown earth ancestry by featuring relic clay coatings partially obscured by sesquioxides (Fisher and Macphail, in press). It may be possible to selectively leach out this secondary iron illuviation to reveal earlier pedogenic events (Bullock et al., 1975). In cases of post burial hydromorphism it should be possible to remove secondary ferro-manganiferous staining by similar methods. Household bleach has already been successfully used to remove organic coatings (Goldberg, pers. comm.).

Textural pedofeatures (Bullock et al., in press) are particularly helpful in identifying pedological events, especially in the instance of superimposed coatings and infills. In western temperate areas the illuviation of 'fine clay' to form the clear or limpid clay coatings present in Bt horizons is considered a natural process (Weir et al., 1971; Slager and van de Wetering, 1977), and one typical of soils developed under woodland (Duchaufour, 1977). However, coarse grained illuvial coatings, sometimes referred to as dusty clay coatings (Figure 1) or impure clay coatings according to the size of inclusions (Bullock et al., 1984), are believed to be associated with deforestation (Slager and van de Wetering, 1977; Kwaad and Mücher, 1977). Very poorly sorted mineral coatings and infills, which contain charcoal and organic matter, are thought to result from soil slaking under a modern agricultural system (Jongerius, 1970), with an extreme product being a low porosity plough pan (Jongerius, 1983). Coatings and infills which show superimposed phases of (a) limpid clay, (b) dusty clay and (c) impure clay and unsorted mineral illuviation, represent a microfabric sequence common to many modern French soils (Federoff, pers. comm.). This has been interpreted as evidence of sequential undisturbed woodland (a), woodland clearance (b), and tillage (c) respectively (Scaife and Macphail, 1983). The recognition of 'dusty' clay coatings formed by *in situ* clay decomposition, or the identification of siltans caused by silt movement during a cold climatic phase (Federoff and Goldberg, 1982) are instances where caution is needed in interpreting such dusty coatings.

Other pedogenic processes such as the development of crystalline and amorphous pedofeatures (Bullock et al., 1984) may also provide data. For example, the presence or absence of apatite in archaeological paleosols from Israel is being used to identify increasing alkalinity in the later Flandrian (Goldberg, pers. comm.); and post-burial hydromorphism beneath monuments as detailed later may result in the preservation of organic pseudomorphs by iron hydroxide.

Where no recognizable organic fragments are present in paleosols microstructural analysis may indicate the character of buried A horizons. Interpretations of fine porosity (Cornwall, unpublished notes) and lamina structure indicating mull horizons have been corroborated by evidence from mollusca (Clay, 1981). At the field level, coarse (1-2 cm) root channels noted in a brown soil beneath the long cairn at Hazleton, Gloucestershire suggests a scrub cover locally in the Neolithic (Figure 2) (Macphail, in prep.), a suggestion supported by soil pollen data indicating one of *Corylus* (Scaife, pers. comm.). As noted below, mor horizons of podzols may be still recognizable after burial under acid conditions. Lastly, an example of exceptional organic matter preservation can be cited from Silbury Hill, Wiltshire (Evans, 1972) where over 40 m of overburden have perfectly preserved the Neolithic

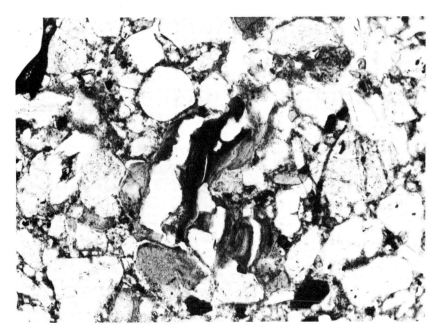

Figure 1. Photomicrograph. Selmeston, Sussex. Lower Greensand: Mesolithic to Saxon brown argillic sand, Bt(g)2 horizon. Very dusty clay coatings (argillans). Plane polarized light. Field of view is 1.348 mm wide. Such coatings are associated with deforestation.

monocotyledonous vegetation of a buried mull horizon, by the exclusion of post-burial oxygen preventing microbiological activity.

Problems

Interglacial and periglacial soil features within buried profiles (Bullock and Murphy, 1979; Catt, 1979) have first to be differentiated from the products of Flandrian pedogenesis if the latter are to be understood properly. Soil processes themselves, such as podzolization and hydromorphic leaching, have a considerable influence on the soil, even as far as obliterating earlier soil horizons. For example, the original soil forming processes associated with a layer of burned Palaeolithic flints, dated by thermoluminescence to the Allerød, at Hengistbury Head, Hampshire (Barton, pers. comm.) cannot be identified because the present soil fabric relates purely to Flandrian podzolization.

Burial may have the affect of changing local drainage characteristics by altering ground water levels of buried soils, so producing iron pans along the junction between the old ground surface and the monument above (Limbrey, 1975). Where water-tables have fluctuated manganese deposits may also obscure soil characteristics, and soil fabrics may become leached of their iron content, both processes hindering interpretation.

Burial may also lead to compaction, so that original soil structure, porosity, faunal channels and excrements are difficult to interpret. 'Ageing' in general may have the same effect on organic components as organic matter oxidizes. In the case of the latter, chemical measurements of organic carbon and nitrogen may not reflect the original levels in the paleosol. The degree of preservation noted in thin section can be related to pH, as this determines intensity of biological activity and rate of organic matter loss (Jenkinson and Rayner, 1977; Cerri and Jenkinson, 1981). In base-rich soils all visible organic matter may disappear (Clay, 1981), whereas in the humus horizons of buried podzols plant remains and excrements are better preserved but may become welded into an amorphous mass, blackened by fungal and possible bacterial melanic products (Fisher and Macphail, in press).

Immediate pre-burial events, such as slaking may cause a loss of structure and features in the paleosol (Macphail and Courty, in press). In addition, when unconsolidated materials bury a soil, soil water may slake and carry soil material into the buried soil, producing a last phase of coatings and infills in the microfabric (Romans and Robertson, pers. comm.). Argillic brown earths and even podzols buried by stone and mortar walls or calcareous mounds, often have alkaline pH's as a result of percolating soil waters. Post-burial contamination, however, may not always be so evident. For example Iron Age soils buried beneath chalk ramparts at Balksbury hillfort, Hampshire were originally described as brown calcareous earths (Macphail, unpub.). Microfabric analysis revealed that these soils had become recalcified from the overlying chalk rampart, which had encouraged post-burial reworking by earthworms so that all the previous void clay coatings within the original Bt horizon of the argillic brown earth had been integrated into the matrix. Micromorphology also has the ability to identify many of the components within barrows or present in anthropogenic deposits when, for example bulk grain size analysis would only provide an 'average' of all the different mineral constituents (Macphail and Courty, in press). Another advantage is that the influence of burning can be recognized. This is important as the resulting charred organic matter in the soil is believed to produce confusingly high C/N ratios (Courty and Federoff, 1982).

In summary, both man's activities and the alteration of the chemical and physical character of the soil during and after burial may cause difficulties in interpretation which have led to erroneous conclusions concerning Flandrian pedogenesis (Fisher, 1982). In addition, single paleosols should not be regarded as totally representative but only as examples of natural variation in a pedologic continuum (Valentine and Dalrymple, 1976). Archaeologists have fallen into the trap of actually dating their monuments on heaths by the degree of podzolization exhibited by the buried soil, with an incipient podzol indicating an early Bronze Age barrow and a well-formed podzol suggesting a late Bronze Age barrow (Cornwall, 1958).

THE PROCESSES OF DECALCIFICATION AND LESSIVAGE, AND MAN'S EFFECTS UPON THEM

On base-rich parent materials, such as calcareous sands, loess and limestone, soils develop by the major processes of decalcification and lessivage. Soils investigated beneath archaeological monuments, and within anthropogenically produced colluvium, show that human activity had an increasing effect on natural pedogenesis. Man may also have destroyed much of the evidence of early Flandrian pedogenesis, accelerated particle translocation (both downprofile and downslope), and in some cases reversed the trend from decalcification to calcification.

THE EARLY FLANDRIAN

Flandrian pedogenesis commenced with the process of decalcification under late Glacial and early Flandrian vegetation communities. Similar rates of calcium carbonate loss from dateable soils can be estimated, from both English sand-dunes (Salisbury, 1925) and calcareous glacial beach sands in Switzerland (van der Meer, 1982), at approximately 0.02−0.04% per year. Even in the instance of finer materials, loess is believed to have been decalcified by early post-Glacial times at Pegwell Bay, Kent (Weir et al., 1971). Low temperature, as in the very early post-Glacial, and high percolation rates, favour decalcification, and it is thought that by the later warmer Holocene the rate of calcium carbonate dissolution decreased (Catt, 1979; van der Meer, 1982). Full decalcification of the profile, however, is not considered necessary for clay translocation (van der Meer, 1982; Aguilar et al., 1983), although the development of a cambic B horizon may predate lessivage in loess soils in southern England (Dalrymple, 1962).

Few Boreal paleosols have been investigated but there is increasing evidence that soils formed during this period may have already been affected by anthropogenic activity (Macphail, in press). Evidence of significant silt inwash during the Boreal has been obtained from studies of alluvium from rivers in Sussex and the Isle of Wight which infer valley side instability, possibly as the result of Mesolithic woodland disturbance (Scaife, 1982; Scaife and Burrin, 1983; Burrin and Scaife, 1984). Experimental studies of soil aggregate stability and erosion in the Luxembourg Ardennes, suggest that the removal of a woodland canopy can lead to soil erosion (Imeson and Jungerius, 1974, 1976), but also that a number of natural agencies under undisturbed broad-leaved woodland may also produce significant quantities of colluvium (Imeson et al., 1980). The investigation of a polycylic soil profile from a Mesolithic site near the Sussex Ouse investigated by Burrin and Scaife (1984), provided some additional evidence to the debate on anthropogenic disturbance of early Flandrian soil development (Scaife and Macphail, 1983). Here, at Selmeston, an argillic brown sand appeared to show only short-lived illuviation of limpid clay prior to much more extensive dusty clay illuviation (Figure 1). From this it was inferred that only minor lessivage occurred under a closed woodland canopy (resulting in sparse limpid clay translocation) before Mesolithic man made his impact on the woodland cover and promoted the formation of dusty clay coatings (Federoff, pers. comm.; Scaife and Macphail, 1983).

Pedogenesis on Shallow Soils

Mature argillic brown soil formation which occurred during the Atlantic period from Pegwell Bay, Kent is often cited (Weir et al., 1971), whereas in Luxembourg and Belgium these soils are believed to have been a mainly Sub-Boreal phenomenon related to a drier climatic regime (Langohr and Van Vliet, 1979). In either case, it is therefore not unexpected to find that many Neolithic (Atlantic/Sub-Boreal) paleosols on the fine superficial deposits on the Chalk of southern England were found to be argillic brown earths (Evans, 1971, 1972). Thus, when unburied modern argillic brown soils are studied the inference is that most of the argillic Bt horizon fabric is inherited from Atlantic and Sub-Boreal pedogenesis. However, few of these Neolithic paleosols have been investigated in detail. Hence the importance of the recent excavation of the long cairn at Hazleton, Gloucestershire, which revealed large areas of the Neolithic landsurface (Figure 2). This allowed a detailed review of argillic brown earth formation on this site up to Neolithic date.

At Hazleton, a number of surface features, archaeological influences and post-burial changes were differentiated (Macphail, unpub.) (Figure 2). One area had a grass turf top preserved by iron hydroxide replacement of the organic matter while another A horizon, as deduced from relic, coarse root holes, had a scrub cover. At two other areas the top of the paleosol had been clearly unprotected by any surface

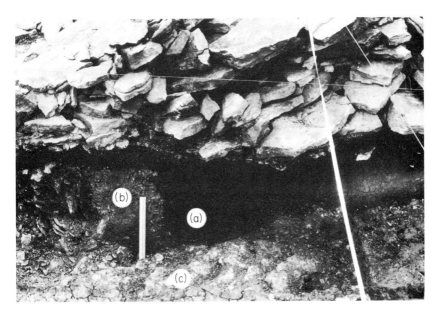

Figure 2. Hazleton Neolithic Long Cairn, Gloucestershire, Section 211. Probable tree throw hollow (a); on the left, upthrow (b) of subsoil and parent material (c) as tree fell to right. Buried soil is an argillic brown earth developed on Hampen Marley Beds, Great Oolite Series, Jurassic.

vegetation cover and had slaked under the impact of prehistoric tillage. This activity gave rise to frequent very dusty coatings which could be traced downprofile along sub-vertical soil fissures. Morphologically distinct, last phase coatings could be related to immediate pre-burial disturbance prior to cairn construction, whereas earlier sequences containing high proportions of fine charcoal could have been associated with pre-cairn occupation. During the latter domestic waste such as charred grain and hazelnuts, were mixed into the soil but there was insufficient time for biological activity and tillage to break them down before burial (Courty and Federoff, 1982).

Sub-soil horizons were also characterized by dusty coatings. At a possible tree-throw feature (Figure 2) these coatings post-dated a heterogeneous fabric comprising fragments of reddish (possible paleo-argillic) soil from long weathered limestone, 'remobilized' matrans from the latter, small pieces of limpid Bt horizon, and massive infills of 'fallen' and 'washed in' A and possibly eluviated Eb horizon material. The latter were identifiable by their less ferruginous but more organic character (Bullock, pers. comm.; Federoff, pers. comm.). Contrary to expectation these shallow (13–21 cm, maximum 46 cm) soils revealed an absence of limpid clay coatings other than in the disrupted relic features described above. Thus the question must be asked, what happened to the argillic fabric with its limpid coatings formed under early Flandrian forests up to the Neolithic period? One possible explanation is that woodland clearance and the pulling up of tree-roots disrupted the whole soil fabric.

Two other shallow Neolithic soils can be highlighted from a review of Dr Cornwall's thin sections (Macphail, in press), which are still available for study (Cornwall, pers. comm.). Thin section material from the Neolithic barrow at Ascott-under-Wychwood (Evans, 1971), Oxfordshire (developed on Jurassic limestone), showed a faunally reworked fabric, but again any *in situ* coatings were of the very dusty type. The soil also included a fragment of possible limpid Bt horizon material. At Kilham, Yorkshire (on Chalk), the Neolithic Bt horizon (Evans and Dimbleby, 1976), had suffered some minor faunal reworking, but the many coatings and infills were all very dusty (Figure 3). It thus appears that on shallow soils, developed on or over calcareous parent materials, Neolithic deforestation may have obliterated much of the evidence of early Flandrian pedogenesis although soil erosion cannot be discounted. Instead, Neolithic activity, in some cases in the form of tillage (Evans, 1972; Dimbleby and Evans, 1974; Evans and Dimbleby, 1976), superimposed its own fabric (Macphail, unpub.). The uniqueness of this kind of evidence may be illustrated at Hazleton, where the Neolithic paleosol is the only known example left of an argillic brown earth in an area surrounded by brown rendzinas and colluvial soils for tens of kilometres (Courtney and Findlay, 1978; Macphail, unpub.).

On deeper soils, as noted earlier a primary phase of limpid clay coatings may still be recognizable, but if the soil is not buried it is difficult to date successive features. At Balksbury hillfort, Hampshire, Iron Age ramparts bury various brown earths developed on clay-with-flints head (Macphail, unpub.). One argillic brown earth had been truncated as far as the Bt horizon, but in this case, both well formed limpid clay and dusty clay illuviation was preserved. One inference is that this horizon, and its limpid fabric, was deep enough not to be affected by disruption during Iron Age and earlier woodland deforestation and activity. Since we know that very few areas of Western Europe were not deforested in prehistoric and historic times, or suffered tree-throw affects, it may argued that much of the top 50 cm and deeper, of present day soils in these areas are not truly representative of full Flandrian soil formation. Features in such horizons may relate to much more recent pedogenesis. Indeed, Fisher (1983) clearly demonstrated major late Flandrian clay translocation in Late Bronze Age colluvium in Wiltshire, and so consequently many argillic features in modern soils elsewhere could relate to this later period of lessivage.

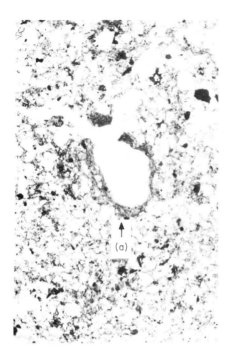

Figure 3. Photomicrograph. Kilham, North Yorkshire. Chalk: Neolithic B horizon: void lined with dusty clay coatings (a). Plane polarized light. Field of view is 1.348 mm wide.

AGRICULTURE, EROSION AND THEIR EFFECTS ON PEDOGENIC TRENDS

Although more experimental work needs to be carried out, there is sufficient evidence to suggest that some of the very dusty clay coatings from Hazleton and Kilham, for example, occurred through tillage which broke up the soil surface, exposing it to slaking under rainsplash impact (Jongerius, 1970; Imeson and Jungerius, 1974, 1976; Romans and Robertson, 1983; Macphail, in press). Clearance and agricultural activity therefore produced increasing quantities of coarse illuvial material which has to be differentiated from limpid clay coatings produced by lessivage (Fisher, 1982).

It has been suggested that Neolithic agriculture consisted of 'slash and burn' episodes (Romans and Robertson, 1975a), and often a scrub woodland or a grass cover occurred after occupation and tillage (Evans, 1971, 1972; Thomas, 1982; Scaife, pers. comm.). With the abandonment of an unburied Neolithic site in the Dordogne, France woodland regeneration followed and limpid clay illuviation occurred (Courty and Federoff, 1982). Although we can only guess at rates of

Neolithic soil erosion, we do know that deforestation and tillage produced colluvium (Kwaad and Mücher, 1977). Nevertheless significant quantities of dry valley colluvial fills have been dated, by included artefacts and charcoal, to the Neolithic (Evans and Valentine, 1974; Burleigh and Kerney, 1982). Evidence for soil erosion in the Bronze Age, including alluvial deposits, is much greater (Shotton, 1978; Bell, 1983), and although a few Bronze Age barrows may bury brown soils the landscape was seemingly being increasingly dominated by a rendzina cover on the Chalk plateaus and by deep colluvial brown calcareous soils in the valleys (Evans, 1972; Macphail, in press). Thus, the early Flandrian pedogenic trend towards progressive decalcification and lessivage was reversed, especially by the intensification of agriculture in the Iron Age. By the later Flandrian in Europe large areas of calcareous soils and limestone pavement had been created (Greig and Turner, 1974; Drew, 1982; Bell, 1983; Macphail, in press).

MEDITERRANEAN EXAMPLES

Other anthropogenic effects may be noted from this warmer and drier regime. In an area of brown mediterranean soils in southern Spain, the development of terracing, because it led to a coarser porosity, gave rise to more rapid leaching and subsequently less fertile soils (de Olmedo Puljol, 1983). Similarly, mediterranean soils formed on impure limestones in the Italian Appenines were decalcified and developed as argillic brown earths by the Copper Age (Macphail, in prep.). However, deforestation and erosion also caused subsoils to be exposed. In these horizons a cracked porosity structure, seen in charcoal fragments, was later occupied by coarse and dusty infills and rounded soil pellets of probable A horizon origin and provide evidence of severe shrinkage and swelling due to this exposure. Evidence from the subsequent colluviums which bury this truncated profile suggests that intermittent lessivage occurred between the late Bronze Age and Iron Age occupations.

PODZOLIZATION AND UPLAND HYDROMORPHISM

This section covers the so-called soil deterioration processes which occur on acid, freely-drained substrates. The processes of podzolization and upland hydromorphism, including peat formation, that characterize later Flandrian pedogenesis in northern latitudes (Franzmeier and Whiteside, 1963; Catt, 1979; Bonneau and Souchier, 1982) have been closely linked with the activities of man in Western Europe (Dimbleby,

1962; Simmons and Tooley, 1981). There are, however, problems in dating and investigating the ancestry of these processes which vary in relative importance according to both local environmental conditions and climatic regime. The latter are related to westerly exposure and altitude (Duchaufour, 1977; Bonneau and Souchier, 1982). The lowland and upland zones will be dealt with separately. Data sources will be outlined so that discussions on whether man or natural causes were responsible for these soil degradation processes (and the non-synchroneity of their onset) can be placed on a sound basis. The ability of prehistoric man to reverse these processes is also noted at the end, to provide a contrast.

Approaches

For the reasons stated above dating pedogenic events may be difficult, especially where a number of periods of soil formation are superimposed. Nevertheless, this was possible at West Heslerton, North Yorkshire, where microfabric analysis of soils formed in dateable archaeological features, of Neolithic to Roman in age, permitted a reconstruction of the site's soil history (Fisher and Macphail, in press; Table 2).

The dating of podzolization has also been attempted by ^{14}C assays on Bh horizons of unburied podzols (Perrin et al., 1964), by comparing the 'mean residence dates' with the known timing for the establishment of *Calluna* heaths (Guillet, 1982). De Coninck (1980) with Righi and Guillet (1977), however, showed that, in France, the spodic (Bh) horizon is comprised of two materials: a biologically active and 'young' polymorphic fabric, and an inert, and 'old' cemented monomorphic fabric, the latter having mean residence dates of 2–3000 years. If Guillet (1982) is right, the latter dates may project podzolization back into Neolithic and Mesolithic times. This is of significance, as podzolization had been accepted as a mainly Bronze Age and Iron Age event (Dimbleby, 1962; Perrin et al., 1964; see Table 1).

Such subjective concepts as 'poorly degraded' and 'strongly degraded' podzols are a problem when characterizing podzolization from single archaeological paleosols. Natural and paleosolic variation is best seen in catenary sequences (Valentine and Dalrymple, 1975, 1976), but these are rare in archaeological situations. One exception are the land boundaries called reaves which are rock and soil banks that extend for kilometres across Dartmoor, Devon (Smith et al., 1981; Maguire et al., 1983). At Shaugh Moor, these provided archaeological paleosols showing a Bronze Age landscape of stagnopodzols on the plateau edge, stagnogleys on the slopes and incipient peats in lower-

Table 2. An outline of soil and environmental history at West Heslerton, based on the pedological evidence

Period	Event	Soil type	Soil process
Modern		Calcareous brown sand	Recalcification (neutralization)
	——Burial by blown sand——		
Saxon (Late Anglian)	Settlement and cemetery (Wolds footslope)		
	Heath?	Podzols	Pozolization 3
	——Burial by blown sand——		
Roman	Agriculture		
	Aeolian erosion		
	Heath?	Podzols	Podzolization 2
	——Burial by blown sand——		
Late Bronze Age –Early Iron Age	Agriculture		
	Aeolian erosion		
	Heath?	Humo-ferric podzols	Podzolization 1
	Agriculture Clearance?		
	Woodland regeneration?	Argillic brown earth	Acidification Decalcification
	——Burial by blownsand——		
	Aeolian erosion		
Early Bronze Age	Barrow construction		
	——Burial by blown sand——		
	Aeolian erosion		
Late Neolithic	Area already cleared?	Calcareous brown sand	

slope sites (Keeley and Macphail, 1981) (Figure 4). Also the excavation of whole barrow cemeteries has permitted the inspection of buried podzols from different landscape positions in upland Wales (Keeley, 1982) and lowland England (Scaife and Macphail, 1983). In the latter, differences in profile morphology were not ascribed to age or to a varying vegetation history, but were related to the influence of slope position on the podzol process (Macphail, 1983; Scaife and Macphail, 1983).

Lowland Podzolization

Man or natural causation
In lowland, low rainfall regimes a classic soil degradation sequence has been suggested for western Europe (Mackney, 1961; Dimbleby, 1962; Duchaufour, 1965) in the order of increasing soil deterioration: argillic brown earth to brown podzolic soil to podzol (ferric or humo-ferric podzol) (Anderson *et al.*, 1982). This sequence and podzolization itself,

Figure 4. Saddlesborough Reave, Shaugh Moor, Dartmoor, Devon. Section AJ. Bronze Age Reave, containing phases of upcast and short-lived organic A horizon (a) formation, buries a stagnohumic-gley soil (b).

which entails the eluviation and illuviation of sesquioxides and organic matter in an acidifying regime, is believed to have occurred through man's interference with the environment, leading to the replacement of broad-leaved woodland by *Calluna* heath (Dimbleby, 1962). Man's actions through woodland clearance, burning and agriculture encouraged the depletion of soil nutrients and progressively acidified, leached and podzolized the soils (Dimbleby, 1962). Decreasing evapotranspiration rates through these Bronze Age woodland clearance phases also initiated gleys and acid mires to form on some parent materials (Moore and Wilmott, 1976).

Pollen investigations linking podzolization and vegetation have, however, shown that podzols have developed purely under forest (Dimbleby, 1962; Guillet, 1975), even as early as the Atlantic period (Valentine and Dalrymple, 1975). Thin section analysis of podzols formed under deciduous woodland at Hengistbury Head, Hampshire (Scaife, pers. comm.) and at the Iron Age Camp at Keston, Kent (Cornwall, 1958; Dimbleby, 1962) reveal a primary illuvial phase of clay translocation followed by organic matter and sesquioxidic deposition (Macphail, unpub.). Presumably increasing acidification, even under woodland, led to clay destruction and the cessation of lessivage.

This was replaced by the chemical translocation processes of podzolization to produce a secondary spodic fabric over an argillic fabric. Comparisons with buried Mor horizons developed under *Calluna* (Scaife and Macphail, 1983; Fisher and Macphail, in press) show that Ah horizons formed under acid oak woodland are more biologically active (Macphail, unpub.). Podzols formed under a *Calluna* heath vegetation are considered more acid, and more strongly degraded (Dimbleby, 1962; Guillet, 1975; Simmons and Tooley, 1981).

Other suggestions have been made which contrast with these classic views and instances of 'natural' podzolization under woodland. For example, the early erosion of unstable, fine, superficial deposits, such as loess in the Boreal period, could also trigger podzolization if their loss exposed a coarse and acid substrate (Scaife and Macphail, 1983). In addition the presence of *Calluna* in the Late Glacial vegetation may infer that acidification and podzolization were already active on particularly poor parent materials from the outset of the Flandrian. From this an alternative soil development sequence could be suggested, of raw soil to podzolic ranker to podzol, without an argillic brown earth phase (Scaife and Macphail, 1983).

Alternative dating
Instances of the presence of *Calluna* and podzolization have been traced back to the Mesolithic at a number of archaeological sites (Dimbleby, 1962; Keef et al., 1965). In addition, a mean residence date of 3770–80 years BP for a predominately monomorphic Bh horizon in a Bronze Age paleosol at West Heath, Sussex, could be linked with pollen and archaeologically dated evidence from the same soil to Mesolithic woodland removal and the appearance of heath vegetation (Scaife and Macphail, 1983).

There is also plenty of evidence to show that some heaths were not podzolized by the Neolithic, and only partially podzolized by the Bronze Age (Macphail, in press). Vagaries in local vegetation sequences, such as an undisturbed broad-leaved woodland lasting until the Iron Age at Keston Camp, Kent (Dimbleby, 1962), and differences in parent materials, caused local variations in the onset of podzolization. The latter can be illustrated from the Lower Greensand of southern England where, on the medium sandy Folkestone Beds, very early podzolization may have occurred (Scaife and Macphail, 1983), whereas on the finer and in places, more base rich Hythe Beds, podzolization may have been resisted until possible Medieval clearances and agricultural expansion (Macphail and Scaife, in press).

Upland Podzolization and Hydromorphism

As in the last section, the examination of a number of archaeological paleosols and the use of other environmental evidence, shows that the onset of these processes, which occur in uplands and westerly and northerly areas, also varies in date across the British Isles. The soil degradation sequence is similar to that on lowland heaths, but the cooler, wetter climatic conditions commonly lead to waterlogging in the upper part of the podzol, leading to the development of a thin ironpan (ironpan stagnopodzol) if the subsoil remains oxidized (Anderson et al., 1982). Increased hydromorphism will also produce stagnohumic gley soils and peats, especially in receiving sites (Crompton, 1952). Ball (1975) has also emphasized the natural inclination of upland soils (cf. soils of the High Latitudes) towards podzolization and hydromorphism. Climatic deterioration in later prehistory (Sub-Atlantic) more plainly affected these soils than those of the lowlands (Keeley, 1982).

Many workers maintain that the major anthropogenic impact occurred during the Iron Age (see Table 1) when metal technology made clearance easier (Atherden 1976), and heath became a major component in the pollen record during this time of climatic deterioration (Dimbleby, 1962: Ball, 1975: Smith et al., 1981: Keeley, 1982). A combination of a wetter climate and a decrease in evapo-transpiration rates as woodland was replaced by grassland or heath after the clearances (Eyre, 1968), is believed to have accelerated leaching, surface waterlogging and resulted in acid soil and peat formation.

It is, however, dangerous to generalize about dates for the initiation of these degradation processes. There are, in fact, a number of examples of possibly much earlier soil deterioration and peat formation which have been ascribed to man's influence on the environment. Firstly, the presence of bleached sand, truncated soils and charcoal at the base of blanket peat deposits of Atlantic age in parts of Dartmoor, the Cleveland Hills and the Pennines has suggested to pollen analysts that soil degradation and peat formation may have been initiated by Mesolithic man (Dimbleby, 1961, 1976; Simmons and Tooley, 1981). However, larger areas were affected by Neolithic activities (Simmons and Tooley, 1981). Major woodland clearances on the Isles of Scilly truncated brown soils which were later podzolized under an open moorland vegetation before preservation by blown sand burial (Dimbleby et al., 1981; Macphail, in Evans, in press) (Figure 5). At Daladies, Scotland in an area of present day podzols developed on fluvio-glacial gravels, a barrow preserves micromorphological evidence of how the Neolithic acid brown forest soil was cleared by 'slash and burn' techniques (Romans and Robertson, 1975a). Once many westerly areas had been cleared, podzolization could occur between periods of tillage. A reconstruction of the pedogenic history of Goodland and Torr Townlands, Northern Ireland, suggested that clearance initiated the podzolization of Boulder Clay soils, a process only interrupted by phases of Neolithic agriculture (Proudfoot, 1958). Peats which had been initiated in the Neolithic, eventually buried this site (Proudfoot, 1958).

As on the lowland heaths, many examples of the progress of Flandrian pedogenesis in upland areas have been detailed from soils buried beneath Bronze Age barrows and other features. Dimbleby (1962) found that in Lancashire and Yorkshire a few Bronze Age paleosols were still little affected by podzolization. In contrast, the Shaugh Moor area of Dartmoor had developed a mature sequence of stagnopodzols, stagnogleys and incipient peats by this period (Keeley and Macphail, in Smith *et al.*, 1982; Figure 4), whereas on nearby Holne Moor such soil deterioration is not thought to have occurred until later prehistoric and Medieval times (Maguire *et al.*, 1983). Indeed, the last major phase of soil deterioration in upland areas caused by massive deforestation and the subsequent spread of grasses,

Figure 5. Bar Point, St. Marys, Isles of Scilly. Romano-British East/West wall (a) with upslope plough soil lynchet to right (b), buries earlier plough soil (c), resting on a podzolic B horizon (d) developed on a probable truncated brown soil. Site preserved by blown 'beach' sand (e).

can be dated to the period of Cistercian monastic sheep husbandry (Roman and Robertson, 1975b). The fact that some parts of the British Isles resisted podzolization and hydromorphism until this Medieval period may indicate that although prehistoric affects on pedogenesis were dramatic they tended to be localized prior to the Iron Age. During this time the combination of technically advanced, expanding populations, and climatic deterioration led to the development of large areas or moorland (Macphail, in press).

Montane Soil Amelioration

Many instances of man's deleterious affects on soil have been described, and so it useful to cite a case of soil improvement through human influence on vegetation regimes. In a sub-montane area of the Western Italian Alps archaeological information can help us date known pedological changes. The zonal soil for sub-montane environments is an iron podzol developed under coniferous woodland (Kubiena, 1953), which if replaced by a herbaceous vegetation (including legumes) may be converted to a brown soil (Duchaufour, 1958, 1977). Few palaeoenvironmental studies have been carried out on alpine archaeological sites, but it is known that areas of woodland over 1000 m were being cleared and terraced for agriculture in the early Iron Age (Nisbet and Macphail, 1983). It may also be possible, through pollen studies of dated rock shelter sediments and investigations of local polycyclic soils, to link the development of pasture (and the conversion of iron podzols to brown soils) to Copper Age activities around 1400 m in Val chisone, Piedmont, Italy (Biagi et al., in press).

CONCLUSIONS

The investigation of paleosols preserved by dateable archaeological features, and present in archaeologically and environmentally understood landscapes, clearly aids our comprehension of Flandrian pedogenesis. Paleosols beneath dateable archaeological features both provide a *terminus antiquem* for soils developing in this period, and also help us accurately appreciate man's role in influencing this development.

In Europe at least, man's effects have been shown to be more decisive locally on pedogenic trends than any climatic change during the Flandrian. By causing massive soil erosion, through woodland clearance and agriculture, he created large areas of shallow calcareous soils, thus reversing the natural trend of decalcification. By accelerating the leaching rates on acid substrates such clearance triggered podzolization. Deforestation leading to decreased evapo-transpiration rates also resulted in progressive upland hydromorphism and peat formation, at a much earlier time in some cases, than the onset of cooler and wetter conditions of the sub-Atlantic phase. Thus limestones could have their decalcified topsoils removed by erosion, and the processes of podzolization and upland hydromorphism could be hastened by the destruction of the natural vegetation. Differences in the soil forming factors, such as parent material and the localized nature of man's affects also meant that there were vast differences in the timing of the onset of these soil deterioration processes. In contrast, man also actively improved the soil and moderated erosion in some Alpine and Andean montane areas.

In addition, archaeological paleosols can provide local and detailed information. The inspection of some British Neolithic brown earths indicated that soil fabric disturbance by natural tree-throw, woodland clearance and anthropogenic activities such as tillage may be underestimated. Equally, they suggested that the importance of late Flandrian lessivage on the formation of present day profiles may be undervalued.

We know far less about Flandrian pedogenesis (and man's effect upon it) from other climatic regions of the world because of the lack of studied archaeological paleosols. Nevertheless this is changing and increasing interest in, for instance, present day arid areas of the Middle East and India, is providing evidence to suggest that the processes of alkalization and salinization were less prevalent in prehistory. These findings, and initial studies from the New World, indicate the widening utilization of archaeological paleosols in this field.

Techniques and interpretive methods for understanding both archaeologically buried soils and soils in archaeologically known contexts, are constantly under review, but micromorphology appears the best single analytical tool. Any conclusions reached using micromorphology are strengthened by the integration of corroborative environmental data. Such approaches may be applicable to the study of paleosols in general. This is because archaeological paleosols being much younger, have often suffered far less from post-burial processes. They have also been unaffected by geomorphological and geological alteration. Thus their pedological features can be more accurately interpreted, especially when the soils themselves are from well-studied environmental contexts. They may, therefore, act as the best- understood paleosols for the study of comparable but older paleosols.

ACKNOWLEDGEMENTS

The author wishes to thank the following for their help: Peter Bullock, Ian Cornwall, Marie-Agnès Courty, Keith Dobney, John Evans, Nick Federoff, Paul Goldberg, Nevile Henderson, Helen Keeley, Jill Macphail, Chris Murphy, John Romans, Alan Saville, and Rob Scaife; and acknowledges the Ancient Monument and Historic Buildings Commission, and the Soprintendenza Archaeologica della Liguria, for pre-publication data.

REFERENCES

Aguilar, J., Guardiola, J.L., Barahona, E., Dorronsoto, C. and Santos, F. 1983. Clay illuviation in calcareous soils, in *Soil Micromorphology 1*, P. Bullock and C.P. Murphy (eds). A.B. Academic Press, Berkhamsted. 541–50.

Allen, M.J. 1983. *Sediment Analyses and Archaeological Data as Evidence of the Palaeoenvironments of Early Eastbourne: The Bourne Valley Excavation*. Unpublished B.Sc. Dissertation, Institute of Archaeology, University of London.

Anderson, H.A., Berrow, M.L., Farmer, V.C., Hepburn, A., Russel, J.D. and Walker, A.D. 1982. A reassessment of podzol formation process. *J. Soil Sci.* **33**, 125–36.

Atherden, M.A. 1976. Vegetation of the North York Moors. *Trans. Inst. Br. Geogr.* **1**, 284–300.

Avery, B.W. and Bascomb, C.L. (eds) 1974. *Soil Survery Laboratory Methods*. Soil Survey Technical Monograph, Harpenden.

Ball, D.F. 1975. Processes of soil degradation: a pedological point of view, in *The Effect of Man on the Landscape: the Highland Zone*, J.G. Evans, S. Limbrey and H. Cleere (eds). Council British Archaeology Res. Rep. 11, 20–7.

Bell, M. 1983. Valley sediments as evidence of prehistoric land use on the South Downs. *Proc. prehist. Soc.* **49**, 119–50.

Biagi, P., Macphail, R., Nisbet R. and Scaife, R.G., in press. Early farming communities and short-range transhumance in the Cottian Alps. (Chisone Valley, Turin) in the Late 3rd millenium B.C., in *Early Settlement in the W Mediterranean Islands and their Peripheral Areas*, J. Lewthwaite (ed.) British Archaeological Reports, Oxford.

Bonneau M. and Souchier, B. 1982. *Constituents and Properties of Soil*. Academic Press, London.

Bullock, P. and Murphy, C.P. 1979. Evolution of a paleo-argillic brown earth (Paleudalf) from Oxfordshire, England. *Geoderma* **22**, 225–52.

Bullock, P., Loveland, P.J. and Murphy, C.P. 1975. A technique for selective solution of iron oxides in thin sections of soil. *J. Soil Sci.* **26**, 247–9.

Bullock, P., Federoff, N., Jongerius, A., Stoops, G. and Tursina, T., in press. *Handbook for Soil Thin Section Description*. Waine Research Publications, Wolverhampton.

Burleigh, R. and Kerney, M.P. 1982. Some chronological implications of a fossil molluscan assemblage from a Neolithic site at Brook, Kent, England. *J. Archaeol. Sci.* **9**, 29–38.

Burrin, P.J. and Scaife, R.G. 1984. Aspects of Holocene valley sedimentation and flood plain development in southern England. *Proc. Geol. Ass.* **1**, 81–96.

Catt, J.A. 1979. Soils and Quaternary geology in Britain. *J. Soil Sci.* **30**, 607–42.

Cerri, C.C. and Jenkinson, D.S. 1981. Formation of microbial biomass during decomposition of 14c labelled rye grass in soil. *J. Soil Sci.* **32**, 619–26.

Clay, P. 1981. Two multi-phase barrow sites at Sproxton, and Eaton, Leicestershire. *Archaeological Report* **2**. Leicester Museum.
De Coninck, F. 1980. Major mechanisms in formation of spodic horizons. *Geoderma* **24**, 101–28.
Cornwall, I.W. 1953. Soil science and archaeology with illustrations from some British Bronze Age monuments. *Proc. prehist. Soc.* **19**, 129–47.
Cornwall, I.W. 1958. *Soils for the Archaeologist*. Phoenix House Ltd, London.
Courtney, F.M. and Findlay, D.C. 1978. Soils in Gloucestershire II. *Soil Survey Record* **52**, Harpenden.
Courty, M.A. and Federoff, N. 1982. Micromorphology of a Holocene dwelling, in *Nordic Archaeometry 2*. PACT 7, 2, 257–277.
Courty, M.A. and Federoff, N., in press. Comparison of present-day and buried soils in a semi-arid region between the Ghagger Valley and the Arawalli mounts of Northwest India. *Geoderma*.
Courty, M.A. and Nørnberg, P., in press. Comparison between buried virgin and cultivated Iron Age podzols on the west coast of Jutland, Denmark. *Nordic Archaeometry 3*. ISKOS, Archaeological Society of Finland, Aarland, Finland.
Courty, M.A., Goldberg, P. and Macphail, R.I. in prep. *Micromorphology in Archaeology*. Cambridge University Press, Cambridge.
Crompton, E. 1952. Some morphological features associated with poor drainage. *J. Soil Sci.* **3**, 277–89.
Dalrymple, J.B. 1958. The application of soil micromorphology to fossil soils and other deposits from archaeological sites. *J. Soil Sci.* **9**, 199–205.
Dalrymple, J.B. 1962. Some micromorphological implications of time as a soil-forming factor, illustrated from sites in SE England. *Zeitshrift fur pflanzenernahrung Dungung und Bodenkunde 98*, (**143**) 3, 232–9.
Dimbleby, G.W. 1961. The ancient forest of Blackamore. *Antiquity* **35**, 123–8.
Dimbleby, G.W. 1962. *The Development of British Heathlands and Their Soils*. Clarendon Press, Oxford.
Dimbleby, G.W. 1976. Climate, soil and man, *Phil. Trans. R. Soc. B*. **275**, 197–208.
Dimbleby, G.W. and Evans, J.G. 1974. Pollen and land snails of calcareous soils. *J. Archaeol. Sci.* **1**, 117–33.
Dimbleby, G.W., Greig, J.R.A. and Scaife, R.G. 1981. Vegetational history of the Isles of Scilly, in *Environmental Aspects of Coasts and Islands*, D. Brothwell and G.W. Dimbleby (eds). Symposium Association Environmental Archaeology, 1. British Archaeological Reports, International Series 94, Oxford. 127–44.
Drew, D.P. 1982. Environmental archaeology and karstic terrains: the example of the Burren, Co. Clare, Ireland, in *Archaeological Aspects of Woodland Ecology*, M. Bell and S. Limbrey (eds). British Archaeological Reports, International Series 146, Oxford. 115–27.
Duchaufour, P.L. 1958. Dynamics of forest soils under the Atlantic climate. *Lectures in Surveying and Forest Engineering*, Quebec L'Institut Scientifique Franco-Canadien.
Duchaufour, P.L. 1965. *Précis de Pedologie*. Masson and Cie, Paris.
Duchaufour, P.L. 1977. *Pedology, Pedogenesis and Classification*. George, Allen and Unwin, London.
Eden, M.J., Bray, W., Herrera, L. and McEwan, C. 1984. Terra Preta soils and their archaeological context in the Caqueta Basin of Southeast Colombia. *American Antiquity* **49** (1), 125–40.
Eidt, R.C., in press. *Advances in Abandoned Settlement Analysis: Application to prehistoric anthrosols in Colombia*, South America. University of Wisconsin, Milwaukee.
Evans, J.G. 1971. Habitat change on the calcareous soils of Britain: the impact of Neolithic man, in *Economy and Settlement in Neolithic and Early Bronze Age Britain and Europe*, D.D.A. Simpson (ed.). Leicester University Press, Leicester.
Evans, J.G. 1972. *Land Snails in Archaeology*. Seminar Press, London.

Evans, J.G., in press. Excavations at Bar Point, St Marys, Isles of Scilly. *Cornish Studies*.
Evans, J.G. and Valentine, K.W.S. 1974. Ecological change induced by prehistoric man at Pitstone, Buckinghamshire. *J. Archaeol. Sci.* **1**, 343–51.
Evans, J.G. and Dimbleby G.W. 1976. In *Excavation of the Kilham Long Barrow, East Riding of Yorkshire*, T.G. Manby (ed.). *Proc. prehist. Soc.* **42**, 111–59.
Eyre, S.R. 1968. *Vegetation and Soils*. Edward Arnold, London.
Federoff N. and Goldberg, P. 1982. Comparative micromorphology of two late Pleistocene paleosols. *Catena* **9**, 227–51.
Fisher, P.F. 1982. A review of lessivage and Neolithic cultivation in southern England. *J. Archaeol. Sci.* **9**, 299–304.
Fisher, P.F. 1983. Pedogenesis within the archaeological landscape at South Lodge Camp, Wiltshire, England. *Geoderma* **29**, 93–106.
Fisher, P.F. and Macphail, R.I., in press. The study of archaeological soils and deposits by micromorphological techniques, in *Palaeoenvironmental Investigations*, N.R.J. Fieller, D.D. Gilbertson and N.G.A. Ralph (eds). British Archaeological Reports, Oxford.
Franzmeier, D.P. and Whiteside, E.P. 1963. A chronosequence of podzols in northern Michigan. *Mich. State Univ. Agr. Exp. Sta. Quar. Bull* **46**, 2–57.
Goldberg, P. 1983. Applications of micromorphology in archaeology, in *Soil Micromorphology*, Bullock and C.P. Murphy (eds). A.B. Academic Press, Berkhamsted. 139–50.
Goudie, A. 1977. *Environmental Change*. Clarendon Press, Oxford. Chap. 4.
Greig, J.R.A. and Turner, J. 1974. Some pollen diagrams from Greece and their archaeological significance. *J. Archaeol. Sci.* **1**, 177–94.
Guillet, B. 1975. Forested podzols and degraded podzols: relationship between vegetation history and podzol development on the Vosges Triassic sandstone. *Revue d'Ecologie at du Biologie du Sol* **12**, 1, 405–14.
Guillet, B. 1982. Study of the turnover of soil organic matter using radio-isotopes (^{14}C), in *Constituents and Properties of Soils*, M. Bonneau and B. Souchier (eds). Academic Press, London. 238–60.
Imeson, A.C. and Jungerius, P.D. 1974. Landscape stability in the Luxembourg Ardennes as exemplified by hydrological and (micro) pedological investigations of a catena in an experimental watershed. *Catena* **1**, 273–95.
Imeson, A.C. and Jungerius, P.D. 1976. Aggregate stability and colluviation in the Luxembourg Ardennes. *Earth Surf. Proc.* **1**, 259–71.
Imeson, A.C., Kwaad, F.J.P.M. and Mücher, H.J. 1980. Hillslope processes and deposits in forested areas of Luxembourg, in *Timescales in Geomorphology*, R.A. Cullingford, D.A. Davidson and J. Lewis (eds). Wiley, Chichester.
Jenkinson, D.S. and Rayner, J.H. 1977. The turnover of soil organic matter in some of the Rothamsted classical experiments. *Soil Sci.* **123**, 298–305.
Jongerius, A. 1970. Some morphological aspects of regrouping phenomena in Dutch soils. *Geoderma* **4**, 311–31.
Jongerius, A. 1983. The role of micromorphology in agricultural research, in *Soil Micromorphology 1*, P. Bullock and C.P. Murphy (eds). A.B. Academic Press, Berkhamsted. 111–38.
Jorda, M. and Vaudour, J. 1980. Sols, morphogenese et actions anthropiques a l'epoque historique S.L. sur les Rives Nord de la Mediterranee. *Naturalia Monspeliensia, Hors Serie*, (Montpellier), 173–83.
Keef, P.A.M., Wymer, J.J. and Dimbleby, G.W. 1965. A Mesolithic site on Iping Common, Sussex, England. *Proc. prehist. Soc.* **31**, 85–92.
Keeley, H.C.M. 1982. Pedogenesis during the later prehistoric period in Britain, in *Climatic Change in Later Prehistory*, A.F. Harding (ed.). Edinburgh University Press, Edinburgh.
Keeley, H.C.M., in press. Soils of Pre-Hispanic Terrace systems in the Cusichaca Valley,

Peru, in *Prehistoric Intensive Agriculture in the Tropics*, I. Farrington (ed.). British Archaeological Reports, International Series, Oxford.

Keeley, H.C.M. and Macphail, R.I. 1981. A soil handbook for archaeologists. *Inst. Archaeol. Bull.* **18**, 225–44.

Keeley, H.C.M. and Macphail, R.I. 1981. A soil survey of part of Shaugh Moor, Devon, in *The Shaugh Moor Project: Third Report*, K. Smith, J. Coppen, G.J. Wainwright and S. Beckett (eds). *Proc. prehist. Soc.* **47**, 240–5.

Kubiena, W.L. 1953. *The Soils of Europe*. Thomas Murby and Co., London.

Kwaad, F.J.P.M. and Mücher, H.J. 1977. The evolution of soils and slope deposits in the Luxembourg Ardennes near Wiltz. *Geoderma* **17**, 1–37.

Kwaad, F.J.P.M. and Mücher, H.J. 1979. The formation and evolution of colluvium on arable land in Northern Luxembourg. *Geoderma* **22**, 173–92.

Langohr, R. and van Vliet, B. 1979. Clay migration in well to moderately well drained acid brown soils of the Belgian Ardennes. *Pédologie* **29**, 367–85.

Limbrey, S. 1975. *Soil Science and Archaeology*. Academic Press, London.

Mackney, D. 1961. A podzol development sequence in oakwoods and heath in central England. *J. Soil Sci.* **12**, 23–40.

Macphail, R.I. 1983. The micromorphology of spodosols in catenary sequences on lowland heathlands in Surrey, England, in *Soil Micromorphology 2*, P. Bullock and C.P. Murphy (eds). A.B. Academic Press, Berkhamsted. 647–54.

Macphail, R.I., in press. A review of soil science in archaeology in England, in *Environmental Archaeology of a Regional Review 2*, H.C.M. Keeley (ed). English Heritage Commission Occasional Paper.

Macphail, R.I. and Courty, M.A., in press. Interpretation significance of urban deposits, in *Nordic Archaeometry 3*, ISKOS, Archaeological Society of Finland, Aaland, Finland.

Macphail, R.I. and Scaife, R.G., in press. The geographical and environmental background, in *The Archaeology of Surrey*, D. Bird and J. Bird (eds). Surrey Archaeological Collections.

Maguire, D., Ralph, N. and Fleming, A. 1983. Early land use on Dartmoor, in *Integrating the Subsistence Economy*, M. Jones (ed.). British Archaeological Reports, International Series. 181, Oxford. 57–106.

van der Meer, J.J.M. 1982. *The Fribourg Area, Switzerland. A Study in Quaternary Geology and Soil Development*. Ph.D. Thesis, Publ. Fys. Geogr. Bodemk. Lab. Univ. Amsterdam, Nr 32.

Moore, P. and Wilmott, A. 1976. Prehistoric forest clearances and the development of peatlands in the uplands and lowlands of Britain. *VII. International Peat Conference*, Poznan, Poland. 1–15.

Mücher, H.J. 1974. Micromorphology of slope deposits, in *Soil Micromorphology*, G.K. Rutherford (ed.). Limestone Press, Kingston. 553–67.

Nisbet, R. and Macphail, R.I. 1983. Organizzazione del territorio e terrazzamenti preistorici nell Italia settentrionale. *Quaderni della Soprintendenza Archeologica del Piemonte* **2**, 43–57.

Oldfield, F., Dearing, J.A., Thompson, R. and Garnet-James, S. 1978. Some magnetic properties of lake sediments and their possible links with erosion rates. *Polskie Archiwum hydrobiologii* **25**, 321–31.

de Olmedo Pujol, J.L. 1983. Anthropic influence on a Mediterranean brown soil, in *Soil Micromorphology 2*, P. Bullock and C.P. Murphy (eds). A.B. Academic Press, Berkhamsted. 583–8.

Parsons, R.B., Scholtes, W.H. and Riecken, F.F. 1962. Soils of indian mounds in north eastern Iowa as benchmarks for studies of soils genesis. *Proc. Soil Sci. Soc. Am.* **26**, 491–6.

Pennington, W. 1975. The effect of Neolithic man on the environment in north-east

England: the use of absolute pollen diagrams, in *The Effect of Man on the Landscape: the Highland zone*, J.G. Evans, S. Limbrey and H. Cleere (eds). Council British Archaeology. 74–85.
Perrin, R.M.S., Willis, E.H. and Hodge, C.A.H. 1964. Dating of humus podzols by residual radio carbon activity. *Nature, London* **202**, 165–6.
Proudfoot, V.B. 1958. Problems of soil history. Podzol development at Goodland and Torr Townlands, Co Antrim, Northern Ireland. *J. Soil Sci.* **9**, 186–98.
Proudfoot, V.B. 1976. The analysis and interpretation of soil phosphorus in archaeological contexts, in *Geo-archaeology*, D.A. Davidson and M.L. Shakley (eds). Duckworth, London. 93–113.
Righi, D. and Guillet, B. 1977. Datations par le Carbone −14 naturel de la matiere organique d'horizons spodiques de podzols des landes du Medoc, in *Soil Organic Matter Studies 2*. Int Atom. Energy Agency, Vienna. 187–92.
Romans, J.C.C. and Robertson, L. 1975a. Soils and archaeology in Scotland, in *The Effect of Man on the landscape: the Highland Zone*, J.G. Evans, S. Limbrey and H. Cleere (eds). Council British Archaeology, 11, 37–9.
Romans, J.C.C. and Robertson, L. 1975b. Some genetic characteristics of the freely drained soils of the Eltrick Association in East Scotland. *Geoderma* **14**, 297–317.
Romans, J.C.C. and Robertson, L. 1983. The environment of North Britain: Soils, in *The George Jobey Conferences: Settlement in North Britain 1000 BC–1000 AD*, J.C. Chapman and H.C. Mytum (eds) British Archaeological reports, British Series 118, Oxford. 55–80.
Salisbury, E.J. 1925. Note on the edaphic succession in some dune soils with special reference to the time factor. *J. Ecol.* **13**, 322–8.
Scaife, R.G. 1982. Late-Devensian and early Flandrian vegetation changes in southern England, in *Archaeological Aspects of Woodland Ecology*, M. Bell and S. Limbrey (eds). British Archaeological Reports, International Series, 146, London. 57–74.
Scaife, R.G. and Burrin, P.J. 1983. Floodplain development in the vegetational history of the Sussex High Weald and some archaeological implications. *Sussex Archaeological Collections* **121**, 1–10.
Scaife, R.G. and Macphail, R.I. 1983. The post-Devensian development of heathland soils and vegetation, in *Soils of the Heathlands and Chalklands*, P. Burnham (ed.). Seesoil, 1, 70–99.
Shotton, F.W. 1978. Archaeological inferences from the study of alluvium in the Lower Severn–Avon valleys, in *The Effect of Man on the Landscape: The Lowland Zone*, S. Limbrey and J.G. Evans (eds). Council British Archaeology, 21, 27–32.
Simmons, I.S. and Tooley, M. 1981. *The Environment in Prehistory*. Duckworth, London.
Slager S. and van der Wetering, H.T.J. 1977. Soil formation in archaeological pits and adjacent loess soils in southern Germany. *J. Archaeol. Sci.* **4**, 259–67.
Smith, K, Coppen, J, Wainwright G.J. and Beckett, S. 1981. The Shaugh Moor Project: Third Report. *Proc. Prehist. Soc.* **47**, 205–73.
Soil Survey Staff 1975. *Soil Taxonomy*. Agriculture Handbook 436, USDA Soil Conservation Service, Washington D.C.
Thomas, K.D. 1982. Neolithic enclosures and woodland habitats on the South Downs in Sussex, England, in *Archaeological Aspects of Woodland Ecology*, M Bell and S. Limbrey (eds). British Archaeological Reports, International Series, 146, Oxford. 147–70.
Valentine, K.W.G. and Dalrymple, J.B. 1975. The identification, lateral variation, and chronology of two buried paleocatenas at Woodhall Spa and West Runton, England. *Quaternary Research* **5**, 551–90.
Valentine K.W.G. and Dalrymple, J.B. 1976. Quaternary buried paleosols: A critical review. *Quaternary Research* **6**, 209–22.
Vita-Finzi, C. 1969. *The Mediterranean Valleys*. Cambridge University Press, Cambridge.

Waterbolk, H.T. 1957. Pollen analytisch onderzoek van twee nordbrabanste tumuli, in G. Beex Twee grafheuvels in Nord Brabant. *Bijdr. Studie Brababantse Heem.* **9**, 34–9.

Waton, P.V. 1982. Man's impact on the Chalklands: some new pollen evidence, in *Archaeological Aspects of Woodland Ecology*, M. Bell and S. Limbrey (eds). British Archaeological Reports, International Series, 146, Oxford. 75–91.

Weir, A.H., Catt, J.A. and Madgett, P.A. 1971. Postglacial soil formation in the loess of Pegwell Bay, Kent, England. *Geoderma* **5**, 131–49.

GLOSSARY

A horizon A mineral horizon containing an accumulation of organic matter formed or forming at or near to the surface and which has lost iron, aluminium or clays, being concentrated in resistant sand or silt size minerals.

Albic horizon An horizon from which clay and free iron oxides have been removed, or in which the oxides have been segregated.

Alfisols These are soils with an umbric or ochric epipedon, an argillic horizon and moderate to high base saturation. Water is held at < 15 bar tension for at least three months a year; USDA system with suborders Aqualfs, Boralfs, Udalfs, Ustalfs and Xeralfs.

Argillans Coatings in the soil composed of clay minerals.

Argillasepic fabric Soil fabrics in which the plasma consists dominantly of anisotropic clay minerals and exhibits a flecked extinction pattern.

Argillic horizon An horizon characterized by an illuvial concentration of clays. It is usually a B horizon.

Argillipedoturbation Mixing of soil components caused by the shrink-swell behaviour of some clays.

Aquents Suborder of Entisols; formed on recent sediments in flood plains, delta plains, lake margins or marshes; formed under wet conditions; USDA system.

Aridisols Soils formed in arid areas with water tension of < 15 bar; may have argillic, calcic, or petrocalcic horizon; USDA system.

Asepic fabric Type of plasmic fabric with a dominantly anisotropic plasma with anisotropic domains which are unoriented with regard to one another, exhibiting a flecked extinction pattern.

B horizon Horizons which exhibit an illuvial concentration of silicates, iron, aluminium, or humus, or residual concentrations of sesquioxides or clays, or coatings of sesquioxides.

Braunification The development of brown colour in a soil.

C horizon A mineral horizon in the soil, which may be unconsolidated or consolidated which retains evidence of original structure and lacks properties of A or B horizons. It may possess accumulations of carbonate or other soluble salts.

Calcan Calcareous coating, see *cutan*; synonomous with calcitan.

Calcic horizon An horizon with an accumulation of alkaline earth carbonate, referred to as a ca horizon.

Calcrete A widely used term which is much abused. Refers to a terrestrial material, composed mainly of calcium carbonate which occurs in a variety of physical states from powdery to nodular to indurated forms (petrocalcic horizons) Synonomous with caliche. Calcrete has a variety of origins from purely pedogenic to cementation in the phreatic zone.

Cambic horizon A subsoil horizon with fine sand or finer, weakly, argillic or spodic.

Catena These are groups of soils with similar parent materials, developed under similar climates but with different characteristics related to variations in relief and drainage.

Channel Tubular-shaped pore, burrow or root mould.

Clay coatings Argillans, see cutans.

Composite profiles Stacked soil profiles in which overlap and overprinting occurs between profiles. cf. compound profiles. Common in ancient alluvial sequences.

Compound profiles Stacked profiles in which the profiles are separated by sediment.

Cutan A concentration of particular soil constituents occurring on or near a surface in the soil such as a crack (plane) or on a rock fragment or nodule. May be composed of clay (clay cutans, clay skins or argillans) or calcite (calcan or calcitan) or gibbsite, gypsum, iron oxides, organic matter or soil material. May become fragmented to form *papules*.

Crystallaria Concentrations of crystals usually calcite or gypsum, filling former voids such as cracks (planes).

Crystic plasmic fabric The soil plasma consisting of recognizable crystals, typically of calcite or gypsum. Calcretes have a crystic plasmic fabric.

Duricrust Terrestrial material formed within the zone of weathering in which either iron or aluminium sesquioxides, silica, calcium carbonate or other materials have accumulated, which may become strongly indurated. Examples include calcrete, silcrete and ferricretes.

Edaphic Of or pertaining to the soil, or resulting from or influenced by soil factors.

Eluviation The removal of soil material from the upper horizons in suspension (lessivage) or solution (leaching).

Entisols Soils with little or no evidence of soil horizon development. There must be some evidence of pedogenic modification such as an ochric epipedon. Other horizons may develop and salts may accumulate deeper in the profile. USDA soil order with five sub-orders including Fluvents and Aquents. Such soils are likely to be very common in ancient continental deposits.

Epipedon The uppermost soil horizons.

Fluvents Entisols developed on flood plain sediments.

Gilgai Surface micro-relief, as ridges or knolls and basins, developed on clay soils with potential for expansion and contraction with wetting and drying, i.e. Vertisols. Associated with pseudo-anticlinal structures in section. See *Argillipedoturbation*.

Glaebules A soil feature, typically a nodule or concretion, which has not formed within a pre-existing void. It is recognizable because it has a greater concentration of some constituent or because it has a different fabric to that of the matrix.

Gleying The reduction of iron forming grey and blue colours, associated with anaerobic conditions. Reflects poor drainage and is associated with hydromorphic soils. Surface water gleys result from poor drainage caused by impervious horizons within the profile. Ground-water gleys result from impervious horizons beneath the soil.

Hardpan A hard cemented soil layer, for example a petrocalcic horizon.

Histosols Soils composed mainly of organic matter, for example peats.

Horizon A layer of soil approximately parallel to the soil surface with characteristics produced by pedogenic processes which distinguish it from other horizons. This is a different usage to that in geology and confusion can result unless this distinction is made when describing paleosols.

Hydromorphic soils Soils developed in the presence of excess water, typified by gleying and build-up of organic matter.

Illuviation The movement and deposition of material from one horizon to another.

Inceptisol These are soils which exhibit some degree of horizon development (unlike Entisols) but do not contain features diagnostic of the other soil orders. They show evidence of leaching but also contain weatherable minerals. They usually have an ochric or umbric epipedon over a cambic horizon; USDA system.

Intercalary crystals Individual large crystals or groups of a few large crystals set in a soil matrix, for example calcite or gypsum crystals.

K horizon A type of strongly indurated calcareous horizon, a petrocalcic horizon or hardpan. Sometimes referred to as a Km horizon (m is a symbol for strongly indurated horizons). Corresponds to mature calcrete horizons.

Laterite A term with a wide meaning but generally used for a soil horizon which is hard or will harden on exposure, composed mainly of oxides or iron and/or aluminium.

Lessivage The removal of clay and silt down the soil profile (see eluviation).

Luvisol Soils having an argillic B horizon and of low base status, similar to Alfisols, FAO/UNESCO system.

Micromorphology Basically the study of the fine morphology of the soil, soil petrography.

Mollic epipedon A dark surface horizon containing > 1% organic matter with base saturation of over 50%.

Mollisols Dark coloured, base-rich soils typically with a mollic epipedon. They may have an argillic, matric or calcic horizon; USDA system.

Mull A mixture of organic and mineral matter with a crumb or granular structure.

Natric horizon An horizon which meets the requirements of an argillic horizon but also has a prismatic, columnar or blocky structure and over 15% saturation with exchangeable sodium.

Neoferrans A neocutan composed of a concentration of iron oxides. Neocutans are concentrations of some soil component which occur immediately adjoining the natural surfaces with which they are associated, but do not actually coat that surface.

O horizon Organic soil horizons.

Ochric epiedon A surface horizon that is light in colour and contains < 1% organic matter.

Oxic horizon An horizon which is at least 30 cm thick and contains > 15% clay, composed of kaolinite and sesquioxides and a very low content of weatherable minerals.

Oxisols Soils with an oxic horizon within 2 m of the surface, or plinthite that forms a continuous phase within 30 cm of the mineral surface of the soil. There is no argillic or spodic horizon above the oxic horizon. Correspond to Ferralsols or Latosols; USDA system.

Paleustalf A great group soil of the Ustalfs suborder of Alfisols. They represent soils of warm humid to semi-arid regions which have a warm, rainy season.

Papules Glaebules composed of clay minerals, which represent disrupted clay skins (argillans). Their presence indicates pedoturbation.

Ped A soil aggregate separated from adjoining peds by surfaces of weakness such as voids or by the occurrence of cutans.

Pedoturbation The mixing of soil components by various processes including bioturbation, and shrink-swell cycles; does not include illuviation.

Petrocalcic horizon An indurated calcic horizon, synonomous with duricrust, hardpan or Km horizon.

Phytoliths Siliceous accumulations in plant tissue which contribute to the soil on the death of the plant.

Placic horizon A single, thin iron or manganese pan within 50 cm of soil surface.

Planes Elongate voids in soils, cracks.

Planosols Soils developed in areas with level or depressed topography with poor drainage, showing hydromorphic properties; FAO/UNESCO system.

Plasma Refers to that part of the soil material which is capable of being moved, concentrated or re-organized by soil processes. This includes material of colloidal size and soluble material.

Plinthite A non-indurated mixture of iron and aluminium oxides, clays and other material; typically red and mottled which undergoes induration on exposure to repeated wetting and drying.

Podzolization A process by which sesquioxides are moved through the soil profile. Acid humus, resulting from slow decomposition, results in solutions capable of breaking down clay minerals and releasing silica, aluminium and iron. Insolubles such as silica accumulate in the upper parts of the profile and result in a grey, bleached horizon (albic horizon or E horizon).

Podzols Soils having a spodic B horizon; correspond to Spodosol of USDA system. They commonly show an iron-cemented placic horizon.

Podzoluvisols Podzols having an argillic B horizon showing deep intertonguing of the E horizon into the B horizon.

Regolith Unconsolidated material consisting of weathered rock, overlying solid rock.

Rendzina These are shallow, organic-rich soils (mollic epipedon) usually developed on calcareous parent materials. The A horizon rests directly on the C horizon. Correspond to Rendolls in USDA system.

Rubifaction Development of reddening in soils typical of soils in tropical areas with a marked dry season with the transformation of hydrated iron oxides into hematite.

S-matrix Material consisting of plasma and skeleton grains which forms the matrix of the soil comprising peds or apedal material.

Sepic Fabrics Plasmic fabrics in which patches or zones of plasma have striated extinction patterns under crossed polars. Various types are recognized.

Sesquioxide Amorphous oxides of aluminium and iron.

Silcrete A variety of duricrust composed of silica.

Skeleton grains Individual grains in the soil, larger than colloidal size which are relatively stable and are not readily translocated, concentrated or reorganized by pedogenic processes.

Soilstone crust A type of finely laminated carbonate that forms a crust over indurated limestones or over petrocalcic horizons. Commonly used for crusts on exposed or soil-covered limestones.

Solifluction The slow flowage of material down sloping ground which is commonly caused by alternating periods of freezing and thawing.

Solum Upper part of the soil profile, generally the A and B horizons.

Spodic horizon Horizon with an illuvial accumulation of free sesquioxides and organic matter.

Spodosol See Podzol; soils which have a spodic horizon; USDA system.

Torrerts Suborder of Vertisols order which are developed under an arid climate.

Translocation General term for the movement of material through the soil in solution, suspension or by organisms.

Ultisols Soils which have an argillic horizon and low base saturation; represent soils of low to mid latitudes and properties reflect leaching in wet season; USDA system.

Umbric epipedon Similar to mollic epipedon but less than 50% base saturated.

Vertisol Clay-rich soils predominantly composed of swelling clays (smectite) in climates with alternating wet and dry seasons. As a result of seasonal shrinkage and swelling the soils undergo argillipedoturbation and commonly show gilgai.

USDA systems refer to the Soil Survey Staff, *Soil Taxonomy*, 1975, United States Department of Agriculture, Agricultural Handbook No. 436, 754 pp. FAO/UNESCO, 1974. Soil Map of the World, Vol, 1. Legend, Paris, 59 pp.

AUTHOR INDEX

Page numbers in italic indicate text pages on which references are quoted; those in upright figures indicate pages in reference lists

Abbott, P.L. *31*, 43, *181*, 203
Abed, A.M. *29*, 43
Adams, A.E. *29*, 43
Adams, J.B. *14*, *15*, 55
Aguilar, J. *271*, 285
Al-Rawi, G.J. *162*, 177
Alexander, C.S. *184*, 204
Alimen, M.H. *40*, 43
Allen, J.R.L. *28*, *29*, 43; *58*, *59*, *60*, *61*, *62*, *67*, *69*, *70*, *72*, *74*, *75*, *77*, *78*, *79*, *80*, *81*, *82*, 83, 85; *139*, *161*, *162*, *169*, *171*, *172*, 177; *181*, *184*, *186*, *199*, 204
Allen, M.J. *267*, 285
Allen, P. *29*, 44; *246*, *247*, *251*, *252*, *257*, 261
Allen, R.H. *256*, 261
Allison, K.R. *184*, 205
Alt, D. *31*, 56
Alvin, K.L. *116*, *132*, *133*, 135
America Commission on Stratigraphic Nomenclature 1961 *180*, 204
Anderson, F.W. *117*, 135
Anderson, H.A. *278*, *281*, 285
Andries, R.R. *35*, 44
Andrews, H.N. *2*, *20*, 54
Andrews, P. *38*, 44, 47, 52
Andriesse, J.P. *89*, 110
Annovi, A. *27*, 44
Arakel, A.V. *128*, 135
Ardrey, R.H. *36*, 50
Aristarian, L.F. *128*, 135
Arkell, W.J. *112*, *113*, *114*, *115*, *117*, *118*, 135
Aronson, J.L. *39*, 44
Atherden, M.A. *281*, 285
Atkinson, C.D. *142*, *144*, *154*, *161*, *173*, 177, 179

Avery, B.W. *247*, *252*, *253*, *254*, *255*, 258; *266*, 285
Baden-Powell, D.F.W. *245*, 258
Badgley, C. *36*, *38*, 44, 52
Badham, J.P.N. *12*, 55
Badham, N. *2*, 46
Bainbridge, R.B. *39*, 57
Baker, C.A. *252*, 258
Baker, V.R. *78*, 83
Ball, D.F. *247*, *249*, 258; *264*, *281*, 285
Ball, H.W. *62*, 83
Banham, P.H. *245*, 261
Banks, H.P. *18*, *23*, 44
Barahona, E. *271*, 285
Barghoorn, E. *14*, 44
Barghoorn, E.S. *15*, 44
Barrell, J. *23*, *25*, 44; *183*, 204
Bascomb, C.L. *266*, 285
Basilevsky, A.T. *3*, 48
Basu, A. *6*, 44
Bathurst, R.G.C. *128*, 135
Batten, D.J. *29*, 44
Beck, C.B. *23*, *25*, 44
Beckett, S. *264*, *265*, *277*, *281*, *282*, 289
Beckmann, G.G. *70*, 84
Beerbower, J.R. *34*, 46; *199*, 204
Behrensmeyer, A.K. *36*, *38*, *39*, *42*, 44, 51; *181*, *195*, 204
Bell, M. *265*, *267*, *276*, 285
Bennacef, A. *22*, 48
Bennett, D.K. *22*, *28*, 51
Bernoulli, D. *29*, 44; *128*, 135
Berrow, M.L. *278*, *281*, 285
Bertrand-Sarfati, J. *13*, 44
Besly, B. *149*, 177
Besly, B.M. *27*, 44

297

Best, M.   *12*, *13*, 46
Biagi, P.   *283*, 285
Bickford, M.E.   *12*, 45
Bidwell, O.W.A.   *36*, 45
Bildgen, P.   *9*, *29*, 52
Birkeland, P.W.   *5*, *7*, *15*, *23*, 45; *98*, 110; *180*, 204
Birrell, K.S.   *225*, *229*, *231*, *233*, *234*, *236*, 239, 240, 241
Bishop, W.W.   *38*, 45; *245*, 261
Blackwelder, E.   *183*, 204
Blades, E.L.   *12*, 45
Blank, H.R.   *29*, 45
Bläsi, H.   *130*, 135
Blaxland, A.B.   *22*, 45
Bledsoe, A.O.   *26*, *29*, 50
Blodgett, R.H.   *29*, 45
Blokhius, W.A.   *162*, 177
Blom, G.J.   *29*, 45
Boardman, J.   *244*, *246*, *247*, *248*, *253*, *257*, 258, 259, 261
Bohn, H.L.   *13*, 45
Bonamo, P.M.   *18*, *20*, 54
Bonneau, M.   *266*, *276*, *277*, 285
Booth, W.E.   *17*, 45
Bosellini, A.   *29*, 45
Boucot, A.B.   *2*, *20*, 54
Boucot, A.J.   *17*, *18*, *19*, *22*, *29*, 45, 49
Boulaine, J.   *129*, 135
Bowen, D.Q.   *244*, *245*, *247*, *249*, *259*, 261
Bown, T.M.   *25*, *28*, *31*, *42*, 45; *59*, 83; *139*, *151*, *161*, *169*, *171*, 177; *182*, *184*, *188*, *196*, *197*, *200*, *201*, 204, 205
Braun, H.M.H.   *39*, 55
Braunagel, L.H.   *31*, 45
Bray, W.   *264*, 286
Bretz, J.H.   *128*, 135
Brewer, R.   *15*, 45; *63*, *75*, 83; *90*, *100*, *105*, 110, 111; *119*, *120*, *123*, 135; *147*, *158*, *159*, 177; *181*, 204
Briden, J.C.   *88*, 111
Bridge, J.S.   *171*, *172*, 177; *184*, *186*, *199*, 204
Bridges, E.M.   *167*, *172*, 177
Brown, G.   *255*, 258
Brown, P.R.   *113*, *120*, *130*, 135
Bruce, J.G.   *213*, *215*, *216*, *238*, 239, 240
Bryan, A.L.   *246*, *247*, *250*, 260
Bryant, I.D.   *244*, 259
Bubb, J.N.   *173*, 179
Buchbinder, B.   *29*, 45
Buchbinder, L.B.   *29*, 45
Buckland, Rev. M.   *113*, 135
Bullerwell, W.   *62*, 85
Bullock, P.   *100*, 111; *243*, *247*, *252*, *253*, *254*, *255*, *256*, 258, 259, 261; *267*, *268*, *269*, *273*, 285
Buol, S.W.   *101*, 111; *123*, 135; *161*, *162*, 177

Burgess, I.C.   *58*, 83
Burgess, I.S.   *139*, 177
Burggraf, D.R.   *39*, 45, 57
Burke, A.S.   *220*, *237*, 240
Burleigh, R.   *276*, 285
Burrin, P.J.   *265*, *272*, *285*, 289
Butler, R.W.H.   *173*, 177
Button, A.   *6*, *8*, *9*, *16*, 45, 46
Butuzova, O.V.   *101*, 111
Buurman, P.   *31*, 46; *101*, 111; *112*, *119*, 135; *139*, *154*, 177

Callen, R.A.   *78*, 84
Calver, M.A.   *102*, *107*, 111
Campbell, F.H.A.   *13*, 46
Campbell, I.B.   *217*, *218*, *222*, 239, 240
Campbell, S.E.   *14*, *15*, 46
Carroll, D.M.   *254*, 259
Castledine, C.J.   *247*, 259
Catt, J.A.   *2*, 46; *242*, *243*, *244*, *245*, *246*, *247*, *249*, *250*, *252*, *253*, *254*, *255*, *256*, *257*, *258*, *259*, 261; *268*, *269*, *271*, *272*, *276*, *285*, 290
Caty, J.L.   *13*, 46
Cecile, M.P.   *13*, 46
Cerri, C.C.   *270*, 285
Chaloner, W.G.   *23*, 46; *117*, 135
Chalyshev, V.I.   *27*, 46; *182*, 205
Chandler, F.W.   *8*, 49
Charpal, O.   *22*, 48
Chartres, C.J.   *256*, *258*, 259
Chayka, V.M.   *8*, *9*, 46
Chernaya, I.M.   *3*, 48
Chernyakhovskii, A.G.   *31*, 46
Chester, D.K.   *247*, 259
Chiarenzelli, J.R.   *12*, *13*, 46
Childs, C.W.   *234*, 239
Chown, E.H.   *13*, 46
Clark, B.C.   *3*, 46
Clark, J.   *34*, 46
Clark, L.D.   *22*, 50
Clarke, M.R.   *243*, *247*, *249*, 259
Clay, P.   *268*, *270*, 286
Clayden, B.   *124*, *131*, 137
Clemmey, H.   *2*, 46
Cloud, P.E.   *14*, *15*, 46
Cohen, A.S.   *39*, 46
Connell, E.R.   *247*, *248*, 259
Conway, B.W.   *247*, *250*, 259
Coope, G.R.   *244*, *245*, *260*, 261
Cooper-Driver, G.   *25*, 55
Cope, M.J.   *117*, 135
Coppen, J.   *264*, *265*, *277*, *281*, *282*, 289
Cornwall, I.W.   *263*, *264*, *266*, *267*, *268*, *271*, *274*, *279*, 286
Cossey, P.J.   *29*, 43
Cotillon, P.   *130*, 135
Cotter, E.   *23*, 46
Coultas, C.L.   *6*, 46
Courtney, F.M.   *124*, 135; *274*, 286

Courty, M.A. *264*, *267*, *270*, *273*, *275*, 286, 288
Cowie, J.D. *218*, *219*, 239
Cox, A.V. *78*, 84
Crandell, D.R. *180*, 204
Crompton, E. *281*, 286
Crook, K.A.W. *181*, 204
Cummings, M.L. *22*, 46
Currant, A.P. *244*, *245*, 260, 261
Curtis, L.F. *124*, 135
Cutler, E.J.B. *220*, *221*, 240

Daghlian, C.P. *30*, 47
Dalland, A.B. *27*, 55
Dalrymple, J.B. ix, xiii; *176*, 179; *233*, 239; *246*, *249*, *250*, 259, 262; *267*, *271*, *277*, 279, 286, 289
Damon, R. *113*, *133*, 135
Daniels, R.B. *100*, 111
Danilov, I.S. *27*, 47
Davaud, E. *128*, *130*, *131*, 137
Davies, H. *245*, 261
De Coninck, F. *277*, 286
De Ford, R.K. *147*, 178
De La Beche, H.T. *113*, 135
Dean, J.M. *199*, 206
Dearing, J.A. *267*, 288
Denham, C.R. *200*, 206
Dennison, J.M. *17*, *22*, 53
Denny, C.S. *168*, 177
Dewey, J.F. *18*, *22*, 45
Dilcher, D.L. *27*, *29*, *42*, 53
Dimbleby, G.W. *254*, 259; *264*, *265*, *266*, *274*, *276*, *277*, *278*, *279*, *280*, *281*, *282*, 286, 287
Dimroth, E. *2*, 47
Dineley, D.L. *18*, *22*, 45; *61*, *62*, *67*, *83*, 84
Dingus, L. *183*, *195*, *196*, *197*, *198*, 205, 207
Dixon, E.L. *61*, *82*, 84
Dixon, J.K. *208*, 239
Donaldson, J.A. *12*, *13*, *46*, 47
Doner, H.E. *13*, 47
Donner, J.J. *245*, 262
Dorn, R.I. *14*, *15*, *17*, 47
Dorronsoto, C. *271*, 285
Downing, R.A. *62*, 85
Doyle, J.A. *132*, 137
Drake, M. *6*, 51
Drew, D.P. *276*, 286
Duchafour, P. *6*, *8*, 47; *124*, 136
Duchafour, P.L. *264*, *268*, *277*, *278*, *283*, 286
Dudal, R. *90*, 111
Dunham, R.J. *29*, 47
Dunne, K.C. *184*, 205
Dunoyer de Segonzac, G. *6*, 47
Dury, G.H. *29*, 47; *182*, 205
Dutton, C.E. *29*, 50

Eager, R.M.C. *102*, *107*, 111
Easton, R.M. *13*, 47
Edelman, M.J. *8*, *17*, 47
Eden, M.J. *264*, 286
Eden, R.A. *102*, 111
Edwards, K.J. *247*, *248*, 259
Edwards, W. *102*, 111
Eidt, R.C. *264*, *266*, 286
Eigen, M. *14*, 47
Eiler, J.P. *184*, 207
Elliott, T. *98*, 111; *154*, 178
Ellison, R.A. *252*, 260
Elston, D.P. *12*, 47
Emry, R.J. *34*, 47
Epstein, J.B. *6*, 47
Ermanovics, I. *8*, 49
Eskola, P. *8*, 47
Eslinger, E.V. *6*, 49
Evans, A.H. *243*, 260
Evans, E.M.N. *38*, 47
Evans, J.G. *246*, *247*, 259; *265*, *268*, *272*, *274*, *275*, *276*, *281*, 286, 287
Evans, L.J. *181*, 205
Eyre, S.R. *281*, 287

Fagerstrom, J.A. *28*, 53
Fahey, B.D. *181*, 206
Faller, A.M. *88*, 111
Fang, R.-S. *29*, 45
FAO-UNESCO 1974 *88*, *89*, *90*, *100*, *101*, 111
Farmer, H.G. *200*, 206
Farmer, V.C. *278*, *281*, 285
Faugeres, L. *29*, 47
Federoff, N. *264*, *267*, *268*, *270*, *272*, *273*, *275*, 285, 286, 287
Feofilova, A.P. *27*, *47*, 48
Ferrer, J. *145*, 178
Fields, R.W. *31*, 56
Findlay, D.C. *274*, 286
Firman, D.C. *40*, 48
Fisher, G.C. *23*, 48
Fisher, P.E. *243*, *247*, *249*, 259
Fisher, P.F. *267*, *270*, *274*, *275*, *277*, *280*, 287
Fitton, W.H. *113*, 136
Fitzpatrick, E.A. *9*, 48; *90*, 111; *129*, *132*, *133*, 136; *154*, *161*, *162*, *167*, 178; *247*, *254*, 259
Flannery, T. *30*, 48
Fleagle, J.G. *184*, 204
Fleming, *277*, *282*, 288
Fletcher, R. *18*, *22*, 45
Florensky, C.P. *3*, 48
Flores, R.M. *199*, 205
Folk, R.L. *29*, 48; *126*, *129*, *130*, 136, 137
Fontana, D. *27*, 44
Frakes, L.A. *124*, 136

Francis, J.E. *114*, *116*, *117*, *120*, *122*, *126*, *129*, *131*, *132*, *133*, 136
Frank, H.J. *39*, 45
Franks, P.C. *27*, 48
Frarey, M.J. *7*, 48
Franzmeier, D.P. *264*, *276*, 287
Freeman, P.S. *29*, 51
French, C. *246*, *247*, 260
Frenguelli, J. *35*, 48
Freytet, P. *29*, 48; *139*, 178; *182*, 205
Friedman, G.M. *29*, 54
Friend, P.F. *142*, *144*, *147*, *150*, 178
Froggatt, P.C. *231*, 239
Frye, J.C. *80*, 85; *180*, *181*, *182*, 205, 206, 207
Funnell, B.M. *244*, 260
Fyson, W.K. *18*, *22*, 45

Galloway, W.E. *29*, 48; *199*, 205
Gamble, E.E. *100*, 111
Gardiner, P.R.R. *59*, 84
Gardiner, W. *14*, 47
Gariel, O. *22*, 48
Garnet-James, S. *267*, 288
Garrido-Megias, *140*, *144*, 178
Gay, A.L. *3*, *4*, *5*, *6*, *7*, *8*, *9*, *15*, *17*, *29*, 48
Gelmini, R. *27*, 44
Gerasimov, I.P. *125*, 136
Gibbs, H.S. *208*, 239, 240
Giele, C. *17*, 50
Gile, L.H. *62*, *63*, *78*, 84; *129*, 136; *161*, *165*, 178
Gillespie, R. *78*, 84
Gillespie, W.H. *25*, *27*, 48
Gingerich, P.D. *195*, *196*, *198*, 205
Glacken, C.J. *36*, 48
Gladfelter, B.G. *246*, 260
Glaeser, J.D. *6*, 47
Goddard, E.N. *147*, 178
Goh, K.M. *216*, *217*, *234*, *237*, 239
Goldberg, P. *264*, *267*, *268*, *286*, 287
Goldbery, R. *29*, 48; *59*, *70*, 84; *112*, 136; *162*, 178; *184*, 205
Goldich, S.S. *29*, 48
Gorgoni, C. *27*, 44
Gossage, D.W. *67*, 84
Goudie, A. *59*, *77*, *78*, *80*, 84; *161*, 178; *182*, 205; *263*, 287
Goudie, A.S. *59*, *62*, *63*, *75*, *77*, *80*, 84
Gould, S.J. *42*, 49
Goulding, M. *184*, 205
Grabert, H. *12*, 49
Grandstaff, D. *15*, 54
Grandstaff, D.E. *3*, *4*, *5*, *6*, *7*, *8*, *9*, *15*, *17*, *29*, *47*, *48*, 50
Grange, L.I. *208*, 239
Graustein, W.C. *6*, 49
Gray, J. *17*, *18*, *19*, *29*, *45*, 49

Gray, W. *113*, 136
Green, C.P. *243*, *244*, *245*, 260
Greenfield, S. *256*, 261
Greig, J.R.A. *276*, *281*, *286*, 287
Grierson, J.D. *18*, *20*, 54
Griffin, J.G. *18*, *22*, 45
Griffiths, E. *215*, 239
Grigor'ev, N.P. *29*, 49
Grossman, R.B. *62*, *63*, 84; *129*, 136; *161*, *165*, 178
Gruhn, R. *246*, *247*, *250*, 260
Guardiola, J.L. *271*, 285
Guillet, B. *277*, *279*, *280*, *287*, 289

Habermann, G.M. *29*, 47
Halffter, G. *22*, 49
Hall, A.M. *247*, *248*, 259
Hallbauer, D.K. *14*, 48
Hallsworth, E.G. *70*, 84
Harland, T.L. *20*, 51
Harland, W.B. *78*, 84
Harmer, F.W. *245*, 260
Harmon, R. *245*, 261
Harms, J.C. *23*, 56
Harris, C.S. *208*, 240
Harris, T.M. *117*, 136
Harris, W.B. *29*, 56
Harrison, R.S. *129*, 136
Harvey, C. *181*, 207
Hassan, K.E.-D.K. *29*, 52
Hattelid, W.G. *173*, 179
Hay, R.L. *39*, *40*, 49; *78*, 84
Hayes, J.M. *13*, 54
Heckel, P.H. *80*, 84
Heine, J.C. *231*, *233*, 239, 240
Heling, D. *6*, 49
Henderson, J.B. *8*, *9*, 54
Hepburn, A. *278*, *281*, 285
Herd, R.K. *8*, 49
Herrera, L. *264*, 286
Heward, A.P. *168*, 178
Hey, R.W. *243*, *246*, *247*, *251*, *252*, 260, 261
Hickox, C.F. *18*, *22*, 45
Hieskanen, K.I. *6*, 55
High, L.R. *151*, 178
Hill, G.J.C. *21*, 55
Hill, R.S. *117*, *131*, 136
Hlustik, A. *29*, 49
Hodge, C.A.H. *277*, 289
Hodgson, L. *208*, 239
Hodson, F. *102*, *107*, 111
Hoffmeister, J.E. *129*, *132*, 136
Hogg, A.G. *237*, 239
Holberg, L. *128*, 135
Hole, F.D. *36*, 45; *101*, 111; *123*, 135; *161*, *162*, 177
Holland, H.D. *2*, *3*, *6*, *8*, *12*, *17*, 49

Holliday, D.W.   *102*, *107*, 111
Hollig, C.S.   *32*, 56
Holyoak, D.T.   *244*, *245*, 259, 260
Hooker, P.J.   *38*, 54
Horne, R.R.   *59*, 84
Horwitz, R.C.   *77*, 84
Howarth, D.T.   *216*, *217*, *234*, *235*, 240
Hower, J.   *6*, 49
Hower, M.E.   *6*, 49
Howitt, F.   *113*, 136
Howorth, R.   *217*, *229*, 240, 241
Hubert, J.F.   *27*, *29*, 49; *139*, 178
Hubert, J.R.   *59*, 84
Huddle, J.W.   *27*, 49
Hull, C.B.   *91*, 111
Hunt, C.B.   *2*, 49
Hunt, J.H.   *98*, 111
Hutton, J.T.   *59*, *77*, 84
Huzayyin, S.   *40*, 49

Illing, L.V.   *128*, *130*, 136
Imeson, A.C.   *272*, *275*, 287
Isaac, K.P.   *29*, *31*, 49
Ivanovich, M.   *244*, *245*, 260
Ives, D.   *216*, *237*, 241
Ives, D.W.   *213*, *215*, *237*, 239, 240

Jahns, R.H.   *184*, 205
James, H.L.   *22*, *29*, 50
James, N.P.   *128*, 136
Jarvis, R.A.   *254*, 259
Jenkinson, D.S.   *270*, 285, 287
Johnson, G.D.   *36*, 50
Jones, D.K.C.   *252*, 258
Jones, D.S.   *20*, 56
Jones, R.L.   *244*, *245*, 260
Jongerius, A.   *267*, *268*, *275*, 285, 287
Jorda, M.   *265*, 287
Jungerius, P.D.   *29*, 50; *182*, 205; *272*, *275*, 287

Kabata-Pendias, A.   *27*, 50
Kalliokosi, J.   *13*, 50
Keef, P.A.M.   *280*, 287
Keeley, H.C.M.   *264*, *265*, *266*, *278*, *281*, *282*, 287, 288
Keen, D.H.   *244*, *245*, 260
Keller, W.D.   *26*, *29*, 50
Kemp, R.A.   *246*, *247*, *250*, *251*, *252*, *254*, *257*, 260, 261
Kennett, J.P.   *32*, 50
Kent, L.E.   *9*, 50
Kerney, M.P.   *246*, *247*, *250*, 260; *276*, 285
Kerr, D.R.   *31*, 52
Kesel, R.H.   *184*, 205
Khosbayar, P.   *31*, 46
Kidston, R.   *22*, 50
Kietzke, K.K.   *34*, 46
Kimberley, M.M.   *2*, *3*, *8*, *15*, *17*, 47, 50, 54
King, W.W.   *60*, 84
Kjellesvig-Waering, E.N.   *20*, 50
Klappa, C.F.   *129*, *132*, 136
Klein, C.   *3*, *6*, 56
Klinge, H.   *89*, 111
Knight, A.E.   *12*, 50
Knight, M.J.   *77*, 84
Kohn, B.P.   *217*, 241
Komarek, E.V.   *131*, 136
Koryakin, A.S.   *13*, 50
Kraus, M.J.   *25*, *28*, *31*, *42*, 45; *59*, 83; *139*, *151*, *161*, *169*, *171*, 177; *182*, *184*, *188*, *197*, *200*, *201*, 204, 205
Kroonenberg, S.B.   *12*, 50
Krumbein, W.C.   *17*, 50
Kryuchkov, V.P.   *3*, 48
Kubiena, W.L.   *25*, *38*, 50; *124*, *131*, 136; *249*, 260; *283*, 288
Kukla, G.J.   *40*, 50
Kusmin, R.O.   *3*, 48
Kwaad, F.J.P.M.   *267*, *268*, *272*, *276*, 287, 288

Laffan, M.D.   *220*, *221*, 240
Lake, R.D.   *252*, 260, 262
Lamey, C.A.   *22*, 50
Lang, W.H.   *17*, *22*, 50
Langohr, R.   *257*, 260; *272*, 288
Laruelle, J.   *162*, 177
Leakey, M.G.   *39*, 51
Leakey, R.E.   *39*, 51
Leamy, M.L.   *213*, *220*, *221*, *237*, *238*, 239, 240
Leeder, M.R.   *58*, *79*, 84; *139*, *161*, *169*, *171*, *172*, 177, 178; *184*, *186*, *199*, 204, 205
Leonard, A.B.   *80*, 85
Leopold, L.B.   *184*, 207
Leslie, D.M.   *220*, 240
Lewan, M.D.   *13*, 51
Lewis, J.S.   *3*, 52
Lewontin, R.L.   *42*, 49
Li, Z.-P.   *29*, 45
Limbrey, S.   *266*, *269*, 288
Lindemann, W.L.   *29*, 51
Lindsay, J.F.   *2*, 51
Lionnet, J.F.G.   *124*, *132*, 136
Lipschutz, B.   *36*, 52
Llewellyn, P.G.   *78*, 84
Lobo, R.C.   *181*, 206
Loughnan, F.C.   *29*, 51
Loveland, P.J.   *267*, 285
Lowe, D.J.   *236*, 240
Lucas, C.   *26*, *29*, 51
Ludwig, D.   *32*, 56
Luterbacher, H.P.   *145*, 178
Lynn, W.C.   *13*, 47

Mabesoone, J.M.   *181*, 206
MacCarthy, I.A.J.   *59*, 84
MacClintock, P.   *247*, *250*, 260
MacDonald, J.R.   *34*, 51
Mackney, D.   *100*, 111; *264*, *278*, 288
Macphail, R.   *267*, 285, 286
Macphail, R.I.   *265*, *266*, *267*, *268*, *270*, *271*, *272*, *274*, *275*, *276*, *277*, *278*, *279*, *280*, *282*, *283*, 287, 288, 289
Madgett, P.A.   *268*, *271*, *272*, 290
Maguire, D.   *277*, *282*, 288
Mahaney, W.C.   *180*, *181*, 206
Maiklem, W.R.   *29*, 51; *128*, *130*, 136
Mann, A.W.   *77*, 84
Marbut, C.F.   *6*, *25*, 51
Margaritz, M.   *29*, 45
Marshall, J.D.   *60*, *62*, *67*, 85
Martin, L.D.   *22*, *28*, 51
Martou, M.   *131*, 137
Massa, D.   *18*, 48
Matthews, B.   *254*, 260
Matthews, E.G.   *22*, 49
Mazzoni, M.M.   *35*, 55
McBride, E.F.   *27*, *29*, *48*, 51
McCracken, P.J.   *101*, 111
McCracken, R.J.   *123*, 135; *161*, *162*, 177
McCraw, J.D.   *237*, 239
McDonald, R.C.   *184*, 205
McEwan, C.   *264*, 286
McFarlane, M.J.   *5*, *12*, *29*, *31*, 51
McGowran, B.   *31*, 51
McGregor, D.F.M.   *243*, *244*, *245*, 260
McKerrow, W.S.   *18*, *22*, *28*, *45*, 57
McNeal, B.L.   *13*, 45
McPherson, J.G.   *25*, 51; *59*, 84; *112*, 136; *184*, 206
Mehlich, A.   *6*, 51
Mencher, E.   *2*, *20*, 54
Meyer, R.   *27*, 51; *112*, 136
Miall, A.D.   *186*, *199*, 206
Middleton, L.T.   *188*, *201*, 205
Miller, A.R.   *12*, *13*, *46*, 51
Miller, J.A.   *38*, 54
Milne, J.D.G.   *218*, *219*, *238*, 239, 240
Milner, A.R.   *28*, 51
Milnes, A.P.   *59*, *77*, 84
Minch, J.A.   *181*, 203
Minch, J.A.K.   *31*, 43
Mitchell, G.F.   *244*, *247*, *248*, 260
Mitchell, J.L.   *27*, 55
Mitchum, R.M.   *173*, 179
Moffat, A.J.   *247*, *252*, 261
Molenaar, N.   *59*, 84
Moore, P.   *279*, 288
Moorlock, B.S.P.   *252*, 260
Morand, F.   *31*, 51
Morey, G.B.   *22*, *29*, 51
Morgan, W.C.   *12*, 50
Morris, S.C.   *20*, 51

Morrison, R.B.   *151*, 178; *180*, *181*, 206
Moseley, K.A.   *244*, 259
Moss, A.J.   *246*, *247*, *250*, 260
Moussine-Pouchkine, A.   *13*, 44
Mücher, H.J.   *29*, 50; *267*, *268*, *272*, *276*, 287, 288
Muir, A.   *89*, 111
Muller, J.   *30*, 51
Multer, H.G.   *129*, *132*, 136
Murphy, C.P.   *243*, *255*, *256*, 259; *267*, *269*, 288
Mutti, E.   *145*, 178

Naraeva, M.K.   *3*, 48
Neall, V.E.   *225*, *228*, *230*, *231*, 240
Nelson, L.A.   *100*, 111
Netterberg, F.   *59*, *77*, 85
Nicholas, J.   *9*, *29*, 52
Nijman, W.J.   *140*, *142*, *143*, *144*, *173*, 178
Nikiforoff, C.C.   *2*, 52
Niklas, K.J.   *17*, *20*, 52
Nikolaeva, O.V.   *3*, 48
Nilsen, T.H.   *31*, 52
Nio, S.D.   *140*, *142*, *143*, *144*, *173*, 178
Nisbet, R.   *283*, 285, 288
Nkedi-Kizza, P.   *31*, 55
Nørnberg, P.   *267*, 286
Norris, G.   *117*, 136
North American Commission on Stratighraphic Nomenclature 1983   *181*, 206
Norton, P.E.P.   *244*, 260
Norton, R.A.   *18*, *20*, 54
Norton, S.A.   *13*, 52
Nozette, S.   *3*, 52
Nunan, W.E.   *29*, 56
Nussinov, M.D.   *13*, 52

Oberlander, T.M.   *14*, *15*, *17*, 47
O'Connor, G.A.   *13*, 45
Oldfield, F.   *267*, 288
de Olmedo Pujol, J.L.   *276*, 288
Opdyke, N.D.   *36*, *50*, 55
Ori, G.G.   *142*, 178
Ortlam, D.   *2*, *27*, 52
Overbeck, R.M.   *147*, 178
Owen, N.   *246*, *247*, 261

Pain, C.F.   *225*, 240
Palmer, F.   *14*, *15*, 55
Pape, T.   *162*, 177
Parker, A.   *77*, 85
Parnell, J.   *29*, *52*; *58*, 85
Parsons, R.B.   *181*, 206; *264*, 288
Parti, G.   *131*, 137
Patel, I.M.   *22*, 52
Paterson, K.   *246*, *247*, 261
Patterson, S.H.   *27*, 49

Pawluk, S. *182*, 206
Payton, R.W. *254*, 261
Pearce, A.J. *222*, 240
Pennington, W. *264*, 289
Penny, L.F. *244*, 260
Percival, C.J. 26, 52; *87*, *91*, *101*, 111
Perrin, R.M.S. *245*, 261; *277*, 289
Perry, E.A. 6, 49
Peterman, R.M. *32*, 56
Peterson, F.F. *62*, *63*, 84; *129*, 136; *161*, *165*, 178
Peterson, G.L. *31*, 43; *181*, 203
Pettijohn, F.J. 22, 29, 50
Pettyjohn, W.A. *31*, 52; *181*, 206
Phillips, C.J. *222*, 240
Phillips, T.L. *17*, *22*, 53
Philobbos, E. 29, 52
Picard, M.D. *151*, 178
Pick, M.C. 58, *62*, *66*, 85
Pickerill, R.K. *20*, 51
Pickford, M. *38*, 52
Pickton, C.A.G. *78*, 84
Pilbeam, D. *36*, *38*, *40*, 52
Pilbeam, D.R. *36*, 52
Pittman, E.D. *100*, 111
Plaziat, J.-C. 29, 48; *182*, 205
Polach, H.A. 78, 86; *182*, 207
Pohlen, I.J. *208*, 240
Pomerol, C. 27, *31*, 52
Pons, L.J. *155*, 178
Poty, E. 29, 53
Power, P.E. 27, 53
Powers, M.C. *129*, 137
Pratt, L.M. *17*, *22*, 53
Price, W.A. *80*, 85
Prior, J.C. *184*, 204
Pronin, A.A. *3*, 48
Pronk, J.W. *147*, *163*, 178
Prothero, D.R. *200*, 206
Proudfoot, V.B. *264*, 266, *282*, 289
Pugh, M.E. *120*, 137
Puigdefabregas, C. *199*, 206
Puigdefabregas, C.T. *154*, *173*, 178
Pullar, W.A. *225*, *229*, *231*, *233*, *234*, *236*, *238*, 239, 240, 241

Raeside, J.D. *208*, *211*, *216*, *217*, *218*, *220*, *237*, 240
Rafter, A.T. *216*, *217*, *237*, 239
Ragg, J.M. *124*, *131*, 137
Ralph, N. *277*, *282*, 288
Ramaekers, P. *12*, 53
Ramsbottom, W.H.C. *102*, *107*, 111
Rankama, K. *8*, 53
Ratcliffe, B.C. *28*, 53
Rayner, J.H. *247*, *252*, *255*, 258; *270*, 287
Read, J.F. 77, 85; *128*, 137
Read, W.A. *199*, 206
Reeder, R.J. *40*, 49; *78*, 84
Reeves, C.C. 59, 77, *80*, 85
Reeves, C.C. Jnr. *161*, 179
Reffay, A. *31*, 53
Reid, C. *246*, *247*, 261
Reid, E.M. *246*, *247*, 261
Reinson, G.E. 29, 56
Rekshinskaya, L.G. 27, 48
Retallack, G.J. 2, *6*, *15*, *18*, *19*, *21*, *22*, *23*, *24*, *25*, *26*, *27*, *28*, *29*, *31*, *33*, *34*, *35*, *36*, *37*, *38*, *42*, 53, 54; *112*, 137; *139*, *161*, *167*, *169*, 179; *181*, *182*, *184*, *200*, *201*, *202*, 206
Rey, P.H. *36*, 50
Richmond, G.M. *180*, *181*, 204, 207
Ricketts, B. *13*, 47
Ricq-Debouard, M. *31*, 53
Riding, R. *128*, *129*, 137
Riecken, F.F. *264*, 288
Righi, D. *277*, 289
Ritzkowski, S. *31*, 54
Ritzma, H.R. *181*, 207
Riveline-Bauer, J. *31*, 51
Robbin, D.M. *128*, *132*, *133*, 137
Robert, P. 29, 47
Robertson, L. *249*, 261; *270*, 275, *282*, *283*, 289
Robinson, J.E. *244*, *245*, 260
Roeschmann, G. 27, 54
Rolfe, W.D.I. *18*, *20*, *28*, 54
Romans, J.C.C. *247*, *248*, *249*, 261; *270*, 275, *282*, *283*, 289
Ronca, L.B. *3*, 48
Roscoe, S.M. *6*, 7, 48, 54
Rose, J. *245*, *246*, *247*, *251*, *252*, *257*, 261
Rose, M.D. *36*, 52
Rossel, J. *145*, 178
Rossi, D.L. 29, 45
Rothwell, G.W. *25*, 27, 48
Rove, O.N. *147*, 178
Rozen, O.M. *9*, 54
Rubin, D.M. 29, 54
Rudeforth, C.C. *247*, *248*, 261
Runge, E.C.A. *216*, *217*, *234*, *235*, *237*, 240, 241
Rushton, A.W.A. *62*, 85
Russel, J.D. *278*, *281*, 285
Russell, E.W. *15*, 54
Ryka, W. 27, 50

Sabine, P.A. *62*, 85
Sadler, P.M. *183*, *195*, *196*, *197*, *198*, 205, 207
Salisbury, E.J. *271*, 289
Salop, L.J. *9*, 54
Sanders, J. *257*, 260
Sangres, J.B. *173*, 179
Sanson, G.D. *30*, 54

Santos, F. *271*, 285
Saveliev, A.A. *13*, 55
Scaife, R.G. *265*, *268*, *272*, *275*, *278*, *279*, *280*, *281*, *283*, 285, 286, 288, 289
Schankler, D.M. *196*, 207
Schau, M.K. *8*, *9*, 54
Scheckler, S.E. *25*, *27*, 48
Schidlowski, M. *3*, *6*, 56
Schindel, D.E. *195*, 207
Scholtes, W.H. *264*, 288
Schopf, J.M. *2*, *20*, 54
Schopf, J.W. *3*, *6*, *13*, *14*, 49, 54, 56
Schulenschko, I.K. *13*, 55
Schultz, C.B. *181*, 207
Schumm, S.A. *17*, *22*, *28*, 54
Schuster, P. *14*, 47
Scott, G.R. *12*, 47
Scrivner, J.V. *22*, 46
Searle, A.B. *26*, 54
Searle, P.L. *229*, 239
Seelye, F.T. *208*, 239
Seguret, M. *140*, *144*, 179
Selivanov, A.S. *3*, 48
Selleck, B.W. *29*, 54
Sevon, W.D. *6*, 47
Shackleton, N.J. *245*, 261
Sharp, R.P. *12*, *22*, 54
Shear, W.A. *18*, *20*, 54
Sheerin, A. *23*, 46
Shipman, P. *38*, 54
Shlemon, F.J. *78*, 85
Shotton, F.W. *244*, *245*, 260, 261; *276*, 289
Siegel, B.Z. *14*, 54
Sieveking, G. de G. *250*, 260
Sighinolfe, G. *27*, 44
Simmons, I.S. *265*, *277*, *280*, *281*, 289
Simons, E.L. *184*, 204
Singer, A. *29*, 54
Singer, M.J. *31*, 55
Singewald, J.T. *147*, 178
Slager, S. *162*, 177; *268*, 289
Slate, M.J. *144*, *147*, *150*, 178
Sloan, R. *29*, 55
Smart, J.G.O. *62*, 85
Smith, A.G. *78*, 84
Smith, E.L. *18*, *20*, 54
Smith, K. *264*, *265*, *277*, *281*, *282*, 289
Smocovitis, V. *17*, 52
Sneed, E.D. *126*, *129*, 137
Sochava, A.V. *13*, *29*, 55
Soil Survey Staff 1960 *89*, *98*, *100*, 111
Soil Survey Staff 1975 *9*, *12*, *13*, *20*, *22*, *23*, *25*, *26*, *27*, *29*, *30*, *31*, *35*, *36*, *38*, *41*, *55*; *125*, 137; *147*, *155*, *162*, *167*, 179; *266*, 289
Sokolov, V.A. *6*, 55
Sombroek, W.G. *29*, *31*, *39*, 55
Souchier, B. *266*, *276*, *277*, 285

Spalleti, L.A. *35*, 55
Speight, J.G. *181*, 204
Speksnijder, A. *147*, 179
Spicer, B.E. *184*, 205
Squirrell, H.C. *60*, *62*, *66*, 85
Stace, H.C.T. *124*, *132*, 137
Staley, J.T. *14*, *15*, 55
Stanley, K.O. *31*, 45
Stanworth, C.W. *12*, 55
Stebbins, G.L. *21*, *29*, *30*, 55
Steel, R.J. *29*, 55; *59*, 85; *139*, *161*, *165*, *176*, 179
Steila, D. *139*, *162*, *167*, *172*, 179
Stephen, I. *255*, 258
Stephens, N. *247*, *248*, *249*, 261
Stevens, G.R. *219*, 240
Stevenson, D.J. *3*, *6*, 56
Stevenson, F.J. *8*, *15*, 55
Stevenson, I.P. *102*, 111
Stipp, J.J. *128*, *132*, *133*, 136
Stoops, G. *267*, *268*, 285
Størmer, L. *20*, 55
Stout, T.M. *181*, 207
Strahan, A. *103*, 111; *114*, 137
Strasser, A. *128*, *130*, *131*, 137
Straw, A. *245*, *247*, 261
Strickland, E.L. *3*, 55
Stringer, C.B. *245*, 261
Strother, P.K. *17*, *22*, 55
Stubblefield, C.J. *102*, *107*, 111
Sturdy, R.G. *246*, *247*, *256*, *257*, 261
Sturt, B.A. *27*, 55
Sugden, W. *128*, *130*, 137
Suggate, R.P. *210*, *211*, *213*, 240
Summerfield, M.H. *13*, 55
Sutcliffe, A.J. *245*, 261
Sutherland, C.F. *208*, 239
Swain, T. *25*, 55
Swineford, A. *80*, 85
Sys, C. *162*, 177
Szabo, I. *131*, 137

Tahirkheli, R.A.K. *36*, 50
Taieb, M. *39*, 44
Tanaka, R.T. *3*, 50
Tanner, L.G. *181*, 207
Tauxe, L. *36*, 44, 55; *181*, 204
Taylor, K. *62*, 85
Taylor, N.H. *208*, 239, 240
Thomas, K.D. *275*, 289
Thomas, R.G. *60*, 83
Thomassen, J.R. *30*, 56
Thomasson, A.J. *252*, 262
Thompson, G.R. *31*, 56
Thompson, I. *20*, 56
Thompson, R. *267*, 288
Thompson, S. III. *173*, 179
Tiffney, B.H. *27*, 56; *184*, 204
Tilley, P.D. *247*, *250*, 261

Todd, R.G. *173*, 179
Tonkin, P.J. *216, 217, 237*, 239, 241
Tooley, M. *265, 277, 280, 281*, 289
Topping, W.W. *232*, 241
Townsend, W.N. *124, 131*, 137
Trask, P.D. *147*, 178
Traverse, A. *17, 22*, 55
Tremblay, L.P. *12*, 56
Trichet, J. *31*, 51
Trotter, F.M. *62, 67*, 85
Trudgill, S. *124*, 135
Tunbridge, I.P. *59, 60*, 85
Turner, C. *245*, 261
Turner, J. *276*, 287
Turner, P. *27, 44*; *139*, 177
Tursina, *267, 268*, 285
Tyler, N. *6, 8, 9*, 46
Tyuflin, Y.S. *3*, 48

Upchurch, G.R. *132*, 137

Vadour, J. *265*, 287
Vail, P.R. *173*, 179
Vakhrameev, V.A. *116*, 137
Valentine, K.W.G. *ix*, xiii; *176*, 179; *246, 247*, 262; *271, 277, 279*, 289
Valentine, K.W.S. *276*, 287
Valeton, I. *9, 29*, 56
Van Couvering, J.A. *30, 38*, 54, 56
Van Couvering, J.A.H. *38*, 47
Van der Meer, J.J.M. *271*, 288
Van der Poouw, B.J.A. *39*, 55
Van der Wetering, H.T.J. *268*, 289
Van Hart, D.C. *3*, 46
Van Kliet, B. *272*, 288
Van Moort, J.C. *6*, 56
Van Vliet, A. *199, 206*; *154*, 178
Van Wormelo, K.T. *14*, 49
Veen, T.B. van der *147*, 179
Vehkov, A.A. *13*, 52
Velbel, M.A. *6*, 49
Visser, C.F. *36*, 50
Vita-Finzi, C. *264*, 290
Vogel, D.E. *8*, 56
Vogl, R.J. *32*, 56
Vondra, C.F. *39, 45, 57*; *184*, 204
Vucetich, C.G., *217, 225, 229, 231*, 241

Waechter, J. de A. *247, 250*, 259
Wagner, C.W. *128*, 135
Wagner, L.W. *29*, 44
Wainwright, G.J. *264, 265, 277, 281, 282*, 289
Walkden, G.M. *29*, 56
Walker, A. *38*, 54
Walker, A.D. *278, 281*, 285
Walker, B.H. *32*, 56

Walker, J.C.G. *3, 6*, 56
Walker, R.G. *23*, 56
Walker, T.W. *216, 217, 234, 235*, 240
Walls, R.A. *29*, 56
Walter, M.R. *3, 6, 13*, 54, 56
Walters, R. *78*, 84
Ward, W.C. *129, 130*, 137
Ward, W.T. *222, 224, 225*, 241
Wardlaw, N.C. *29*, 56
Wasser, G.G.M. *147*, 179
Wasson, R.J. *78*, 84
Waterbolk, H.T. *264*, 290
Watts, N.L. *29*, 56; *70, 78*, 85
Weaver, C.E. *6*, 56
Webb, S.D. *30, 34, 35*, 56
Webster, T. *113, 126*, 137
Weir, A.H. *2, 46*; *247, 252, 255*, 258; *268, 271, 272*, 290
Welch, F.B.A. *62, 67*, 85
Wescott, J.F. *26, 29*, 50
Wesley, A. *124*, 137
West, I.M. *29*, 57; *113, 114, 117, 124, 126, 130*, 138
West, R.G. *243, 244, 245, 246, 260*, 261, 262
Wheeler, H.E. *183, 184*, 207
White, D.E. *60, 66*, 85
White, E.I. *62*, 85
White, H.J. *39*, 45
White, H.T. *39*, 57
Whiteman, A.J. *169*, 179
Whiteman, C.A. *246, 247, 257*, 261
Whiteside, E.P. *264, 276*, 287
Whyte, F. *38*, 45
Widmeir, J.M. *173*, 179
Wieder, M. *158, 161*, 179
Wier, K.L. *29*, 50
Wilde, R.H. *220*, 241
Willard, B. *22*, 57
Willams, B.P.J. *59, 60, 61, 62, 67, 69, 72, 74, 75, 77, 79, 81, 82, 83*, 85
Williams, G.D. *140, 142, 173*, 179
Williams, G.E. *6, 9, 11, 12, 13*, 57; *58, 78*, 85, 86; *151*, 179; *182*, 207
Williams, L.A.J. *39*, 57
Williams, M.A.J. *40*, 57
Williams, R.C. *144, 147, 150*, 178
Willis, E. *277*, 289
Wilmott, A. *279*, 288
Wilson, D. *252*, 262
Wilson, D.G. *246*, 262
Wilson, J.L. *129, 130*, 137, 138
Wilson, M.D. *100*, 111
Wilson, R.B. *102, 107*, 111
Wing, S.L. *184*, 204
Winkler, D.A. *31, 42*, 57
Winkler, H.G.F. *6*, 57
Winkler-Oswatitsch, R. *14*, 47
Witzke, B.J. *80*, 84

Wolfe, J.A. *32*, 57
Wolman, M.G. *184*, 207
Wopfner, H. *29*, 57
Wright, V.P. 28, *29*, 57; *70*, 86; *112*, *119*, *128*, *129*, *131*, *133*, 137, 138, *161*, 179
Wymer, J. *245*, 261
Wymer, J.J. *280*, 287

Yaalon, D.H. 2, 57; *158*, *161*, 179; *255*, 258

Yam, O.-L. *23*, 48
Yang, X-C. *29*, 45
Young, D.J. *208*, *213*, *214*, 241

Zang, N. *29*, 45
Zaviyaka, A.I. *8*, 46
Zeigler, A.M. *18*, *22*, *28*, 45, 57
Zeuner, F.E. *247*, *250*, 262
Zimmerle, W. *27*, 52
Zonneveld, I.S. *155*, 178

SUBJECT INDEX

A horizon 291
  acid oak woodland 280
  Great Dirt Beds 121-3, 129
  microstructure analysis 268
  see also Podzolization
Abdon Limestone Formation 62, 66-7, 68
  geomorphological record 76, 80
Albic horizon, E/A2 paleosol profile 87, 291
  definition 88-9
  modern soils 88-90
Alfisols 41, 291
  earliest 25
  Oligocene 32-3
  Oxic Haplustalf 37-8
Algae, evolution of land plants 21-2
Alkalinity, identification, apatite 268
Alleröd interstadial 265
  soil processes 269
Alluvial formations
  Capella 142
  depositional model 145
  sedimentary facies 144
Alluvial stratigraphy 167-72
  analysis 201
  simulation studies 171-2
  time resolution, section completeness and fossil record 194-9
    sediment accumulation rates 199-201
    sedimentation and erosion 183-8
    palaeotopography 188-94
    summary 201-3
Alston Block see Firestone Sill
Alton Marine Band 107-9
Aluminium, indicator, weathering status 234

Amino acid levels in dating 234-5, 243
Andic Ustochrept, columnar section 33
Anglian 244, 247
  humic gley soil 246
Animals see Land animals
Antian stage 244
Apatite, use in identifying alkalinity 268
Apes, Dryopithecine, Kenya 38
Aquents 291
  Capellan paleosols 167
  and flood plain drainage 168, 171
Araucaria, *Araucarioxylon* 117, 118
Archaeopteris halliana, evidence for forest 23, 25
Archaeology
  Flandrian, dates 265
  micromorphology 267
  soil dating 263
Argillans
  argillic horizons 291
  and paleo-argillic 252-5
    yellow argillans, mineralogy 253, 256, 258
  Capella paleosols 153, 159
  definition 151, 291
  Pebbly Clay Drift 252
  pre-Flandrian sequence 248
  see also Paleo-argillans
Argillasepic fabric 291
Argillopedoturbation 161, 291
Aridisols 41, 291
  desert ecosystems 30, 32-3
  Old Red Sandstone see Calcretes
Arikareean, transition to savannah 31-3
Asepic fabric 291
Ash Formations, New Zealand 224-30
Ashgillian
  Dunn Point Formation 18

308 SUBJECT INDEX

Ashgillian *continued*
  Juniata Formation 18
Atmosphere, early 2–13
Augite weathering 248
Avulsion
  channel, alluvial fans 145
  sediment-filled scours 188

B horizon 161, 291
  Hazleton long cairn 272–4
  illuviation, fine clay coatings 268
  mean residence date, woodland
    removal 280
  spodic paleosols 191
'Backshedding', clastic, Capella
    Formation 142
Badlands National Park
  grassland development 31–5
  paleosol sequence, Scenic
    Member 200–2
  superimposed paleosols 31
  columnar sections 33
Barham soil 246–7
  and Valley Farm soil 251
Basin deposits, extra- and intrabasinal
    controls 199, 203
Bauxites 41
  examples 9
  formation 9
  as indicators 29
Baventian stage 244
Beestonian
  pedogenesis 244, 246, 247
  podzolization 246
  Valley Farm soil 251
Bennettitalean cycads 117
Biarritzian transgression 173
Biotite weathering 248
Bighorn Basin, Wyoming
  faunal resolution 198
  Willwood paleosols 198
Biopedoturbation 161
Boreal paleosols 265, 271
  erosion of loess 280
Bramertonian stage 244
Braunerde fabric 249, 250
  braunification 294
  erde fabric 249
'Braunlehm' paleosol 38
Brick earths, paleo-argillic soils 255
Brimpton interstadial 244
Britain
  stratigraphy, Quarternary 243–5
    buried pre-Flandrian soils 245–52
    non-buried pre-Flandrian
      soils 252–7
British Quarternary stages 244
Bronze age
  human activity 265
  podzolization 280

soil burial 267, 271
soil erosion 276
stagnogleys and stagnopodzols 277–8
woodland clearance and
  podzolization 279
Brown earths
  missing phase, podzolization 280
  sub-Boreal formation 272–4
  see also Argillans
Burning *see* Fire

C horizon 292
  Great Dirt Bed 121–3, 129
C/N ratios, burning by man 270
Caesar's Camp gravels, Pleistocene
    soil 249
Calcan, calcitan 292
Calcareous paleosols
  analysis 124
  geology 112–17
  Great Dirt Bed 118–20, 121
    discussion 130–4
    interpretation 122–4
    origins 128–30
    pebbles 126–8
  length of exposure 131
  Lower and Basal Dirt Beds 120–2
    interpretation 125
Calciorthids 32–3
Calcium carbonate nodules 158, 160
  lacustrine carbonates 144, 146
  see also Calcretes: Limestone
Calcium, pedotranslocation 235
Calcretes 41
  alluvia 81–2
  bioturbation 68–70
  climate changes 13, 40, 80
  definitions 128, 161, 292
  field characteristics 62–6
  fracture systems 70–1
  glaebules 63–7
  Great Dirt Beds 12
  lateral extent 67–8
  paleogeomorphology 79–80
  pedogenic model 75–8
  profile truncation 71–4
    eroded 73
    weathered 72
  stage III caliches 165
  stratigraphy 59–62
  time-scales 78–9
Caldey Island, Moor Cliffs Formation,
  calcretes 64–5, 73–5
Caliche *see* Calcrete
Calluna, podzol indicator 279–80
Cambic horizon 292
Campodarbe Group
  depositional model 145
  sedimentary facies 143, 144

SUBJECT INDEX

stratigraphy 146
Capella Formation xii
  different sedimentation rates 172–4
  field and laboratory procedures 147
  geology 140–4
  origin of paleosol variation 167–72
  sedimentation and paleosols 147
    Laguarres-Capella sequence 147–55
    Torsal Gross sequence 162–7
    Valturo sequence 155–62
  stratigraphy 144–6
  study localities 147
  summary 174–6
  time-scale 145
Carbon, evidence for soil microbes 15
Carbon-14 dating
  loess 216–17
  and micromorphology 266
  Quarternary deposits 181
Carbon dioxide calcretes, climate changes 13
Carboniferous, Upper
  climate indications 110
  Firestone Sill 91–102
  paleosols with albic horizons 87–109
  Pot Clay Ganister 102–7
  Sheffield Blue Ganister 107–9
  soil formation, factors affecting 87–8
  modern soils 88–90
Castle rock *see* Tanzania
Catenas
  and calcretes 59
  definition 139, 292
  methodology, in paleosols 176
Chadronian Paleustalfs 31–3
Chalazoidites, correlations 229
Channel migration, sediment-filled scours 188
Cheirolepidiaceae, Purbeck 132
Chelford interstadial 244
Chinle Formation, Arizona
  palaeotopography 188
  paleosol development 191
  scouring 189–90, 193
  Petrified Forest Member 193
  time section 195
  unconformities 201
Classopollis (pollen) 116
Clay
  differences, in identification 234, 236
  dusty clay illuviation and lessivage 272–3
    Neolithic deforestation, effect on pedogenesis 274–5
Clay-with-flints drift deposits 255
Climate, Flandrian 265
Cold climate
  accumulation 222
  gully dissection 223
Conata clay paleosols 33

Coniferous forest, Purbeck 116–17
Connecticut River, floods, restriction of scouring 184
Correlation of paleosols 236–8
  loess sequences 215
Corton interstadial 244
Crevasse channels, alluvial fans 145
Cretaceous extinction of dinosaurs 41
Cromerian stage 244, 247
  paleo-argillic horizons 258
  tills, Eastern England 257
  Valley Farm Soil 246–7, 251–2
Crystallaria, in calcrete 63, 292
Crystic plasmic fabric 292
Cupressinocladus valdensis 116
Cutans, clay 292
  argillans, similarity 100
  and water table 101
Cyanobacteria, microfossil 14–15

Dartmoor podzols 282
Decalcification, rate of loss 271
  see also Lessivage
Deforestation
  Bronze Age woodland clearance 264, 268
  Mesolithic clearance 271–2, 180
  Neolithic clearance 274–5, 276
  'slash and burn' 275, 281
Denison profile, Huronian 3–5, 7–8, 15
Devensian stage 244, 247
  loess, reworked 249
Dhok Pathan Formation, Siwaliks 36–8
Dirt Beds, Purbeck
  Great 117–21
  Lower and Basal 120–2

E/Az (albic) horizon *see* Albic horizon
Earthworms 20
Egg yellow papules, major pre Devensian processes 253, 255
Emsian, red-bed succession 60–2
Entisols 41, 293
  Capella formation 155
  Red Fluvents, relict bedding 32
  zeolitic paleosols 39
Eoastrion and soil microbes 14
Eocene
  Capella Formation xii, 139
  sea-level fluctuations 173
  Tremp-Gauss Basin 143
  Willwood Formation, Wyoming 183
Eohostimella, Early Silurian 19–20
Epipedon 293
  mollic, Great Dirt Bed 121
  umbric 296
Epsilon bedding 150
Erde fabric, definition 249

310 SUBJECT INDEX

Eurypterids, Ordorician 20
Eutrandepts, Ustollic 32
Eutrochrepts, Fluvaquentic 32

Faecal pellets, absence, Great Dirt Beds 129
Fammennian disconformity 60
  fluvial rocks 62
Ferns, indications 117
Ffynnon Limestones 62, 66
Fire, effect on C/N ratios 270
  evidence for 34, 117
  'slash and burn' 275, 281
Firestone Sill sandstone paleosol 91–102
  composition and description 91–7
  discussion 101–2
  interpretation 98–101
Flandrian stage 244, 247
  climatic changes, desertification, Jordan 264
  soil erosion, Mediterranean 264
  pedogenesis, agriculture and erosion 275–6
    decalcification and lessivage 271–4
    effects of man 273–6, 283
    history 263–6
    methodology 266–71
    NW Europe, archaeology 265
    podzolization 276–83
    summary 283–4
  processes, Plateau drift 255
Flood plains
  aggradation, soil reoccurrence rates 168–9, 169–70
  degradation 188–93
  sedimentation and erosion 184–6, 201–3
  stasis, pedogenic modification 203
Fluvaquentic Butrochrepts 32
Fluvents, Red (Entisols) 32, 293
  Capellan paleosols 167–72
  Vertic-Fluvents 162
Fluvial
  lithologic units, hypothetical 185
  regimes, aggradation and degradation, model 188–9
  increase 172–3
Foradada faultline 141, 143
  marine limits 168, 173, 175
  stratigraphy 146
Forests
  coniferous forests 116–17
  distribution 30–1
  earliest trees 23–9
  stabilizing effect 28–9
  see Deforestation
Fragipans, prismatic structure 215, 217
Frasrian, early forests 23, 24
  Upper Old Red Sandstone 60

Freshuraler West Formation, calcretes 74
Fusain (fossil charcoal) 117

Ganisters
  definition 87
  earliest Spodosols 26
  stratigraphy 103
Gedinnian mudstones, Prisoli 60–5, 67
  pseudoanticlinal fractures 71
  smectite clays 78
Gelifluction processes 242–3
Geosol, definition 180
Gilgai, warm-dry regions 70
  compared with calcretes 82
  definition 293
Givetian, early forests 23
Glaciation
  Britain 243–5
  colluvial deposits, paleosols 218–21
  destruction of soils 209
  Devensian glacial limits 243, 252
  drift deposits, paleo-argillan formation 255
  gelifluction processes 242–3
  inference from till deposits 243
  interglacials, between Ipswichian and Hoxnian 244–5
    climate differences 253–4
    relict characteristics 253
  New Zealand 208–9, 213
  periglacial activity 209–11, 218–22
    loess sequences 218–19
    soil horizons 254
  stadials and interstadials 244–5
  soils 247
  temperatures, pre-Devensian and Flandrian 253
Glaebules, calcretes 63–7, 293
  calcite displacement 77
  in fracture systems 70–1
  and pedoturbation 161
Gleys 293
  gley and pseudogley soils 154
  gleysols 9
  Troutbeck soil 248
Grasses
  early grasslands 30–5
  evidence 30
Great Dirt Bed, Purbeck Formation 117–21
'Green Clays' 9, 41
  characteristics 9
Gypsum pseudomorphs 126

Halloysite, volcanic buried soil 225, 234, 236
Haplusalfs, Oxic (Alfisols) 37–8

oxic horizon 294
Hazleton long cairn, brown earth 272–4
Hematite rubification 251, 254, 295
Histosols 41, 293
　oldest 22, 27
Holocene tephra paleosols 231–3
Horizonation 293
　A, B and C horizons 161
　Entisols 155
　times for development, Capella 169
Hornblende weathering 248
Hoxnian stage 244, 247
　interglacial 244–5
　podzols 248
　raised beach, Slindon 249
　Swanscombe stratigraphy 250
Human activity, effects
　artificial soils 265
　Boreal times 271
　deforestation 264, 268
　plough pans 268
　recalcification 270
　soil burial 263, 267, 268–9
　compaction 270
　woodland disturbance, early
　　Flandrian 271–2
　effect on pedogenesis 272–5
Human evolution 35–40
　associated paleosols 36–40
Huronian Supergroup
　Denison profile 3–5, 7–8
　Pronto profile 5–6, 7–8
Hydromorphism 265, 281–3
　post-burial 267–8
　soil characteristics 154–5, 293
　sub-Atlantic period 264

Illerdian transgression 173
Illuviation, and lessivage 272
　contrasted with agricultural causes 275
　definition 293
Inceptisols 41, 293
　Andic Ustochrepts 32
　Lehigh Gap 21
　Fluvaquentic Eutrochrepts 32
　Peas Eddy 23–5
　Sheigra 11
　Ustollic Eutrandepts 32
Ipswichian
　buried podzols 247, 248
　interglacial 244–5, 249
　Llansantiffraid soils 248–9
　Wolstonian loess soil 249–50
Iron
　brunified rendzina 124
　ferruginized paleosols 12
　hematite, rubification 251, 254, 295
　hydroxide, podzols, conversion to
　　brown earth, agriculture 283

preservation, organic
　pseudomorphs 268
laterites 29, 41, 294
lepidocrocite 248
'neoferrans' 151, 153
oxidation, Precambrian 3, 8, 10
pans, altered drainage 269
　stagnopodzols 281
podzoluvisols 90
and potassium metasomatism 7
redistribution 17
secondary ferro-manganiferous
　staining 267
secondary illuviation 267
see also Gleys
Iron Age
　human activity 265
　soil burial 267, 270, 274
　woodland clearance 281

Jebel Qatrani Formation, Egypt,
　stratigraphy 188
Jurassic Upper, Purbeck Dirt Beds xii,
　112–117

K horizon 294
Kakabekia and soil microbes 14
Kaolinite, weathering indicator 12, 105
Karst 41
Kenya, paleosols and hominid
　evolution 36–40
Kesgrave Sands and Gravels 251

Lacustrine carbonates 144, 146
Laddray Wood Soil 248
Laguarres-Capella sequence 141,
　147–55
Lal clay paleosol, Dhok Pathan 37
Land animals, earliest evidence 18, 22
　increasing diversity 27–8
　and modern grasslands 34
Land plants, earliest incidence 19–20
　evolution from algae 21–2
　increasing diversity 27–8
Lapilli, correlations 229
Lateral ramps (thrust surfaces) 173
Laterites 41
　as indicators 29, 294
Lehigh Gap Inceptisol 21
Lehm fabric, definition 249
Lepidocrocite, iron oxide 248
Lessivage
　Atlantic period 265
　decalcification and 271–6
　definition 294
　effect of Mesolithic human activity 272

SUBJECT INDEX

Lessivage *continued*
   effect of Neolithic activity 272–5
   Mediterranean examples 276
   reversal, late Flandrian 276
Levee deposits
   alluvial fans 145
   overbank flooding, suggested 155
Limestones
   Ffynnon 62, 66
   'Psammosteus' 60–2, 65, 68
     features 76
     sedimentation 80
   Purbeck formation 114
Lithologic time units, hypothetical
   fluvial 185
   paleosol 185
Llansantffraid soils, Ipswichian 248–9
Loess deposits, New Zealand
   C-14 dating 216–17
   correlations, with chemical
     properties 215
     with tephras 219
     with weathered units 214
   North Island 218
   periglacial slope and colluvia 218–23
   South Island 211–18
Ludhamian stage 244
Lutetian, Mid, marine horizon 163
Lutetian, Upper, Capella
   Formation 142, 145
   flood plains 161
   marker horizon 148
Luvisols 89–90, 294
   Firestone Sill 101
Luzas fault zone, Capella 142–3, 146
   transitional role 173
Lycopods, indications 117

Maastrichtian Ultisol development 31
Man, evolution *see* Human evolution
Manganese
   extractible, identification 217, 234
   water table fluctuations 269–70
   zone (placic horizon) 36, 295
Mediano anticline, Capella 175
Mediterranean soils, effect of
     lessivage 276
Mesolithic activity 265
   soil burial 269
   woodland clearance 271–2, 280
Metallogenium and soil microbes 14
Microbes, evidence, Precambrian soils
   carbon analysis 15
   predating of marine life? 13–14
   resistance to soil erosion 17
Micromorphology 294
   in archaeology 267
Millipedes, earliest 18, 20, 21

Miocene, Late, Himalayas, vegetation and
     fauna 37–8
Moder fabrics 129
Mollisols 30, 38, 41, 294
   epipedon, mollic 121, 294
   Qquolls 35
   Rendolls 124
   Udolls 35
Mollusca
   buried sequences 265
   dating use, Late Devensian 246
   mull horizons 268
Montanana Group 141, 143
   alluvial fans, model 145
   sedimentary facies 144
Montsec thrust sheet 173–4
Mull horizons, lamina structure 268, 294
Munsell Color Chart System 147

Nematophytes, oldest 17–18
'Neoferrans' 151, 294
   Capella Paleosols 153, 159
Neolithic activity 265
   deforestation, evidence 276
   slash and burn 275, 281
   Hazleton long cairn 272–3
   podzolization 280
   Silbury Hill 268–9
   soil burial 267
   West Heslerton site 277–8
   woodland regeneration 275

New Zealand paleosols 208–39
   identification 233–8
   correlation 236–8
   weathering 233–6
   glacial deposits 209–22
     North Island Loess 218
     periglacial slope 218–22
     South Island Loess 211–18
   volcanic deposits 222–33
     tephra, Holocene 231–3
     older 224–7
     younger 228–31

O horizon 131, 294
   see also Stromatolitic limestones
Ochric epipedon 294
Olduvai Gorge *see* Kenya
Oligocene transitional vegetation 31–5
Oligochaetes 20
Onychophorans 20
Ordovician, Late, earliest land
   animals 18–19
Orellan, soil variety 32
Origins of life 13–14
   evidence, resistance to erosion 15, 17

SUBJECT INDEX 313

Orogenic uplift, New Zealand map 210
Overbank flooding
  frequency 184
  instability in rivers 171
  mudstones, Willwood 194
Oxisols 12-13, 22, 41, 294
Oxygen, Precambrian atmosphere 2-13
  origins of life 17
  oxidizing action 3, 6, 9-13
  reducing environments 6, 7
  texture of soil 7-8

Paleo-argillic horizons
  brown earths, distribution 254, 255
  dating of rubification 255-6
  definition 252-3
  and normal argillic 253
  distribution 254 7
  Soil Survey classification 252-3
  weathering trends 256
Paleosols 294
  correlations, loess sequences 215
  spatial significance 236-8
  hypothetical units, with time 185
  identification, analysis 217
    C-14 dating 181, 216-17, 266
  oldest 3-6, 8-9
  oldest ferruginized 12
  profiles, Capella 152, 156, 159
  total iron 3, 6
Paleudalf volcaniclastic alluvium 31
Paleustalfs, woodland vegetation 32-3, 294
Palynomorphs, early Palaeozoic 18
Pastonian stage 244, 247
  pre-Pastorian, till horizons 257
  Valley Farm soil 251
Pebbles
  analysis 126-8
  origins 128-30
Pebbly Clay Drift, equivalence, Valley Farm soil 252
Pedoderm
  definition 181
  and geosol 181
Pedogenesis
  clacretes 75-8
  time-scales 78-9
Pedotranslocation 235-6
Pedoturbation 161, 295
Perrarva Formation, transitional marine sediments 146
Petrified Forest Member, Chinle, Arizona, superposed scours 193
Phosphorus, soil identification 217
  human occupation 266
  phosphate retention, ash soils 229-30

total, in loess columns, Timaru 235
'Piggy-back' culmination structure 142, 173
Pitstone soil, Windermere 246-7
Placic horizon (Mn zone) 36, 295
  identification 217, 234
  water table fluctuations 269-70
Planosols 90, 295
Plateau Drift, brown earths 255
  uncertain age 255
Pleistocene succession, Hertfordshire and East Anglia 252
Podzols, podzolization 12, 89, 295
  Beestonian 246
  Calluna as indicator 279-80
  cold environment 246
    interglacial pedogenesis 248
    radiocarbon dating 247
  Devensian 247
  faulty dating 271
  and human activity 276 80
  interruption, Neolithic agriculture 282
  iron, conversion to brown earth 283
  leaching, soil ignition tests 266
  in Michigan 264
  in NW Europe 264
  organic matter preservation 270
  peat formation, man-initiated 281
  pre-Flandrian 248
  recalcification by artefacts 270
  soil degradation sequence 278-9
  stagnopodzols 264, 277, 282
  sub-Boreal 265
  uplands and NW areas 281-3
  West Heslerton sequence 277-8
Podzoluvisols 90, 295
  Firestone Sill 101
Pollen
  analysis, in acidic deposits 267
    first evidence for human activity 264
    peat formation, human initiation 281
  distribution, stratigraphy 243
Pongola System, oldest Paleosol 8
Potassium
  metasomatism 7
  removal by land plants 6
Pot Clay Ganister Paleosol 102-7
  description 102-5
  interpretation 106
  stratigraphy 103, 105
Precambrian microfossils, soil 13-17
  evidence 15-17
Pridoli, red-bed succession 60, 61, 65
  geomorphological model 79
  pseudoanticlinal features 71
  smectite clays 78
Protocupressinoxylon purbeckensis 116
Pronto profile, Huronian 4-6, 7-8

'Psammosteus' limestones 60–2, 65, 68, 72
  microscopic features 76
  sediment starvation 80
Puente de Montanana, Capella Formation 141, 143
Pumice, correlation 229
Purbeck Formation 112–17
  Great, Lower, Basal Dirt Beds 117–22
  limestone 114
Puy de Cinca Formation, true marine sediments 146

Qquolls *see* Mollisols
Quaternary
  cooling and warming cycles 238
  paleosols 208–9, 219
  radio-carbon dating 243
  stratigraphy in Britain 243–5
    buried pre-Flandrian soils 245–52
    non-buried pre-Flandrian soils 252–7
  subdivisions, Late, New Zealand 213

Raglan Marl group 72
Rainfall
  identification of paleosols 211, 213
  lateral variation 215
  and pedogenic trends, Flandrian 265
Ramapithecine fossils, and vegetation 36–8
Reading beds 255
Regolith, late Quaternary 222, 295
Relict soils
  pre-Flandrian 242, 243
  and paleosols 208
Rendzinas 295
  Bronze age plateaus 276
  calcareous paleosols, Great Dirt Bed 124
  Late Devensian 246
  Windermere Interstadial 246
Rhynie Chert, histosol 22
Rhyniophytes, Early Siluria 19–20
Rhyolitic paleosols 233–4
Ribagorzana fault-zone, Capella 142–3, 146
  transitional role 173
Ribbon bodies 149, 151
Roots, fossil
  diagnostic use 112, 119
  mycorrhiza 122
  rhizobrecciation 127
  waterlogging 132
Rubification 248, 295
  active, post-Anglian 256
  hematite formation 251, 254, 295
  implication of activity since Anglian 256
  pre-Devensian 253, 256

St Maughan's Group, calcretes 62
Sandstone, Old and New Red *see* Calcretes
Scenic Member, Badlands
  sedimentation rate 200–2
  time restored sequence 201, 202
Scouring, effect on sedimentation
  Chinle Formation 188–90, 193
  Willwood Formation 188, 192
    scourfill deposits 191–4
Sea-level fluctuations
  'Biarritzian transgression' 173
  'Illerdian transgression' 173
Sedimentation rates
  compared with palaeomagnetic data 200
  compared with radiometric data 200
  and erosion, stratigraphy 183–8
  time sections 185, 187, 195
  see also Flood plains: Fluvial aggradation
Sepic plasmic fabrics 253, 295, 296
Sesquioxides 267, 296
Shaugh Moor, Dartmoor 277, 282
Sheffield Blue Ganister paleosol 107–9
Sheigra paleosol 9–12
Shower bedding 228, 230
Shrink-swell cycles, relict soils 253
Siderite nodules, Spodosols 26
Siegenian calcretes 62, 66
Silbury Hill, organic preservation 268–9
Silcretes 41, 296
  climate changes 13, 29
Siltans, identification 268
Silurian, earliest land fossils 17–18
Silurian, Late, Old Red Sandstone *see* Calcretes
Siwaliks, Dhok Pathan Formation 36–8
Skull Cap, Great Dirt Beds 114
Slindon, Hoxnian raised beach 249
Soil development, deposition and erosion rates 169–72
Soil stratigraphy, definition by paleosols 238
Soil Survey of England and Wales, paleo-argillic horizon, definition 252–3
Soil Survey Staff 1975, USDA soil orders 296
Solum
  definition 296
  pedoturbation 161
Speciation, Willwood Formation, Wyoming 196
Spodic horizon 296
  see also Podzols

SUBJECT INDEX 315

Spodosols   9, 41, 296
  definition   26–7
  earliest   25
Stadials and interstadials *see* Glaciation
Stewarts Claim formation   213
Stromatolitic limestones, Purbeck
  'Caps'   113, 116–18, 120
Swanscombe, Thames sequence   250

Tamm's reagent, Aluminium   234, 236
Tanzania, hominids and
  vegetation   38–40
Tardigrada   20
Tchornozems   113
Tectonic activity, New Zealand   208, 210
Temeside Shale Formation   60
  base-level change   81
Temperate soils
  buried   247
  Flandrian, pedogenic trends   265
Tephra
  andesitic tephras   231–4
    organic C levels   234
  chronology   209
  correlations, paleosols   219
  deposits, New Zealand   212
  glass, occurence in loess   217
  principal formations, table   226–7
  rhyolitic tephras   232–3
  in truncated paleosols   223
  weathering profile   228
Terminus antiquem dating   263–4
Terra fusca soil   249
Terra preta   264
Thelon Sandstone, oxisol   12
'Thrust-sheet-top' basin   142
Thrust-slice stacking   173, 175
Thurnian stage   244
Time sections, sediment aggradation and
  degradation   185, 187, 195
Torrert soils   167, 196
  see also Vertisols
Torsal Gros sequence, Capella
  Formation   162–7
  stratigraphy   146
Trees *see* Deforestation: Forests
Tremp-Graus basin, paleo-catena xii
  cross section   143
  sea-level, Lutetian   173
  see also Capella Formation
Triassic, Upper, Chinle Formation,
  Arizona   183
Tropaquept *see* Inceptisols
Troutbeck soil, preservation   248
Tuffs, airfall   67–8, 69

Udolls *see* Mollisols
Ultisols   41, 296

earliest   25–6
forestation   31
United States Department of Agriculture
  soil orders   41, 296
Upton Warren interstadial complex   244
Ustollic Eutrandepts   32
Ustochrept paleosols, Oligocene   32
Ustropept *see* Inceptisols

Valley Farm soil   246–7, 251–2
Valturo sequence, Capella
  Formation   146, 155–62
Venus, atmosphere and soil   2–3
Vertisols   41, 296
  calcrete formation   162
  Old Red Sandstone *see* Calcretes
  oldest   9
  organic matter   8
  Vertic-Fluvents   162, 167
Volcanism, New Zealand   208–10
  link to Quaternary climate   222
  principal tephra formations   226–7
  Taupo Volcanic zone   210
  volcanic centres, map   224

Wales, Old Red Sandstone,
  stratigraphy   61
  see also Calcretes
Waterval Onder paleosol, evidence for
  microbial life   9, 15–16
Weathering cycles, Quaternary   209–11
Weischelian, clay illuviation   257
West Angle Formation, calcretes   62, 64,
  66, 73
White River Group, South Dakota,
  sediment accumulation rates   200
Whitneyan, soil variety   32
  biostratigraphy   196, 198
  lithostratigraphy   196, 198
  paleosols, number   197
  paleotopography   188
    scourfill deposits   191–4
    scouring   189–192
  sediment accumulation   184–5
  stratigraphy   187
  time of formation   197
  unconformities   201
Windermere interstadial   246–7, 251
  Pitstone soil   246
Wolstonian interstadial   244, 247
  loess, Ipswichian shelly deposits   250
Wretton interstadial   244
Woodland *see* Deforestation: Forest

Ypresianxxxx
  Formation   142, 145

Zeolite paleosols   39